D1811040

Lecture Notes in Electrical Engineering

Volume 311

Board of Series editors

Leopoldo Angrisani, Napoli, Italy
Marco Arteaga, Coyoacán, México
Samarjit Chakraborty, München, Germany
Jiming Chen, Hangzhou, P.R. China
Tan Kay Chen, Singapore, Singapore
Rüdiger Dillmann, Karlsruhe, Germany
Haibin Puan, Beijing, China
Gianluigi Ferrari, Parma, Italy
Manuel Ferre, Madrid, Spain
Sandra Hirche, München, Germany
Faryar Jabbari, Irvine, USA
Janusz Kacprzyk, Warsaw, Poland
Alaa Khamis, New Cairo City, Egypt
Torsten Kroeger, Stanford, USA
Tan Cher Ming, Singapore, Singapore
Wolfgang Minker, Ulm, Germany
Pradeep Misra, Dayton, USA
Sebastian Möller, Berlin, Germany
Subhas Mukhopadyay, Palmerston, New Zealand
Cun-Zheng Ning, Tempe, USA
Toyoaki Nishida, Sakyo-ku, Japan
Federica Pascucci, Roma, Italy
Tariq Samad, Minneapolis, USA
Gan Woon Seng, Nanyang Avenue, Singapore
Germano Veiga, Porto, Portugal
Haitao Wu, Beijing, China
Junjie James Zhang, Charlotte, USA

About this Series

"Lecture Notes in Electrical Engineering (LNEE)" is a book series which reports the latest research and developments in Electrical Engineering, namely:

- Communication, Networks, and Information Theory
- Computer Engineering
- Signal, Image, Speech and Information Processing
- Circuits and Systems
- Bioengineering

LNEE publishes authored monographs and contributed volumes which present cutting edge research information as well as new perspectives on classical fields, while maintaining Springer's high standards of academic excellence. Also considered for publication are lecture materials, proceedings, and other related materials of exceptionally high quality and interest. The subject matter should be original and timely, reporting the latest research and developments in all areas of electrical engineering.

The audience for the books in LNEE consists of advanced level students, researchers, and industry professionals working at the forefront of their fields. Much like Springer's other Lecture Notes series, LNEE will be distributed through Springer's print and electronic publishing channels.

More information about this series at http://www.springer.com/series/7818

Marie-Minerve Louërat · Torsten Maehne
Editors

Languages, Design Methods, and Tools for Electronic System Design

Selected Contributions from FDL 2013

 Springer

Editors
Marie-Minerve Louërat
Torsten Maehne
Sorbonne Universités
UMR 7606, LIP6
Paris
France

and

CNRS
UMR 7606, LIP6
Paris
France

ISSN 1876-1100   ISSN 1876-1119 (electronic)
ISBN 978-3-319-06316-4   ISBN 978-3-319-06317-1   (eBook)
DOI 10.1007/978-3-319-06317-1

Library of Congress Control Number: 2014945351

Springer Cham Heidelberg New York Dordrecht London

© Springer International Publishing Switzerland 2015
This work is subject to copyright. All rights are reserved by the Publisher, whether the whole or part of the material is concerned, specifically the rights of translation, reprinting, reuse of illustrations, recitation, broadcasting, reproduction on microfilms or in any other physical way, and transmission or information storage and retrieval, electronic adaptation, computer software, or by similar or dissimilar methodology now known or hereafter developed. Exempted from this legal reservation are brief excerpts in connection with reviews or scholarly analysis or material supplied specifically for the purpose of being entered and executed on a computer system, for exclusive use by the purchaser of the work. Duplication of this publication or parts thereof is permitted only under the provisions of the Copyright Law of the Publisher's location, in its current version, and permission for use must always be obtained from Springer. Permissions for use may be obtained through RightsLink at the Copyright Clearance Center. Violations are liable to prosecution under the respective Copyright Law. The use of general descriptive names, registered names, trademarks, service marks, etc. in this publication does not imply, even in the absence of a specific statement, that such names are exempt from the relevant protective laws and regulations and therefore free for general use.
While the advice and information in this book are believed to be true and accurate at the date of publication, neither the authors nor the editors nor the publisher can accept any legal responsibility for any errors or omissions that may be made. The publisher makes no warranty, express or implied, with respect to the material contained herein.

Printed on acid-free paper

Springer is part of Springer Science+Business Media (www.springer.com)

To Serge Scotti and all our other colleagues from the SoC research community.

Preface

This book in the *Lecture Notes in Electrical Engineering* (LNEE) series presents the most innovative contributions to the Forum on specification and Design Languages (FDL) 2013, which took place in September 2013 at the Université Pierre et Marie Curie (UPMC), Paris, France. These contributions were selected by leading experts from research and industry. Their authors improved the originally presented work based on the feedback received during the conference and recent results.

The increasing integration and complexity of electronic system design requires a constant evolution of the used languages and associated design methods and tools. The FDL is a well-established international forum devoted to the dissemination of research results, practical experiences, and new ideas in this domain. Addressed are in particular the application of specification, design, and verification languages to the modeling, design, and verification of integrated circuits, complex hardware/software embedded systems, and mixed-technology systems. The thus made possible new modeling and specification concepts push the development of new design and verification methodologies to the system level thus providing means for model-driven design of complex information processing systems in a variety of application domains.

This book presents the newest results in five thematic areas:

Part I "Applications of Formal Methods for Specification and Verification": The efficient design space exploration and system verification calls for the use of formal methods to raise the confidence in the taken design decisions and thus accelerate the time to market. Chapter 1 tackles energy component selection and proposes a heuristic algorithm to solve this NP-hard problem. Chapter 2 presents a refinement-based design approach for Systems-on-Chip, supported by model checking technology.

Part II "Embedded Analog and Mixed-Signal System Verification": Tackling this task with classic approaches based on SPICE simulation and visual waveform inspection has reached its limits with today's more and more complex AMS systems. While waiting for a much anticipated revolution in this field, assertion-based approaches are very promising. Two are presented in Chaps. 3 and 4.

Part III "Embedded Analog and Mixed-Signal System Design": Efficient modeling and simulation approaches for analog and mixed-signal systems are at the heart of the design process. Chapter 5 proposes a model to describe the nonlinear behavior of memresistors and presents its implementation in VHDL-AMS. Chapter 6 presents a way to generate homogeneous (SystemC AMS or C++) code for heterogeneous systems by using a formal intermediate representation.

Part IV "Digital Hardware/Software Embedded System Design": SystemC needs to continuously evolve through core library extensions and new methodology-specific libraries to offer modeling capabilities and simulation performance that keep up with the user needs. Chapter 7 proposes the concept of combining SystemC events with TLM transactions so as to simplify and systematize synchronization in TLM models. Chapter 8 shows how predictable platforms can be characterized to form a basis for virtual prototyping of real-time systems.

Efficient simulation, static analysis, and model transformation are key techniques to enable design validation and design space exploration. Chapter 9 combines simulation with analytical techniques to provide estimates that guide the design space exploration of real-time systems. Chapter 10 introduces model transformations and validation methods that open an automated path from algorithm design to ESL design. Chapter 11 supports software allocation in networked automotive system platforms. Chapter 12 shows a method for switching between models of different abstraction levels and its application to trade-off speed and accuracy in network-on-chip simulation.

Part V "Model-Driven Engineering for Embedded Systems Design": The increasing complexity of embedded software requires the application of modern software engineering approaches like model-based software development enabled by standard modeling languages and associated tools. Chapter 13 presents the use of MARTE and its real-time modes specification for the development of cross-layer self-adaptive real-time embedded systems. Chapter 14 investigates the difficult task of design space exploration for the allocation of UML composite structures in the modeling of distributed systems. Chapter 15 explores the use of MARTE on an autonomous robot use case for the application of MAST schedulability analysis tools for model-based performance analysis.

The 2013 edition of FDL was organized by ECSI in technical cooperation with the IEEE Council on Electronic Design Automation (CEDA), the French chapters of the IEEE Computer Society (CS) and IEEE Circuit and System Society (CAS) as well as the International Federation for Information Processing (IFIP). We would like to thank the large number of contributors to FDL 2013: the authors of the submitted papers as well as the 62 members of the technical program committee. We would like to acknowledge the organizational support provided by our colleagues from the Laboratoire d'Informatique de Paris 6 (LIP6), the Université Pierre et Marie Curie (UPMC), and the Centre National de la Recherche Scientifique (CNRS), in particular Roselyne Chotin-Avot. We would like to thank Adam Morawiec and Jinnie Hinderscheit from ECSI for their administrative support.

Paris, May 2014

Marie-Minerve Louërat
Torsten Maehne

Contents

15 ProMARTES: Performance Analysis Method and Toolkit

Contributors

Mohamed Abid CES Laboratory, ENIS, University of Sfax, Sfax, Tunisia

Rabéa Ameur-Boulifa Institut Télécom, Télécom ParisTech, CNRS-LTCI, Sophia-Antipolis, France

Seyed Hosein Attarzadeh Niaki KTH Royal Institute of Technology, Stockholm, Sweden

Ghada Bahig Mentor Graphics Egypt, Cairo, Egypt

Nils Ballmann Hardware/Software Co-Design, Department of Computer Science, Friedrich-Alexander-Universität Erlangen-Nürnberg (FAU), Erlangen, Germany

Faik Baskaya Department of Electrical and Electronics Engineering, Bogazici University, Istanbul, Turkey

Nader Ben Amor CES Laboratory, ENIS, University of Sfax, Sfax, Tunisia

Mouna Ben Said CES Laboratory, ENIS, University of Sfax, Sfax, Tunisia

Egor Bondarev Eindhoven University of Technology, Eindhoven, The Netherlands

Arnaud Cuccuru Laboratory of Model Driven Engineering for Embedded Systems, CEA LIST, Gif-sur-Yvette, France

Peter H.N. de With Eindhoven University of Technology, Eindhoven, The Netherlands

Marcus Eggenberger Embedded Systems Engineering Group, University of Stuttgart, Stuttgart, Germany

M. Watheq El-Kharashi Computer and Systems Engineering Department, Ain Shams University, Cairo, Egypt

Emmanuelle Encrenaz-Tiphene Sorbonne Universités, Université Pierre et Marie Curie Paris 6, Paris, France; CNRS, UMR 7606, LIP6, Paris, France

Andrew N. Fisher University of Utah, Salt Lake City, USA

Franco Fummi EDALab s.r.l., Verona, Italy; University of Verona, Verona, Italy

Sébastien Gérard Laboratory of Model Driven Engineering for Embedded Systems, CEA LIST, Gif-sur-Yvette, France

Michael Glaß Hardware/Software Co-Design, Department of Computer Science, Friedrich-Alexander-Universität Erlangen-Nürnberg (FAU), Erlangen, Germany

Önder Gürcan Laboratory of Model Driven Engineering for Embedded Systems, CEA LIST, Gif-sur-Yvette, France

Joachim Haase Design Automation Division EAS, Fraunhofer Institute for Integrated Circuits IIS, Dresden, Germany

Yessine Hadj Kacem CES Laboratory, ENIS, University of Sfax, Sfax, Tunisia

Bastian Haetzer Embedded Systems Engineering Group, University of Stuttgart, Stuttgart, Germany

Brahim Hamid IRIT, University of Toulouse, Toulouse, France

Sherif Hammad Computer and Systems Engineering Department, Ain Shams University, Cairo, Egypt

Ulrich Heinkel Chair for Circuit and System Design, Technische Universität Chemnitz, Chemnitz, Germany

Fernando Herrera KTH Royal Institute of Technology, Stockholm, Sweden

Thomas Horn Chair for Circuit and System Design, Technische Universität Chemnitz, Chemnitz, Germany

Mickaël Kerboeuf MOCS Team, Lab-STICC, University of Brest, Brest, France

Dhanashree Kulkarni Intel Corporation, Hillsboro, USA

Ajay Kumar Mentor Graphics Corporation, Wilsonville, OR, USA

André Lange Design Automation Division EAS, Fraunhofer Institute for Integrated Circuits IIS, Dresden, Germany

Michele Lora University of Verona, Verona, Italy

Erik Markert Chair for Circuit and System Design, Technische Universität Chemnitz, Chemnitz, Germany

Marcus Mikulcak KTH Royal Institute of Technology, Stockholm, Sweden

Hocine Mokrani Institut Télécom, Télécom ParisTech, CNRS-LTCI, Sophia-Antipolis, France

Chris J. Myers University of Utah, Salt Lake City, USA

Klaus-Holger Otto Alcatel-Lucent AG, Nuremberg, Germany

Ansgar Radermacher Laboratory of Model Driven Engineering for Embedded Systems, CEA LIST, Gif-sur-Yvette, France

Martin Radetzki Embedded Systems Engineering Group, University of Stuttgart, Stuttgart, Germany

Hans-Werner Sahm Alcatel-Lucent AG, Nuremberg, Germany

Ingo Sander KTH Royal Institute of Technology, Stockholm, Sweden

Matthias Sauppe Chair for Circuit and System Design, Technische Universität Chemnitz, Chemnitz, Germany

Alper Sen Department of Computer Engineering, Bogazici University, Istanbul, Turkey

Yasser Shoukry Computer and Systems Engineering Department, Ain Shams University, Cairo, Egypt

Francesco Stefanni EDALab s.r.l., Verona, Italy

Jürgen Teich Hardware/Software Co-Design, Department of Computer Science, Friedrich-Alexander-Universität Erlangen-Nürnberg (FAU), Erlangen, Germany

Konstantinos Triantafyllidis Eindhoven University of Technology, Eindhoven, The Netherlands

Dogan Ulus Verimag, Centre Equation, Gieres, France

Sara Vinco University of Verona, Verona, Italy

Liyuan Zhang Hardware/Software Co-Design, Department of Computer Science, Friedrich-Alexander-Universität Erlangen-Nürnberg (FAU), Erlangen, Germany

Figures

Tables

Algorithms

Listings

Acronyms

AADL	Architecture Analysis and Design Language
ABV	Assertion-Based Verification
AC	Alternating Current
ACET	Average-Case Execution Time
ADC	Analog-to-Digital Converter
ADL	Architecture Description Language
AHB	AMBA High-performance Bus
ALU	Arithmetic Logic Unit
AMBA	Advanced Microcontroller Bus Architecture
AMM	Adaptive Module Manager
AMS	Analog/Mixed-Signal
API	Application Programming Interface
ARM	Allocation by Recursive Mincut
ASL	Analog Specification Language
ASM	Automotive Simulation Model
AT	Approximately Time
ATLM	Arbitrated Transaction Level Model
AUTOSAR	AUTomotive Open System ARchitecture
AVC	Advanced Video Coding (MPEG-4 part 10, H.264 standard)
BCA	Bus Cycle Accurate
BCM	Boundary Condition-based Model
BDD	Binary Decision Diagram
BW	BandWidth
CAAM	Combined Algorithm and Architecture Model
CAN	Controller Area Network
CARAT	Component ARchitectures Analysis Tool
CAS	Circuit and System Society
CATLM	Cycle-Accurate Transaction Level Model
CBML	Component-Based Modeling Language
CBSE	Component-Based Software Engineering
CB-SPE	Component-Based Software Performance Engineering

CCATB	Cycle Count At Transaction Boundaries
CCD	Charge-Coupled Device
CCL	Component Composition Language
CCM	CORBA Component Model
CEDA	Council on Electronic Design Automation
CEPM	Configurable and Executable Performance Model
CIF	Compositional Interchange Format
CISC	Complex Instruction Set Computing
CLI	Command Line Interface
CNF	Conjunctive Normal Form
CNRS	Centre National de la Recherche Scientifique
COMPAS	Component Performance Assurance Solutions
CORBA	Common Object Request Broker Architecture
CP	Constraint-based Programming
CPI	Cycles Per Instruction
CPU	Central Processing Unit
CS	Computer Society
CSP	Constraint Satisfaction Problem
CTL	Computational Tree Logic
D&C	Deployment and Configuration of distributed component-based applications
DAC	Digital-to-Analog Converter
DC	Direct Current
DCT	Discrete Cosine Transformation
DDS	Data Distribution Service
DEAP	Distributed Evolutionary Algorithms in Python
DM	Deadline Monotonic scheduling algorithm
DMA	Direct Memory Access
DNA	DesoxyriboNucleic Acid
DRAM	Dynamic RAM
DRE	Distributed Real-time Embedded (system)
DSE	Design Space Exploration
DSP	Digital Signal Processing
ECSI	Electronic Chip and Systems design Initiative
ECU	Electronic Control Unit
EDA	Electronic Design Automation
EDF	Earliest Deadline First scheduling algorithm
EJB	Enterprise JavaBeans
ELN	Electrical Linear Network
EQN	Extended Queuing Network
ESL	Electronic System Level
ESS	Efficient Safe Solution

ET	Exploration Tool
FCM	Flex-eWare Component Model
FFT	Fast Fourier Transform
FIFO	First In, First Out
FLI	Foreign Language Interface
FPGA	Field-Programmable Gate Array
FSM	Finite State Machine
GIS	Generic Interaction Support
GL	Gate Level
GPU	Graphics Processing Unit
GRM	Generic Resource Modeling
HAL	Hardware Abstraction Layer
HDL	Hardware Description Language
HdS	Hardware-dependent Software
HRM	Hardware Resource Modeling
HRT	Hard Real-Time
HSDF	Homogeneous Synchronous Dataflow
HSDFG	HSDF Graph
HW	HardWare
I/O	Input/Output
IC	Integrated Circuit
ICT	Information and Communications Technology
IDE	Integrated Development Environment
IEC	International Electrotechnical Commission
IEEE	Institute of Electrical and Electronics Engineers
IFIP	International Federation for Information Processing
IP	Intellectual Property
ISO	International Organization for Standardization
ITU	International Telecommunication Union
JAS-DSE	Joint Analytical and Simulation-based Design Space Exploration
JPEG	Joint Photographic Experts Group
JVT	Joint Video Team
KLAPER	Kernel LAnguage for PErformance and Reliability analysis
KPN	Kahn Process Network
LAMP	Language for Analog/Mixed-Signal Properties
LE	Logic Element
LEMA	LPN Embedded Mixed-Signal Analyzer
LIN	Local Interconnect Network
LPF	Low-Pass Filter
LPN	Labeled Petri Net
LQN	Layered Queuing Network
LSF	Linear Signal Flow

LT	Loosely Timed
LTI	Linear Time-Invariant
LTL	Linear Temporal Logic
LTS	Labelled Transition System
MAPE	Monitor, Analyze, Plan, Execute
MARTE	Modeling and Analysis of Real-Time and Embedded systems (UML profile)
MAST	Modeling and Analysis Suite for real-Time applications
MBD	Model-Based Design
MCL	Model Checking Language
MCS	Mixed-Criticality System
MDE	Model-Driven Engineering
ME	Motion Estimation
MITL	Metric Interval Temporal Logic
MoC	Model of Computation
MOEI	Maximum Observed Executed Instructions
MOET	Maximum Observed Executed Time
MPA	Modular Performance Analysis
MPSoC	MultiProcessor System-on-Chip
MTL	Metric Temporal Logic
MuT	Module under Test
NFP	Non-Functional Properties
NHRT	Non-Hard Real-Time
NICTA	National ICT Australia Ltd
NoC	Network on Chip
OMG	Object Management Group
OS	Operating System
OSCI	Open SystemC Initiative
OTA	Operational Transconductance Amplifier
OWL	Web Ontology Language
OWL-S	Ontology of services built on top of the OWL
PB	Processing Block
PBD	Platform-Based Design
PC	Personal Computer
PCM	Palladio Component Model
PE	Processing Element
PECT	Prediction-Enabled Component Technology
PEQ	Payload Event Queue
PGA	Programmable Gain Amplifiers
PI	Phase Interpolator
PLL	Phase-Locked Loop
PMU	Performance Monitor Unit

PSL	Property Specification Language
PSNR	Peak Signal-to-Noise Ratio
QoS	Quality of Service
QP	Quantization Parameter
RAM	Random Access Memory
ReRAM	Resistive RAM
RSS	Reduced set of Safe Solutions
RISC	Reduced Instruction Set Computing
RM	Rate Monotonic scheduling algorithm
ROBOCOP	Robust Open component Based sOftware architecture for COnfigurable devices Project
ROM	Read-Only Memory
ROS	Robot Operating System
RT	Real Time
RTA	Response Time Schedulability Analysis
RTL	Register Transfer Level
RTE	Real-Time Embedded
RTES	Real-Time Embedded System
RTOS	Real-Time Operating System
RT-SVA	Real-Time SystemVerilog Assertions
SA-SDF	Scenario-Aware SDF
SAS-DSE	Scenario-Aware Simulation-based Design Space Exploration
SAT	SATISFIABILITY
SAV	Simulation-Aided Verification
SDF	Synchronous Data Flow
SFDR	Spurious Free Dynamic Range
SIL	Safety Integrity Level
SMT	Satisfiability Modulo Theory
SNDR	Signal-to-Noise and Distortion Ratio
SOA	Service-Oriented Application
SoC	System on a Chip
SPI	System Property Interval
SPICE	Simulation Program with Integrated Circuit Emphasis
SR	Synchronous Reactive
SRAM	Static RAM
SRM	Software Resource Modeling
SS	Safe Solution
STL	Signal Temporal Logic
STL/PSL	Signal Temporal Logic Property Specification Language
SVA	SystemVerilog Assertions
SW	SoftWare
SWC	SoftWare Component

SysML	Systems Modeling Language
TDF	Timed Data Flow
TDM	Time Division Multiplex
TE	Transaction Event
TEAM	ThrEshold Adaptive Memristor Model
TFL	Time-Frequency Logic
THD	Total Harmonic Distortion
TL	Transaction Level
TLM	Transaction-Level Modeling
TML	Task Modelling Language
TTA	Time Triggered Architecture
TV	Torque Vectoring
UCM	Unified Component Model
UI	User Interface
UML	Unified Modeling Language
UML-RT	UML for Real-Time profile
UML-SPT	UML profile for Scheduling, Performance, and Time
URL	Uniform Resource Locator
UVM	Universal Verification Methodology
VCO	Voltage-Controlled Oscillator
VHDL	VHSIC Hardware Description Language (IEEE Std 1076)
VHDL-AMS	VHDL Analog/Mixed-Signal (IEEE Std 1076.1)
VHSIC	Very High Speed Integrated Circuit
VLIW	Very Long Instruction Word
VLSI	Very Large Scale Integration
VP	Virtual Prototype
WCA-DSE	Worst-Case Analytical Design Space Exploration
WCCT	Worst-Case Communication Time
WCEI	Worst-Case Executed Instructions
WCET	Worst-Case Execution Time
WCW	Worst-Case Workload
XMI	XML Metadata Interchange
XML	eXtensible Markup Language

Part I
Applications of Formal Methods for Specification and Verification

Chapter 1
Optimal Component Selection for Energy-Efficient Systems

Matthias Sauppe, Thomas Horn, Erik Markert, Ulrich Heinkel, Hans-Werner Sahm and Klaus-Holger Otto

Abstract Microelectronics have developed very fast in the past. The design process of those systems is getting more and more complex and new design methods have to be applied continuously. One main observation is the increasing design level over time, hence, the re-use of components is getting more important. A key challenge of IC design is the selection of a system architecture which fulfills all requirements in terms of data throughput, area, timing, power and cost. We present a problem class for the optimal component selection in order to assist in selecting the best available alternatives. We will show how to express top-level constraints and optimisation targets including dependencies between components. In addition, heuristic solving algorithms will be presented. The evaluation section shows that the presented algorithms perform well on typical problem sets. Using a framework for evolutionary algorithms results in additional speedup.

Keywords Component selection problem · Design Space Exploration (DSE) · System level · High abstraction level · System integration · Energy efficiency · Heuristic algorithm · Optimization · Evolutionary algorithm · Local search · NP-completeness

M. Sauppe (✉) · T. Horn · E. Markert · U. Heinkel
Chair for Circuit and System Design, Technische Universität Chemnitz, Chemnitz, Germany
e-mail: saum@hrz.tu-chemnitz.de

T. Horn
e-mail: hort@hrz.tu-chemnitz.de

E. Markert
e-mail: erma@hrz.tu-chemnitz.de

U. Heinkel
e-mail: heinkel@hrz.tu-chemnitz.de

H.-W. Sahm · K.-H. Otto
Alcatel-Lucent AG, Nuremberg, Germany
e-mail: hans.sahm@alcatel-lucent.com

K.-H. Otto
e-mail: klaus-holger.otto@alcatel-lucent.com

© Springer International Publishing Switzerland 2015
M.-M. Louërat and T. Maehne (eds.), *Languages, Design Methods,
and Tools for Electronic System Design*, Lecture Notes in Electrical Engineering 311,
DOI 10.1007/978-3-319-06317-1_1

1.1 Introduction

In the last decades, microelectronics have been evolving rapidly fast. The structure sizes of microsystem technologies shrank and their performance increased. Moore's law still holds—the integration density still grows exponentially. The same situation can be observed when considering the worldwide internet bandwidths: The average link capacity grows exponentially as well [9] and network routers have to be able to handle a constantly increasing data throughput.

From a designer's perspective, the growing performance and integration density of microelectronics forces the application of new design methods. This is often called the *productivity gap*, which mainly states that available technologies evolve faster than appropriate design methods. Therefore, the challenge is not only to develop faster and smaller hardware, but also to develop adequate IC design methods which allow for the best possible hardware usage.

Due to this situation, the abstraction level on which microsystems are designed constantly rose during the last decades. Nowadays, microsystems are often implemented on a very high structural level or even on system level. On these abstraction levels it is easier to design a complex system. However, various problems arise.

On high abstraction levels, multiple implementation alternatives are often available. For example, if a microcontroller core is needed in an integrated system, several Intellectual Property (IP) cores will be suitable for the architectural requirements. This might also be the case for other (IP) cores like RAMs or network interfaces. Typically, several constraints will apply as well: The die size of a System on a Chip (SoC) will be limited and the energy consumption should be as low as possible. Additionally, there will be several inter-component requirements, which have to be resolved, too.

When designing a complex microelectronical system, all of these constraints have to be fulfilled while optimising a target function at the same time. The task is now to choose a set of components from a given component database in such a way that all top-level constraints and inter-component dependencies are met. If several solutions exist, the solution which optimises a target function should be chosen.

Now, we formalise the problem description of the component selection problem. Moreover, we prove its NP-hardness and we adapt several well-known heuristic algorithms for this problem class.

Formulating a component selection problem can help to design a complex microelectronical system, because critical design decisions can be made on an objective basis. For example, it is possible to choose system components in such a way that the maximum chip area will not be exceeded. This is a hard constraint. The overall energy consumption could be an auxiliary condition which is optimised while ensuring that the hard constraints will be met. To define a component selection problem, the designer now has to provide a component database including all necessary data and interdependencies. Then, the designer can create top requirements which will describe the target system and its constraints. The algorithm will then return a solution (if it exists) and optimise it, if an optimisation target has been specified.

In Sect. 1.2, previous work on similar problems will be evaluated. Section 1.3 will define the component selection problem formally, a simple example will be provided and the NP-hardness will be proven. In Sect. 1.4, we present an adaptation of common heuristic algorithms which perform well on typical problem input sets. The integration into the specification tool SpecScribe is shown in Sect. 1.5. Finally, we will evaluate the algorithms on larger input sets in Sect. 1.6.

1.2 Related Work

In the past, several types of combinatorial optimisation problems have been discussed and investigated and efficient heuristic algorithms have been provided.

The basic problem class of the component selection problem is the so-called *Constraint Satisfaction Problem* (CSP) class. One of the first works on Constraint Satisfaction Problems (CSPs) was published in 1985 by Mackworth [6]. There, the basic CSP class was introduced: A CSP contains a set of variables and a set of constraints which define the variable value ranges and dependencies or relations between variables. To solve a CSP, at least one variable assignment has to be found for a given set of variables and relations while all constraints are met. The author uses well-known algorithms to solve these CSPs. However, the basic CSP approach is not applicable for the component selection problem since variables cannot be dynamically instantiated.

In the past years, the CSP problem class has been extended in different ways. The *dynamic* CSP class was introduced in 1985 by Mittal [7]. The basic extension is to allow that the set which is relevant for a solution may change during the search, e.g., when certain criteria are met. Nevertheless, this CSP extension is not sufficient to model dynamically instantiable components properly.

An approach where each variable of a CSP may be existentially or universally quantified (*quantified* CSP) was presented by Benhamou [1]. This allows for a broader problem class whose constraints can take advantage of the expressiveness of predicate logic. Since the component selection problem requires dynamic instantiations of variables, the quantified CSP cannot be used to solve this problem class.

The *mixed* CSP class was presented by Gelle and Faltings [5]. The authors introduce a distinction between *controllable* and *uncontrollable* variables. Uncontrollable variables are variables whose value is neither known nor controllable by the solving algorithm. Therefore, this problem class deals with uncertainty. It is also not applicable to the component selection problem because the maximum solution size has to be defined as part of the problem.

An approach to solve *conditional and composite* CSPs was presented by Mouhoub and Sukpan [8]. In this problem class, conditions for every variable are defined, i.e., a variable may only be used in a solution if its according conditions are met. A *composite variable* allows for the disjunction with another variable, so that at most one of them may be used in a solution. Like in the previous approaches, all variables have to be defined during problem definition. For the component selection problem, this approach cannot be used because the number of needed variables is not predictable due to the dynamic structure of the component selection problem.

An approach to formalise and solve the component selection problem class was introduced by Pande et al. [10]. This work focuses on component selection for software components, however, the basic approach is also applicable to hardware systems. The authors define a top-level set of requirements and a set of components. To solve the problem, a subset of components has to be chosen in such a way that all requirements are met. A model which allows for the optimisation of auxiliary conditions is also provided. Nevertheless, it is not possible to represent requirements of component instantiations, only global requirements can be formulated. Additionally, components can only be instantiated once and there is no possibility to express the need of multiple components of the same type. In our work, we introduce hierarchical requirements modeling which allows for arbitrary system complexities.

Another approach facing the component selection problem was shown by Proß et al. [12], where interdependencies between components are modeled. The maximum number of component instantiations is not restricted. Therefore, this approach handles the dynamic problem structure adequately. Moreover, a mechanism is provided to combine requirements of several components. However, this approach was focused on modeling the relationships between components, but the problem class was not defined formally, which complicates implementation work. Additionally, no efficient algorithm for solving this problem class was provided. Our contribution tries to fill this gap.

1.3 Problem Definition

We define the component selection problem Φ as a quadruple:

$$\Phi = (C, F, T, t),$$

where

$$C = \{(name, \text{provides}, \text{requires})\}$$
$$\mid \text{provides}, \text{requires are cost relations},$$
$$F = \{(name, \text{combOp})\}, \text{combOp} \in Op,$$
$$T = \{r\} \mid r \text{ is a cost relation},$$
$$t = (f, op), f \in F, op \in \{\text{min}, \text{max}\}, t \text{ is optional}.$$

Moreover, we define the following (static) sets:

$$R = \{=, \neq, <, >, \leq, \geq, \text{Range}\}$$
$$Op = \{\text{add}, \text{mult}, \text{min}, \text{max}\}.$$

Each *name* can be an arbitrary literal string, which must be unique. It is used for the identification of components and cost factors. An *expression* may be any mathematical expression which can instantly be evaluated. A trivial form of an expression is a number.

A *cost relation r* is defined as:

$$r = (\text{costfactor, relOp, } expression)$$
$$| \text{ costfactor} \in F, \text{relOp} \in R$$

Then, the following functions are defined:

$$\text{provides} ((name, \text{provides, requires})) := \text{provides}$$
$$\text{requires} ((name, \text{provides, requires})) := \text{requires}$$
$$\text{combOp} ((name, \text{combOp})) := \text{combOp}$$

provides and *requires* return the *provides* and *requires* properties of a component. *combOp* returns the combination operation of a cost factor.

Moreover, the evaluation of a cost factor f regarding a set of given properties S is defined:

$$\text{val} (f, S) = \text{combOp} (f) (\{S\})$$

This will combine all elements in S using the associated combination operation of the cost factor f. The result is a number.

A solution Ψ of a problem Φ is defined as:

$$\Psi = \{(c, n)\}, c \in C, n \in \mathbb{N}$$

Ψ is a set of components, where a number is associated to each component. The number will state how often this particular component has to be instantiated.

Informally, C is a set of components a target system may be composed of. A component $c \in C$ may be instantiated several times in a solution if necessary.

Each component $c \in C$ may have several *provides* and/or *requires* properties. Each *requires* property states that if this component is going to be used in the solution, the property has to be covered by *provides* properties, which may be defined for other components.

F is a set of so-called cost factors. Each cost factor $f \in F$ has a unique *name* and an associated combination operation combOp $\in Op$, which defines how to accumulate this cost factor.

Each property $p \in P$ defines a relation on one cost factor. To determine whether the property holds, the term "val (costfactor, P) relOp *expression*" must be evaluated.

T are the top requirement properties, which have to hold if the solution Ψ shall be valid. To test for validity, each cost factor has to be evaluated using its combination operation regarding both the solution components' *provides* and *requires* properties. Since the combination operation, applied to several properties, returns a number, the test is a simple arithmetic comparison for each cost factor. The *provides* property evaluation must result in a number which correlates to the *requires* property evaluation, according to the relation operator that was defined for each particular *requires* property.

t is an optional optimisation target. It is a cost factor, which has either to be minimised or to be maximised.

1.3.1 Example

Consider the following problem: A computer system has to be set up and there are a couple of requirements, but the budget is limited. The system shall provide at least 1 CPU, 8 GB of RAM and an ethernet port. The budget is limited to 600 USD. These are the top requirements T of our problem Φ. The task is to instantiate components from C so that 1. each top requirement is fulfilled and 2. the requirements of each selected component is fulfilled. This problem is now defined formally as $\Phi(C, F, T, t)$.

For simplicity, we only define the following cost factors:

$$F = \{ (price, \text{add}),$$
$$(energy, \text{add}),$$
$$(cpus, \text{add}),$$
$$(ramGb, \text{add}),$$
$$(cpuSockets, \text{add}),$$
$$(ramSockets, \text{add}),$$
$$(ethPorts, \text{add})\}$$

Using these cost factors, the top requirement can be defined as follows:

$$T = \{ (price, \leq, 600),$$
$$(cpus, \geq, 1),$$
$$(ramGB, \geq, 8),$$
$$(ethernetPorts, \geq, 1)\}$$

Next, the component database C has to be defined. It contains every component which may be selected to create a target system. In this example, we define C as follows:

$$C = \{(cpu1, \{(price, =, 100), (energy, =, 800), (cpus, =, 1)\}, \{(cpuSockets, =, 1)\}),$$
$$(mainboard1, \{(price, =, 149), (energy, =, 250),$$
$$(cpuSockets, =, 1), (ramSockets, =, 4)\}, \{\}),$$
$$(mainboard2, \{(price, =, 199), (energy, =, 300), (ethPorts, =, 1),$$
$$(cpuSockets, =, 1), (ramSockets, =, 2)\}, \{\}),$$
$$(ethCard, \{(price, =, 39), (energy, =, 130), (ethPorts, =, 1)\}, \{\}),$$
$$(ramModule1, \{(price, =, 129), (energy, =, 80),$$
$$(ramGb, =, 4)\}, \{(ramSockets, =, 1)\}),$$
$$(ramModule2, \{(price, =, 49), (energy, =, 60),$$
$$(ramGb, =, 2)\}, \{(ramSockets, =, 1)\})\}$$

Note that the *mainboard1* provides 4 RAM sockets, while *mainboard2* provides only 2 RAM sockets, but also an ethernet port. The price of *mainboard2* is higher than the price of *mainboard1*. Up to now, we haven't defined a global optimisation target t for this sample problem.

There are several solutions of this problem. Two sample solutions are:

$$\Psi_1 = \{(cpu1, 1),$$
$$(mainboard1, 1),$$
$$(ethCard, 1),$$
$$(ramModule2, 4)\}$$

and

$$\Psi_2 = \{(cpu1, 1),$$
$$(mainboard2, 1),$$
$$(ramModule1, 2)\}$$

Both solutions are valid, i.e., the top requirements and all implicit requirements are fulfilled. We now evaluate the cost factor *energy* for each solution and result in $energy_1 = 1{,}420$ and $energy_2 = 1{,}260$.

To select the most energy-efficient solution only, our problem definition has to be enhanced by an optimisation target:

$$t = (energy, \min).$$

After applying this restriction, the only possible solution will be Ψ_2.

1.3.2 NP-Hardness

Up to now, no efficient exact algorithms are known for NP-hard problems (though this is not proven). In general, solving an NP-hard problem takes at least exponential time, compared to the input size. Therefore, non-exact heuristic and optimisation algorithms, which are faster, are often preferred for this problem class.

In the following paragraphs, the NP-hardness of the component selection problem will be proven.

We show that the component selection problem is NP-hard by transforming the well-known SATISFIABILITY (SAT) problem into a component selection problem in polynomial time and space. The SAT problem is known to be NP-hard [4] and it will be shown that the component selection problem is at least as hard as the SAT problem by representing an arbitrary SAT instance using the component selection problem.

C, F and T of the problem Φ will be introduced implicitly in the following, the optional parameter t will be left empty.

Let $P = (a_{11} \vee \cdots \vee a_{1m}) \wedge \cdots \wedge (a_{n1} \vee \cdots \vee a_{nm})$ be an arbitrary SAT formula in Conjunctive Normal Form (CNF) with n clauses and m literals $l_1 \ldots l_m$, where $a_{ij} \in \{l_j, \overline{l_j}, 0\}$. $a_{ij} = 0$ means that clause i contains neither l_j nor $\overline{l_j}$. For each clause, we introduce a component $c_1 \ldots c_n \in C$.

Each clause must be fulfilled. Therefore, our top requirement must ensure that each clause is instantiated exactly once. We do this by creating a cost factor $(clauseCount, add) \in F$ and we define $\forall i \in \{1, \ldots, n\} : (clauseCount, =, 1) \in$ provides (c_i) and $(clauseCount, =, n) \in T$.

Next, each literal in the CNF must be either instantiated as negated or non-negated variable. For each literal l_i, we create two components c_{l_i} and $c_{\overline{l_i}} \in C$, representing a non-negated and a negated literal. Each literal must be instantiated exactly once. Therefore, we create a cost factor f_i for each literal: $f_i = (literalInst_i, add) \in F$, define $\forall i \in \{1, \ldots, m\} : (literalInst_i, =, 1) \in T$, $(literalInst_i, =, 1) \in$ provides(c_{l_i}), and $(literalInst_i, =, 1) \in$ provides$(c_{\overline{l_i}})$.

Finally, each clause must evaluate to true. We do this by creating appropriate *requires* properties for each clause: $\forall i \in \{1, \ldots, n\} : (literalTrue_i, \geq, 1) \in$ requires(c_i). Moreover, we consider every literal component c_{l_j} and $c_{\overline{l_j}}$: For every clause i, we check whether it contains the non-negated literal l_j. If yes, we add $(literalTrue_i, =, 1) \in$ provides(c_{l_j}). If the clause i contains the negated literal $\overline{l_j}$, we add $(literalTrue_i, =, 1) \in$ provides$(c_{\overline{l_j}})$. If the clause i doesn't contain l_j or $\overline{l_j}$, we don't extend any *provides* set.

If P is satisfiable, the resulting solution will contain either c_{l_i} or $c_{\overline{l_i}}$ for each literal, which represents the variable assignment. If P is not satisfiable, there will be no solution.

Because the component selection problem could be used to represent an arbitrary SAT problem in CNF, the component selection problem is NP-hard. Knowing this, we also know that there is no efficient exact solution algorithm. Therefore, we will focus on non-exact heuristic algorithms in the following section.

Algorithm 1.1: Random solution generation

---

1 $\Psi = \emptyset$;
2 $j := 0$;
3 **repeat**
4 | add a random component to Ψ or increase the component count of a random component which is already $\in \Psi$;
5 | $j :=$ number of unfulfilled cost factors in Ψ;
6 **until** j *has not decreased for q loop iterations*;

1.4 Algorithms

As the component selection problem is NP-hard, there is no exact efficient algorithm for it, i.e., there is no algorithm which guarantees to finish in polynomial time depending on the input size. Therefore, an algorithm which guarantees to provide the best solution can only be realised using exhausting search techniques like backtracking.

However, the structure of most component selection problems is similar to the example presented in the previous section. In particular, for most component selection problems, only few cost factors are restricted (only *price* is restricted in the example above) compared to the number of components which have to be included in the system. This leads to the situation that it is easy to generate a huge amount of solutions which are *almost* valid, only the restricted cost factors are not fulfilled.

In the next paragraphs, several common heuristic approaches are adapted to face this problem.

1.4.1 Random Solution Generation

Random solution generation is a commonly used technique to generate an initial set of potential solutions, which is altered during the solution search process afterwards. In terms of evolutionary optimisation algorithms, this set is often called *initial population*.

For the component selection problem, we start with an empty solution and add components iteratively until the number of cost factors which are not fulfilled doesn't decrease for a given number of iterations (Algorithm 1.1). q should be set to the expected solution size, multiplied by 1.2 to 2.

As long as j is not zero, we call Ψ a *partial solution*. A set of several partial solutions may form an initial solution set for the local search and evolutionary approach, which are described below.

Algorithm 1.2: Local search

---
1 Generate a set $\Theta = \{\Psi_1, \ldots, \Psi_n\}$ of partial solutions using random solution generation;
2 **foreach** $\Psi \in \Theta$ **do**
3   **while** *one of the following steps is possible* **do**
4   Remove one component of Ψ so that $dist(\Psi)$ does not increase;
5   Add one component $c \in C$ to Ψ so that $dist(\Psi)$ decreases;

6 **return** Θ;

1.4.2 Local Search

In optimisation algorithms, a common approach for locally improving the solution quality is *local search* [2]. If a near local optimum of the solution quality has not yet been reached, the solution will be altered in small steps with the goal that the solution quality will increase. This is done until the quality cannot be improved anymore. Then, a local optimum has been reached.

To use this approach for the component selection problem, we define a distance function $dist(\Psi)$ which indicates the quality of a partial solution. A good partial solution will be assigned a low distance value, whereas a bad partial solution will be assigned a high distance value. The distance function may be as simple as:

$$dist(\Psi) := \sum_i \left[\text{if } f_i \text{ fulfilled: } 0, \text{ else: } diff(f_i) \right],$$

where $diff(f_i)$ is the difference between the accumulated *requires* and *provides* values of the cost factor f_i.

The local search algorithm, adapted to the component selection problem, works as shown in Algorithm 1.2. If the algorithm terminates and there is at least one $\Psi \in \Theta$ with $dist(\Psi) = 0$, a valid problem solution was found using local search.

The main advantage of local search is that it will optimise an existing solution by removing unnecessary components and adding new components if they improve the solution quality directly.

Local search cannot provide global optimisation at all. For example, there might be a partial solution where adding or removing one single component won't increase the solution quality. The local search algorithm would stop then. The solution quality would only improve if multiple operations would be performed. For this reason, the evolutionary approach might be considered.

Algorithm 1.3: Basic evolutionary algorithm

---

1 Generate a set $\Theta = \{\Psi_1, \ldots, \Psi_n\}$ of partial solutions using random solution generation;
2 **for** *generationCount times* **do**
3     $\Theta^* := \Theta$;
4   *crossover*(Θ^*);
5   *mutate*(Θ^*);
6     $\Theta := select(\Theta, \Theta^*, n)$;

7 **return** Θ;

1.4.3 Evolutionary Approach

In contrast to the local search approach, the evolutionary approach does not only optimise locally by adding or removing one single component. Instead, a population (representing a set of potential solutions) is modeled and simulated using evolutionary methods like *crossover*, *mutation*, and *selection* [2]. These methods ensure that the algorithm will not get stuck at one local optimum.

Algorithm 1.3 shows a basic evolutionary algorithm. In the next paragraphs, our adaptations of the functions *crossover*(), *mutate*(), and *select*() will be presented.

crossover(Θ) will randomly take two elements Ψ_1 and $\Psi_2 \in \Theta$ and *cross* them. This is done for several times. For the component selection algorithm, the crossing process is done as follows: First, Ψ_1 and Ψ_2 are randomly selected from Θ. Then, both, Ψ_1 and Ψ_2, are each randomly split into two component subsets so that $\Psi_1 = c_{11} \cup c_{12}$ and $\Psi_2 = c_{21} \cup c_{22}$. Now, two new partial solutions are created: $\Psi_3 := c_{11} \cup c_{22}$ and $\Psi_4 := c_{21} \cup c_{12}$. Then we remove Ψ_1 and Ψ_2 from Θ and add Ψ_3 and Ψ_4 instead. This crossover function should emulate the natural process that an individual, which lives in a certain generation i and which is derived from the generation $i - 1$, has properties of both of its parents.

mutate(Θ) is a function which randomly modifies some elements of Θ. This is done by first selecting a subset of Θ. In each selected element, several components are removed and several components are added from the components database (C) randomly. Mutation is done in order to leave local optima. This mutation process should implement the natural phenomenon that DNA is often not reproduced exactly. Instead, it is randomly mutated at several positions.

select(Θ, Θ^*, n) is a function which returns a set of n elements from $\Theta \cup \Theta^*$. The selection process should work randomly, but *good* elements should be chosen with a higher probability than *bad* elements. To determine the quality of one element, the previously-defined function *dist*(Ψ) is used. This selection function should imitate the evolutionary process that the fittest individuals have the highest chance to survive.

In the last paragraphs, the behaviour of the functions *crossover*(), *mutate*(), and *select*() was presented for the component selection problem. However, important probability constants have not been defined yet. Experiments have shown that for best results the crossover rate should be set to 30 to 35 % and the mutation rate should be around 25 %. When mutating a partial solution, at least 50 % of the solution's components should be replaced by new, randomly chosen components. When using a lower value than 50 %, the probability that local optima are left will decrease rapidly.

The partial solution set size $|\Theta|$ should be, depending on the expected solution size, very large in order to allow for proper algorithm operation. For best results, we set $|\Theta|$ to at least $10 \cdot |\Psi|$.

Both methods, local search and the evolutionary approach, should be used in conjunction and called alternately on a large solution space. *generationCount* should be set to 2 to 3 for best results.

1.5 Integration into a Specification Tool

A basic backtracking algorithm and the heuristics presented in the previous section have been implemented in the tool *SpecScribe*.

SpecScribe [11] is a tool for system specification and requirements management purposes, which has been developed at our chair. SpecScribe allows for *formal* system specification on a high abstraction level. A system may be modeled in different ways, including a hierarchical system description and behaviour modeling. Spec-Scribe offers interfaces to several simulation, verification, synthesis and test tools. If completely covered by the system specification, the design can be exported to several description languages.

In SpecScribe, it is possible to model requirements on the one hand and components and their instantiations on the other hand. These can be linked. In particular, it is possible to formally define an abstract requirement that can be fulfilled by different components, which may also define dependencies in turn. For the integration of the component selection algorithms, the user can add components to the component database and create appropriate *provides* and *requires* properties. The components can be instantiated by the algorithm then. Figure 1.1 shows SpecScribe after the calculation of the above example has finished.

1.6 Evaluation

For evaluation purposes, test sets targeting several problem input domains were created. These are described in the following.

The first and most important problem domain is the problem size, which is directly influenced by the available components C and the top-level requirement set T. Test cases with an expected solution size $|\Psi|$ of 5, 8, 14, 24, 41, 70 and 120 were created.

Next, the density of the inherent dependency graph determines the problem complexity. The density grows as the number of inter-component requirements increases. When creating the dependencies between components, it had to be ensured that the resulting dependency graph is acyclic, as a dependency cycle would lead to an

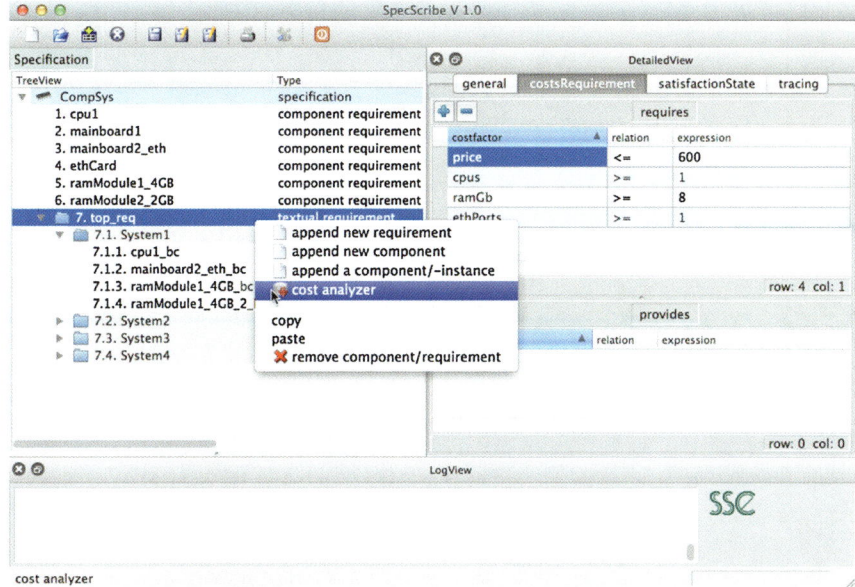

Fig. 1.1 SpecScribe user interface of the component selection algorithm

empty solution. Three graph density domains have been defined: Sparse (at least $\frac{1}{5} \cdot |\Psi|$ unconnected subgraphs exist), normal (the total number of requirements is 1 to 2 times $|\Psi|$), and dense (number of requirements is 3 to 4 times $|\Psi|$).

The last problem domain under consideration is the number of non-functional constraints, which are defined as top requirements, e.g., a limited chip area. The problem complexity increases when adding such constraints. In typical problem sets, the number of non-functional constraints which are part of the optimization goal is low. Therefore, we defined three cases: 1, 2, or 3 constraints.

For all test cases, minimisation of the energy consumption was defined as optimisation target (a random energy consumption value was assigned to each component).

These problem domains were combined, resulting in 63 test cases in total. Every test case was processed by the heuristic algorithms presented in Sect. 1.4.

1.6.1 Results

All test cases and algorithms were implemented in the Python programming language. The heuristic algorithms were able to solve every test case when algorithm parameters have been defined appropriately.

Due to the non-exact approach of the local search and evolutionary algorithms, the algorithm runtime correlates almost linearly to the expected solution size. For finding

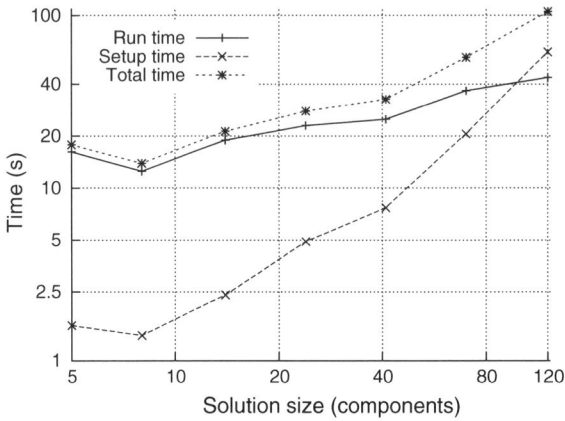

Fig. 1.2 Test case setup time and run time of the presented heuristic component selection algorithms. Both, the algorithm setup time and runtime, increase linearly in relation to the solution size. The algorithm setup time, however, dominates the total runtime for larger problem sets. Measurements were taken on a dual-core 2.53 GHz CPU and 8 GB of RAM

a set of valid solutions, at most four iterations of the local search and evolutionary algorithms were needed.

An important aspect of the overall runtime is the algorithm setup time. When setting $|\Theta|$ to at least 10 times of the expected solution size (as suggested in sect. 1.4), large internal data structures have to be created. For larger problem sets, the setup time dominates the total execution time. This leads to the situation that the total execution time, divided by the solution size, even decreases while the problem size increases. In future algorithm implementations, the long setup time might be avoided. Algorithm runtimes are shown in Fig. 1.2.

Test cases with expected solution sizes greater than 120 have not been considered since total runtime increased too fast.

No significant differences in algorithm runtime could be pointed out regarding the dependency graph densities. This is probably due to the problem creation method. As the test cases were created by targeting a specific expected solution size, the corresponding problems are larger for sparse graphs and smaller for dense graphs, respectively. However, the heuristic algorithm runtime is mainly determined by the solution size $|\Theta|$ and not by the problem size itself.

For the same reason, the number of non-functional top requirement constraints also hardly influences the algorithm runtime.

1.6.2 Using DEAP for Performance Improvements

In the previous paragraphs, we evaluated algorithm performance with a self-written algorithm framework only. To handle problems of larger input sizes, we investigated the usage of existing frameworks that provide functionality of evolutionary algorithms. Since our own framework and test cases have been written in Python, we focused on *Distributed Evolutionary Algorithms in Python* (DEAP) [3], which is a framework that provides highly configurable algorithm components that can be used for most applications of evolutionary algorithms.

We configured DEAP in such a way that it used a similar configuration compared to our own framework. In particular, we used the same population sizes and the same conditions for crossover, mutation and fitness. In contrast to our own model, we could model termination conditions dynamically. Therefore, it was not necessary to use a fixed generation count, leading to improved overall performance.

In addition, evolution strategies were tested. Compared to basic evolutionary optimisation, evolution strategies are more sophisticated methods which do not only focus on the optimisation target, but also on the algorithm parameters like mutation rate. Using these methods, parameters are altered by a given probability distribution during runtime.

Using evolutionary strategies lead to a significant speedup while solving the test cases described in the previous section. The distinction between algorithm setup time and runtime still had to be made since data is initialised outside the DEAP framework. Therefore, only the runtime could be sped up by DEAP.

In Fig. 1.3, performance measurements of the sample test cases are shown. When evolution strategies are used, a speedup of up to 2.43 in runtime could be reached. Achieved speedup increased with rising expected solution sizes, however, no significant speedup was noticable for small test cases. Note that the achieved speedup does not affect setup time, which is the vast amount of total time for large problems.

1.6.3 Application on an FPGA Framer Component

The presented algorithms have also been applied to a framer component of an optical network node by Alcatel-Lucent. The component is an FPGA design, which is synthesised to an Altera Stratix IV device. Its task is basically to interface with several high-speed optical converters bidirectionally and to process data on a low level.

For this device, 1,110 different subcomponents are defined on Register Transfer Level altogether. The subcomponents are tightly linked and form a hierarchical dependency tree. Several components are instantiated many times (e. g., bus systems) and others only once (e. g., the top-level component).

There are several device configuration options for this design, each leading to a specific design structure and subcomponent tree. A typical device configuration leads to 1,376 component instantiations. The global dependency tree, including all configuration options, has been modeled as described in Sect. 1.3. Furthermore, each component was assigned its corresponding chip area on the FPGA device. Due to the data-path-driven design, the chip area is approximately proportional to the energy consumption. For the chip area cost factor, the "add" cost operation was used. How-

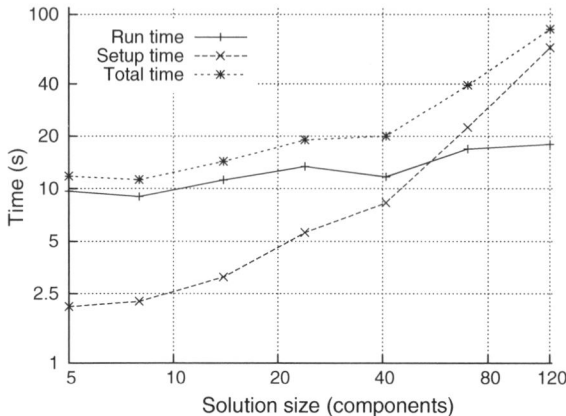

Fig. 1.3 Test case setup time and run time using DEAP

ever, this cost factor only provides a basic estimation, because placement, routing, and optimisation are not considered in this model.

The proposed component selection algorithm can be applied in several ways: If the complete device configuration is given as a set of top level requirements, the solution will contain all necessary component instantiations, and the area cost factor is calculated for this configuration. If the device configuration is not given completely, the component selection algorithm will be able to choose components in such a way that the partial device configuration and all of the hierarchical requirements are met. In this case, several solutions are possible and an optimal configuration regarding chip area can be chosen.

One configuration option targets the so-called timestamp mode of the device. There are two modes: The original timestamp mode, where data frames are processed in parallel, and the Time Division Multiplex (TDM) timestamp mode, where data frames are processed sequentially. The original timestamp mode requires 1,376 component instantiations in total, whereas the TDM timestamp mode requires only 1,330 component instantiations and results in a 2 % lower chip area.

We created a component selection problem for this evaluation example, incorporating an incomplete device configuration regarding the timestamp mode. The problem was extended by the definition of an optimisation target function for the chip area cost factor. The problem size of this example is much larger than the previous examples, which only contained up to 120 components. However, due to the tight component relations, this problem is easier to solve. The proposed component selection algorithm was applied to this problem and was able to find the optimal solution in both cases.

We set $|\Theta|$ to 10 times of the expected solution size (we used a value of 13,000), the generation count was set to two and the other algorithm options were set as proposed above. Like in the previous example, the algorithm setup was most time-consuming. When running this algorithm configuration, the setup needed 550 s. On the contrary, the algorithm execution time was significantly faster and only needed 19 s to complete. When removing either the local search or the evolutionary approach from the global algorithm, the expected solution could not be found in feasible time (we cancelled the algorithm after 1,800 s). Therefore, the local search and evolutionary approach benefit from each other in this example. Nevertheless, the algorithm setup time is still a major concern regarding both our own implementation and the usage of the DEAP framework. This has to be improved in the future.

1.7 Conclusion

We formally introduced the component selection problem class, which a designer can use to generate design decisions in an objective way. The problem class provides the possibility to define top-level requirements, an optimisation target and interdependencies of components. If modeled adequately, the problem solution will return a set of component instantiations which will meet all constraints and therefore can be used to build a target system.

The NP-hardness of the problem class was shown. Therefore, no exact algorithm exists which operates efficiently on all problem cases. We presented heuristic algorithms which guarantee to work efficiently but might only return an approximate solution. The integration of these algorithms into the specification tool SpecScribe was presented.

Finally, we evaluated the heuristic algorithms using test cases of different sizes and structures. It has been shown that a solution could be found for problem solution sizes of up to 120 components, even for hard problems. The algorithm has been applied to an industrial FPGA design and was able to select components in such a way that the chip area was minimised. Algorithm runtime could be improved by using an existing framework for evolutionary algorithms.

For future work, the heuristic algorithms should be investigated in greater depth. Up to now, most algorithm parameters have been defined statically, but the result quality could be improved by deriving optimal parameter values dynamically, since the algorithm performance highly depends on the problem size and problem structure.

Acknowledgments This book chapter is part of the ENERSAVE research project, which is funded by the German ministry of research, BMBF, under the registration number 16BE1100.

References

1. Benhamou F, Goualard F (2000) Universally quantified interval constraints. In: Proceedings of the 6th international conference on principles and practice of constraint programming, Springer, pp 67–82
2. Engelbrecht AP (2007) Computational intelligence: an introduction, 2nd edn. Wiley, New York

3. Fortin FA, De Rainville FM, Gardner MA, Parizeau M, Gagné C (2012) DEAP: Evolutionary algorithms made easy. J Mach Learn Res 13:2171–2175
4. Garey MR, Johnson DS (1979) Computers and intractability: a guide to the theory of NP-completeness. W. H. Freeman, New York
5. Gelle E, Faltings B (2003) Solving mixed and conditional constraint satisfaction problems. Constraints 8(2):107–141. doi:10.1023/A:1022394531132
6. Mackworth AK (1985) Constraint satisfaction. Technical report, University of British Columbia, Vancouver, BC, Canada, http://www.ncstrl.org:8900/ncstrl/servlet/search?formname=detail&id=oai%3Ancstrlh%3Aubc_cs%3Ancstrl.ubc_cs%2F%2FTR-85-15
7. Mittal S, Falkenhainer B (1990) Dynamic constraint satisfaction. In: Proceedings of the 8th national conference on artificial intelligence, pp 25–32
8. Mouhoub M, Sukpan A (2007) Solving conditional and composite constraint satisfaction problems. In: Proceedings of the 2007 ACM symposium on applied computing (SAC), ACM, New York, USA, pp 336–337. doi:10.1145/1244002.1244082
9. Nielsen J (1998) Nielsen's law of internet bandwidth. http://www.useit.com/alertbox/980405.html
10. Pande J, Garcia CJ, Pant D (2013) Optimal component selection for component based software development using pliability metric. SIGSOFT Softw Eng Notes 38(1):1–6. doi:10.1145/2413038.2413044
11. Pross U, Markert E, Langer J, Richter A, Drechsler C, Heinkel U (2008) A platform for requirement based formal specification. In: Proceedings of the forum on specification and design languages (FDL) 2008, ECSI, IEEE, pp 237–238. doi:10.1109/FDL.2008.4641453
12. Proß U, Kröber K, Heinkel U (2010) Abhängigkeitsanalyse und Parameterberechnung auf Spezifikationsebene. In: Dietrich M (ed) Methoden und Beschreibungssprachen zur Modellierung und Verifikation von Schaltungen und Systemen (MBMV), Fraunhofer, pp 197–206

Chapter 2
Assisting Refinement in System-on-Chip Design

**Hocine Mokrani, Rabéa Ameur-Boulifa
and Emmanuelle Encrenaz-Tiphene**

Abstract With the increasing complexity of systems on chip, designers have adopted layered design methodologies, where the description of systems is made by steps. Currently, those methods do not ensure the preservation of properties in the process of system development. In this paper, we present a system on chip design method, based on model transformations—or refinements—in order to guarantee the preservation of functional correctness along the design flow. We also provide experimental results showing the benefits of the approach when property verification is concerned.

Keywords System on a Chip (SoC) · Architecture exploration · Platform-Based Design (PBD) · System modeling · Formal verification · Communication refinement · Property-preservation checking

2.1 Introduction

The System on a Chip (SoC) design faces a trade-off between the manufacturing capabilities and time to market pressures. With the increasing complexity of architectures and the growing number of parameters, the difficulty to explore a huge design space becomes harder to address. An approach to overcome this issue is to use abstract models and to split the design flow into multiple-levels, in order to guide the designer in the design process, from the most abstract model down to a synthesizable model.

H. Mokrani (✉) · R. Ameur-Boulifa
Institut Télécom, Télécom ParisTech, CNRS-LTCI, Sophia-Antipolis, France
e-mail: hocine.mokrani@eurecom.fr

R. Ameur-Boulifa
e-mail: rabea.ameur-boulifa@telecom-paristech.fr

E. Encrenaz-Tiphene
Sorbonne Universités, Université Pierre et Marie Curie Paris 6,
UMR 7606, LIP6, Paris, France

CNRS, UMR 7606, LIP6, Paris, France
e-mail: emmanuelle.encrenaz@lip6.fr

© Springer International Publishing Switzerland 2015   21
M.-M. Louërat and T. Maehne (eds.), *Languages, Design Methods,
and Tools for Electronic System Design*, Lecture Notes in Electrical Engineering 311,
DOI 10.1007/978-3-319-06317-1_2

The use of abstraction levels in the SoC design gives another perspective to cope with design complexity. Indeed, the design starts from a functional description of the system, where only the major function blocks are defined and timing information is not captured yet. During the SoC design process, the system description is refined step by step and details are gradually added. At the end, this process leads to a cycle accurate fully functional system description at Register Transfer Level (RTL).

Furthermore, the verification of complex SoCs requires new methodologies and tools, which include the application of formal analysis technologies throughout the design flow. Indeed, in contrast to simulation technique, formal verification can offer strong guarantees because it explores all possible execution paths of a system (generally in a symbolic way); in the case of model checking, the verification can be automated but has to face the state explosion problem. This approach is applicable for the first steps of the design process or on elementary blocks of the refined components; it can also help in proving the refinement between two successive steps of the design process. This paper proposes a method for assisting the process of refinement along the design flow. The approach is based on a set of transformation rules, representing a concretisation step; the transformation rules are coupled with formal verification techniques to guarantee the preservation of stuttering linear-time properties, hence alleviating the verification process on the last steps of the design and paving the way to a better design space exploration.

This chapter is structured as follows. Section 2.2 summarizes the related techniques in the literature. Section 2.3 describes the major steps of our method for architectural exploration. Section 2.4 details the transformation rules associated with each refinement step. Section 2.5 presents a case-study illustrating the use and benefits one can expect from our approach, concerning behavioral property verification. Section 2.6 concludes and sketches some perspectives.

2.2 Related Works

Nowadays many design methodologies involve formal verification methods to assist the design; generally, verification tools are plugged into the standard (SystemC or SystemVerilog) design flow. These tools are appropriate to perform formal verification at a high level of abstraction, or to derive test-benches generally used for assertions checking on lower design levels. However, there is a lack of design methodologies to assist a designer in the refinement tasks and that offer guarantees about functional properties preservation along the design process. Several frameworks offering design-space exploration facilities have been proposed [3, 9, 15, 18]. However, these frameworks are mostly simulation-oriented and do not formally characterize the relationships between successive abstraction levels. Moreover, formal verification of global functional properties is hard to accomplish when components are described at a low level of abstraction, where many implementation details are provided.

Among design methodologies oriented towards refinement, the B method [2] is one of the most famous, due to its rigorous definition, its (partial) mechanization in Atelier-B, and several success stories for transportation devices. This approach is general and could be applied in the context of SoC design [6]. Although large part of proof obligations can be automatically discharged, the refinement steps are left to the user. Abdi and Gajski [1] defines a model algebra that represents SoC designs at system level. The authors define the refinement as a sequence of model transformations, that allow to syntactically transform a Model Algebra expression into a functionally equivalent expression. The refinement correctness proof is based on the transformation laws of model algebra. Functional equivalence verification is used to compare the values of input and output variables within the models at different levels. Marculescu et al. [11] presents a framework for computation and communication refinement for multiprocessor SoC designs. Stochastic automata networks are used to specify application behavior which allows performance analysis and fast simulations. Our approach is complementary to these last works since we provide transformation rules, representing the introduction of architectural constraints in the design in order to describe more precisely its behavior. Our rules are tuned to be understandable by the designer, who can select which combination of rules to apply in order to perform its refinement; at each step, the refinement can be proven by applying automated verification tools, hence guaranteeing the preservation of a large class of functional properties from abstract levels to more concrete ones.

2.3 Our Method

Our approach for design space exploration of SoCs is based on the Y-chart design scheme [14] as shown in Fig. 2.1. We focus on dataflow applications, modelled as a set of abstract concurrent tasks. Application tasks and architectural elements making up the underlying execution support (e.g., major features of CPU, memory, bus) are first described independently and are related in a subsequent mapping stage in which tasks are bound to architectural elements.

The application is mapped onto the architecture that will carry out its execution: a first platform is available (see Fig. 2.1). The models derived for both applications and architectures may come with some low-level information from designers.

They are analyzed to determine whether the combination of application and architecture satisfies the required design constraints introduced at the initial stage. If the design constraints are not met, then the mapping process is reiterated with a different set of parameters until achieving a satisfactory design quality. Once the desired platform is obtained, it is possible to perform communication refinement for optimizing the communication infrastructure. This makes effective the communication mechanism, and takes into account constraints imposed by the available resources. Referring to Fig. 2.1, this process leads to Platform2. This process is well established in the simulation-based design exploration tools. The boundedness of the execution support and the synchronizations, which it induces, imposes structural constraints.

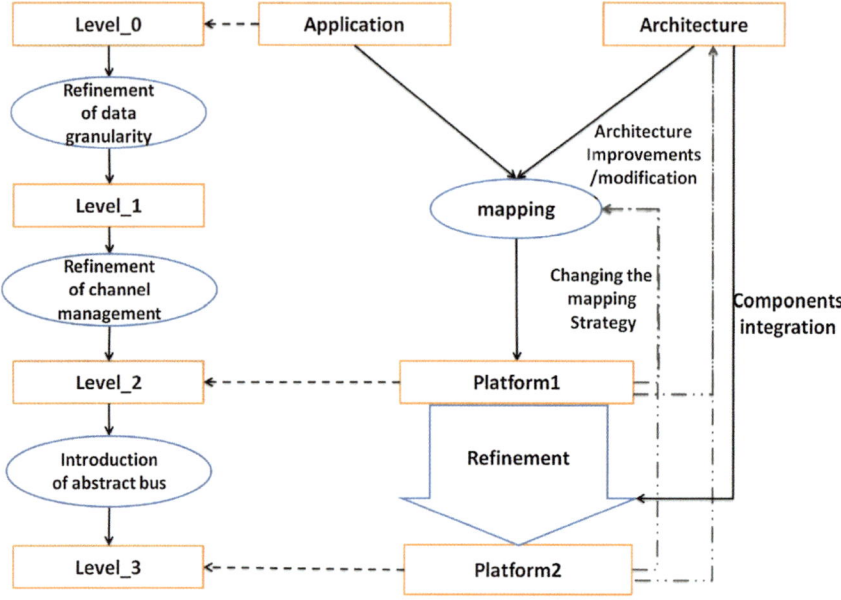

Fig. 2.1 Refinement steps in the design flow

For this reason, the initial set of execution traces of the application is modified along the mapping and refinement process. This means that functional properties that were fulfilled by the initial description of the application may no longer hold once the application has been mapped. For example, deadlocks or livelocks may have appeared, or some good ordering of events may not be respected anymore. These changes are difficult to capture with simulation-only engines, hence formal analysis is required. In order to ensure the preservation of the functional behavior of the application being analysed along the mapping and refinement process, our approach consists of splitting the whole process in defined steps with clearly defined abstraction level (see the left side of Fig. 2.1): Level-0 (application without constraint), Level-1 (application with a defined granularity of the stored data and the transferred data), Level-2 (application with synchronization mechanisms for communication) and Level-3 (application with synchronization mechanisms for communication transiting through a shared bus). Moreover, we provide formal transformation rules as guidelines for the derivation of a concrete model from an abstract one. Then we can prove the preservation of stuttering linear-time functional properties from two successive representation levels by comparing the set of traces of the two descriptions with a formal verification tool.

The remainder of this section gives some precision on the initial application and architecture modeling.

2.3.1 Application

The functional behavior of the application is written in Task Modelling Language (TML) [3]. The model of computation of TML is close to the Kahn networks model [8], however TML supports non-determinism and offers different communication styles. A TML model is a set of asynchronous tasks, representing the computations of the application, and communicating through channels, events or requests. In TML, each task executes forever, i.e., the first instruction is re-executed as soon as the last one finishes.

The main feature of TML models is *data abstraction*. TML models are built to perform design space exploration from a very abstract level; they capture the major features of the application to be mapped, without describing precisely the computation of the application and data value being involved on it. Within tasks, precise computation is abstracted by an action EXEC whose optional parameter represents the amount of time the computation should take. Channels do not carry data values, but only the amount of data exchanged between tasks. Data are expressed in terms of *samples*. A sample has a type which defines its size. Communications are expressed by actions READ or WRITE whose parameters are the channel being accessed and the amount of (typed) samples to be read or written. Other constructs are provided to perform conditional loops, or alternatives (the guarding condition may be non-deterministic, abstracting a particular computation value).

Channels are used for point-to-point unidirectional buffered communication of abstract data, while *events* are used for control purpose and may contain values. *Requests* in their turn can be seen as one-to-many events. A channel may have a maximal capacity or may be unbounded, and is accessed through READ or WRITE actions performed by the emitter and receiver tasks. Channel's type describes its access policy and the type of samples it stores. A channel can be either "Blocking-Read/Blocking-Write" (BR-BW), mimicking a bounded queue (its maximal capacity is defined in its declaration). "Non-Blocking-Read/Non-Blocking-Write" (NBR-NBW) to represent a memory element or "Blocking-Read/Non-Blocking-Write" (BR-NBW) to represent an unbounded queue. A simple example of a TML application is depicted in Fig. 2.2a. It shows two tasks, named TASK1 and TASK2, communicating by FIFO channels, named C1 and C2. TASK1 performs infinite amount of computations and writing actions of a single sample on the channel C1. For each component of the application an abstract model is derived. It captures the component key behavior including both computation and communication aspects. We rely on the *Labelled Transition System* (LTS) formalism [4] for encoding the models.

2.3.2 Execution Platform

The architecture consists of a set of interconnected hardware components, on which the application will be executed. For each *processing element* (e.g., processor or co-processor), the designer provides its number of cores, number of communication

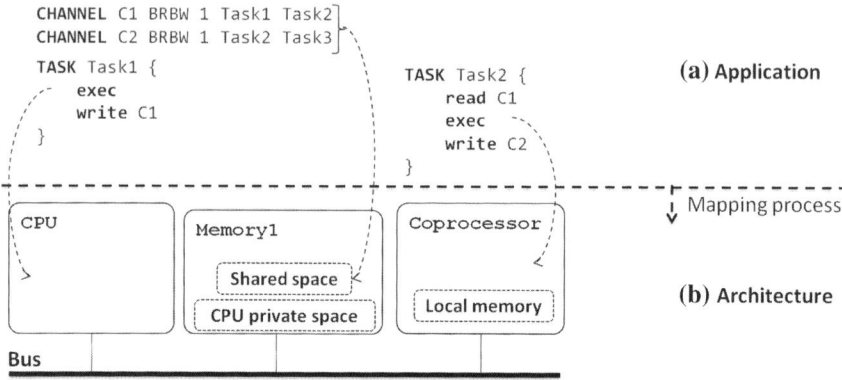

Fig. 2.2 Example of mapping of an application onto an architecture

interfaces, size of local memory. In case of multitask scheduling, the scheduling policy is specified (fixed-priority, random, round-robin). For each storage element (e.g., RAM, ROM, buffer), the size of the storage element and access policy (random access, FIFO) are given. For each *interconnection* or *interface element*, the designer specifies the type of interconnection (e.g., dedicated buffered line, shared bus, full-crossbar, bridge), transfer granularity, arbitration policy. Referring to the architecture in Fig. 2.2b, it consists of a CPU and a dedicated coprocessor, both connected to a bus and a memory.

2.3.3 Mapping and Partitioning

The mapping process distributes application tasks and channels over hardware elements. The mapping determines over which processing elements the tasks will be executed and which memory regions will store data. The *allocation* is static and is described by the designer. The model of the obtained system represents the *combination* of the behavioral models of the application components integrating the constraints imposed by the architecture. Consider the application and the architecture given in Fig. 2.2, TASK1 (resp. TASK2) is mapped see dashed arrows over CPU (resp. CO-PROCESSOR) nodes and the channels C1 and C2 are mapped over a shared memory and communicate through the bus. A more complex example of application, architecture, and mapping is presented in Sect. 2.5, Fig. 2.10.

From a formalism's point of view, this combination can be seen as a product of Labelled Transition Systems (LTSs). However, in order to perform this product, one has to adapt the communication granularity, the interface protocols and to manage the shared resources. This is the purpose of the transformation rules described in the following section.

2.4 Transformation Rules

To assist the designer in developing models from Level-0 to Level-3, we provide guidelines for formally refine the tasks and communication medium from the simple channels to concrete infrastructures. After generating the initial model, the guidelines suggest three steps: 1. Refinement of data granularity, 2. Refinement of channel management, and 3. Introduction of an abstract bus. These transformations manipulate orders and substitutions between elementary actions labelling the LTS of the initial model. In Mokrani [13], these transformations have been formalized with partially ordered multisets referred to in the term *pomset* [16] and then translated into LTS formalism. This section presents an intuitive description of the transformations required by the three steps. A more formal and complete description of these transformations is presented in Mokrani [13].

To begin at Level-0, we build behavioral models of TML applications in terms of a set of interacting LTSs. For each task, we build an LTS, in which the transitions are the atomic actions executed by the TML task. For each channel, we generate an LTS, which encodes its specific behavior and captures the parameters of interest such as maximal capacity and access policy. For instance, the behavior of TASK2 and channels may be modeled by the LTSs presented in Fig. 2.3a, b, respectively.

2.4.1 Refinement Steps

2.4.1.1 First Step: Refining Granularity of Data

The first refinement step considers the capacity (or size) of memory elements allocated to each communication channel during the allocation phase. This capacity may be lower than the size of the TML sample to be transmitted, which imposes a rescaling of the granularity of data transfer and may also impact the granularity of the computation. The granularity of data measures, both, the atomic amount of computations associated to each EXEC statement and the atomic amount of data associated to each READ or WRITE statement (i.e., the amount of data carried away by a channel). The refinement of data granularity converts the unit of data from the coarse-grained unit into the finer-grained one (e.g., from IMAGE into PIXEL) with a given granularity scaling factor of n (so that total size of IMAGE $= n \times$ size of PIXEL). The models of channels are refined by the transformation 1 (denoted by [T1]). The latter associates

Fig. 2.3 Initial LTSs for TASK2 and channels

to each channel a size bounded by the number of samples *of the new granularity*, which it can transfer, and the maximal memory size of the architecture allocated for the channel given by the *MEMSIZE* function.

Transformation 1 *A channel C is transformed into a channel C′ with a granularity scaling factor n such that:*

$$Type(C') = \begin{cases} BR\text{-}BW & \text{if } Type(C) = BR\text{-}BW \lor BR\text{-}NBW \\ NBR\text{-}NBW & \text{otherwise} \end{cases}$$

and

$$size(C') \leq min(MEMSIZE(C), n \times size(C))$$

Models of tasks are also impacted by the rescaling of the data granularity. Each initial action is transformed into an ordered set of micro-actions, according to the granularity of a scaling factor. These ordered sets are gradually built and combined by taking into account the parallelism between actions, data dependency and data persistency. The result of this transformation leads to the transformations [T2]–[T6], which are described in the subsequent paragraphs.

Maximal Parallelism Between Actions

For each action of the models derived at Level-0, the order of the corresponding micro-actions depends on the associated data granularity, as well as on the maximal parallelism that the architecture offers. As channels are point-to-point communication media, the generated order for the micro-actions of communication (READ and WRITE) should be *total*. Whereas for the micro-actions of computation (EXEC), it is defined by the maximal parallelism degree *p* offered by the processing unit executing it (e.g., number of cores within processors). Initially, for each action, an order is built by applying transformation 2 *Expansion of actions* (denoted by [T2]).

Transformation 2 *Consider the model of a task characterised by the set of associated actions S. Given a parameter n (granularity scaling factor), and a parameter p (maximal parallelism degree), the transformation* expansion of actions *consists in replacing each action of S by a (n, p)-ordered group of actions with the same label.*

In the case of TML model $S = \{READ, WRITE, EXEC\}$, it contains primitives of TML model. Once each group of (micro-)actions being locally ordered, the order of the tasks is reinforced by introducing linear order between the terminal elements of each group. This is performed by the transformation 3 *Global order of actions* (denoted by [T3]).

Transformation 3 *Consider the model of a task and the associated set of the groups of ordered micro-actions. This transformation consists of identifying the terminal element of each group of micro-actions, and building an order relating these terminal actions which respects the order of the initial model.*

Data Dependency and Data Persistency

A consequence of data abstraction in TML is the loss of information about the data dependency. This information, which can be expressed as a relation between reading-writing, reading-execution, execution-writing, and execution-execution actions, is required for an optimal management of the memory space of the architecture. In fact, this relation can be restored from the algorithm targeted by the application, or provided by the designer. Thereby, this relation strengthens the order between actions by the transformation 4 *Data dependency introduction* (denoted by [T4]).

Transformation 4 *Consider the model of a task and the relation R between all its actions. Given a data dependency relation D between its actions, the* data dependency introduction *transformation consists in building the order resulting from the transitive closure of R ∪ D, so that it produces an order (i.e., no cycle is introduced).*

Moreover, the models of tasks have to ensure that the data in the local space remains available until its use has been completed. This is taken into account by transformation 5 *Data persistency* (denoted by [T5]). It enforces that the actions consuming data are executed before the ones removing it.

Transformation 5 *Consider the model of a task. Given a data dependency relation and a size of local memory, the* data persistency *transformation consists in strengthening the order of the model of the task such that for any execution, a sequence of actions producing data never exceeds the memory size.*

Shared Resources

Another kind of material constraint taken into account along the refinement process is the number of interfaces and the number of cores which are included in the processing unit onto which the task is executed; this is done by transformation 6 (denoted by [T6]).

Transformation 6 *Consider the model of a task. Given the number of interface and cores of the processing element associated to the task, the transformation forces the order between the actions of the task so as to guarantee a mutual exclusion of shared resources.*

Actually, transformations [T5] and [T6] are performed through a symbolic traversal of the model of a task. They are constrained by the order of micro-actions already computed by way of transformations [T2]–[T4]. At each transformation step compatible with the input order, one has to ensure that both data persistency and resource exclusion conditions are satisfied. If these conditions are not met, the order is strengthened, which forces some sequentialization of concurrent actions, up to the satisfaction of persistence and exclusion conditions.

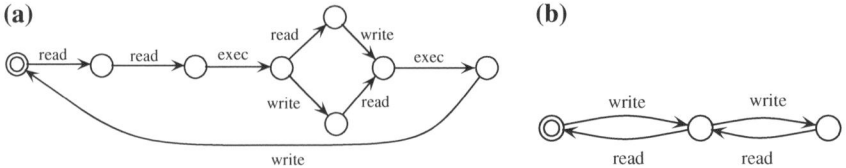

Fig. 2.4 Behavioral models of TASK2 and channels resulting from the first refinement step

The result obtained after applying these transformations is a set of possible execution traces of the considered application, which satisfies imposed design constraints. Referring to the example in Fig. 2.2, we suppose that the written data (resp. read) are refined with a scaling factor equal to 2 (resp. 3). Suppose also that the space allocated to each channel does not exceed two storage compartments. The models of the TASK2 and the channels obtained at the end of this refinement step are shown in Fig. 2.4.

At this stage, the number of atomic execution steps and data transfers are fixed at the granularity offered by the size of memories storing channels, and ordered according the maximal parallelism allowed by the architectural description. However, the communication have to be refined to reflect the access policy of the TML channels. The following step produces a detailed view of these actions.

2.4.1.2 Second Step: Refining Channel Management

In the second step, channels are replaced by communication media equipped with an abstract protocol respecting the blocking-read/blocking-write semantic. The selected protocol is inspired from [10]. The reading and writing primitives are expressed by a series of operations that 1. stall a process until data or memory space (named room) in the shared memory is available, 2. transfer data, and then 3. signal the availability of data or room. This protocol uses six primitives: CHECK-ROOM and SIGNAL-ROOM to test and inform a room is available for writing; STORE-DATA and LOAD-DATA to perform the transfer; and CHECK-DATA and SIGNAL-DATA to check and inform the availability of a data to read. The actual transfer of data are the primitives STORE-DATA and LOAD-DATA, the other operations are synchronization primitives. Transformation 7 (denoted by [T7]) replaces the models of channels by the ones depicting this protocol.

Transformation 7 *The model of channel of type* BR-BW *is replaced by the model encoding the selected protocol of communication.*

Furthermore, the introduction of a communication protocol impacts as well the models of tasks. The communication primitives within each task (READ and WRITE) are changed to support the protocol. The transformation is performed in two phases (transformations [T8] and [T9]): the first replaces all the communication (transfer) actions to a channel of type BR-BW by a communication scheme, which respects the protocol.

Transformation 8 *Consider the model of a task and the selected communication protocol, the actions of transfer to a channel of type* BR-BW *are transformed by the following rules:*

$$
\begin{aligned}
\text{WRITE} &\equiv \text{CHECK-ROOM} \rightarrow \text{STORE-DATA} \rightarrow \text{SIGNAL-DATA} \\
\text{READ} &\equiv \text{CHECK-DATA} \rightarrow \text{LOAD-DATA} \rightarrow \text{SIGNAL-ROOM} \\
\text{EXEC} &\equiv \text{EXEC}
\end{aligned}
$$

In the second phase, the orders between the primitive of transfer and the primitive of synchronization are calculated according to the order established in the abstract model, in a way to preserve a maximal parallelism.

Transformation 9 *Consider the model of a task and the selected communication protocol. The transformation* introduction of the protocol *consists in restoring the orders between actions according to the rules given in* Table 2.1.

The patterns in Table 2.1 are modeled with pomsets. The pomset formalism is a compact representation of concurrent actions without expliciting interleavings. The ordering of the operations on each pattern reflects the happens-before relationship between the actions. For instance, the first pattern specifies that the system tests the availability of data before its loading and then issues the room-release after the load operation; all operations of the first instance of reading precede the corresponding ones of the second instance:

$$
\begin{array}{ccc}
\text{CHECK-DATA} \rightarrow \text{LOAD-DATA} \rightarrow \text{SIGNAL-ROOM} \\
\downarrow \qquad\qquad \downarrow \qquad\qquad \downarrow \\
\text{CHECK-DATA} \rightarrow \text{LOAD-DATA} \rightarrow \text{SIGNAL-ROOM}
\end{array}
$$

This representation leads to the interleaving interpretation, so that a system executing actions concurrently is no different from one that executes them in arbitrary sequential order. With this interleaving interpretation, the system modeled by the pomset above represents five linear traces:

1. CHECK-DATA   LOAD-DATA   SIGNAL-ROOM   CHECK-DATA   LOAD-DATA
   SIGNAL-ROOM
2. CHECK-DATA   LOAD-DATA   CHECK-DATA   SIGNAL-ROOM   LOAD-DATA
   SIGNAL-ROOM
3. CHECK-DATA   LOAD-DATA   CHECK-DATA   LOAD-DATA   SIGNAL-ROOM
   SIGNAL-ROOM
4. CHECK-DATA   CHECK-DATA   LOAD-DATA   SIGNAL-ROOM   LOAD-DATA
   SIGNAL-ROOM
5. CHECK-DATA   CHECK-DATA   LOAD-DATA   LOAD-DATA   SIGNAL-ROOM
   SIGNAL-ROOM

Table 2.1 Patterns of transformation of the actions orders. Each line of the table presents a replacement pattern: the *left pattern* is replaced by the one on the *right*

No	Action pattern	Replacement pattern		
(1)	READ → READ	CHECK-DATA → LOAD-DATA → SIGNAL-ROOM ↓ ↓ ↓ CHECK-DATA → LOAD-DATA → SIGNAL-ROOM		
(2)	WRITE → WRITE	CHECK-ROOM → STORE-DATA → SIGNAL-DATA ↓ ↓ ↓ CHECK-ROOM → STORE-DATA → SIGNAL-DATA		
(3)	WRITE → READ	CHECK-ROOM → STORE-DATA → SIGNAL-DATA ↓ ↓ ↓ CHECK-DATA → LOAD-DATA → SIGNAL-ROOM		
(4)	READ → WRITE	CHECK-DATA → LOAD-DATA → SIGNAL-ROOM ↓ ↓ ↓ CHECK-ROOM → STORE-DATA → SIGNAL-DATA		
(5)	READ → EXEC	CHECK-DATA → LOAD-DATA → SIGNAL-ROOM ↓ EXEC		
(6)	WRITE → EXEC	CHECK-ROOM → STORE-DATA → SIGNAL-DATA ↓ EXEC		
(7)	EXEC → READ	EXEC ↓ CHECK-DATA → LOAD-DATA → SIGNAL-ROOM		
(8)	EXEC → WRITE	EXEC ↓ CHECK-ROOM → STORE-DATA → SIGNAL-DATA		
(9)	EXEC → EXEC	EXEC → EXEC		

Back to our example, the abstract channel is transformed into a shared buffer, which separates data-transfer and synchronisation (see Fig. 2.5).

Because of the interleaving of actions of different operations, the resulting LTS for the TASK2 is large. This makes it highly unreadable. We shall give then its pomset representation (see Fig. 2.6).

2.4.1.3 Third Step: Introduction of Abstract Bus

Once the Level-2 models are available, we introduce information of sharing communication infrastructures. We define an abstract protocol for bus management. The proposed protocol targets a wide family of centralized buses, it contains an arbitration component, interface modules (to depict initiator and target interfaces). It provides a

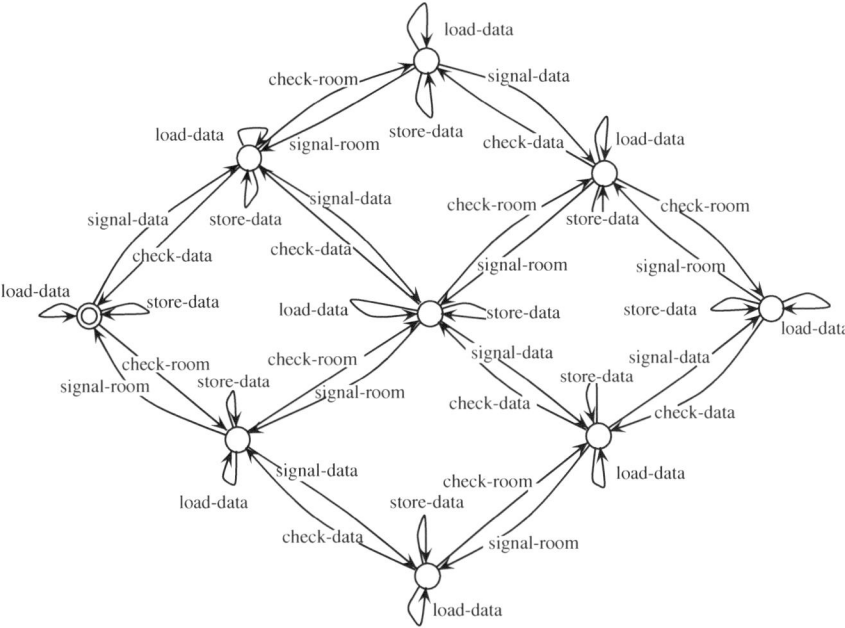

Fig. 2.5 Behavioral model of a two-element channel resulting from the second refinement step

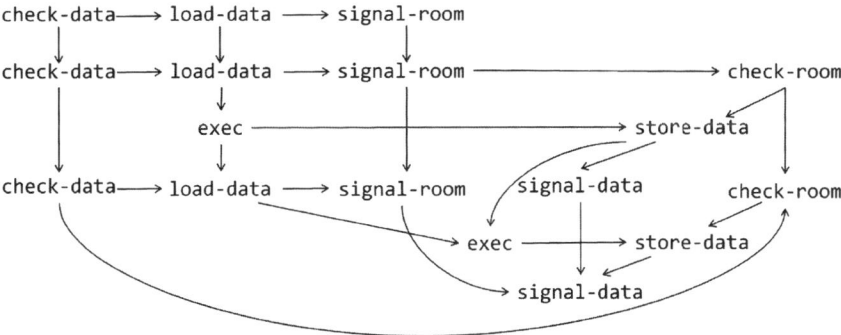

Fig. 2.6 Pomset representation of TASK2 resulting from the second refinement step

transfer policy abstraction which does not distinguish between atomic, burst or split transfers. At this stage, the models of channels remain unchanged, whereas, those of tasks are transformed to incorporate the interface modules. Moreover, a generic model of bus arbiter is introduced. This is performed by transformations [T10] and [T11], which we shall not detail here.

2.4.2 Generation of Models for Level-1, Level-2, and Level-3

As usual in the setting of distributed and concurrent systems, we give behavioral model of the application and the application-architecture combination in terms of a set of interacting finite state machines, called Labelled Transition Systems (LTSs). An LTS is a structure consisting of states with transitions, labelled with actions, between them. The states model the system states; the labelled transitions model the actions that a system can perform.

At each level i, we build an LTS for each component (LTS_t^i and LTS_c^i for resp. task and channel) of the system obtained at this level and by synchronous product (denoted by $||$) of elementary LTSs we build the global model of the overall system (M^i):

$$\forall i \in \{0, 1, 2\}. \ \mathrm{M}^i = ((||_{t \in Task} \ \mathrm{LTS}_t^i) \ || \ (||_{c \in Channel} \ \mathrm{LTS}_c^i))$$

The LTS models of the highest level (Level-0) are generated automatically from the source code TML application. The intermediate models (of Level-1 and Level-2) up to the most concrete one (Level-3) are generated by applying the transformations of the channel models and the task models. The models of Level-1 are built by applying transformation [T1] to each channel and transformations from [T2]–[T6] to each task:

$$\forall c \in Channel : \mathrm{LTS}_c^1 = \mathrm{T1}(\mathrm{LTS}_c^0) \ \text{and} \ \forall t \in Task : \mathrm{LTS}_t^1 = \mathrm{T6}(\mathrm{T5}(\mathrm{T4}(\mathrm{T3}(\mathrm{T2}(\mathrm{LTS}_t^0)))))$$

The models of Level-2 are built by applying transformation [T7] to each channel and transformations from [T6]–[T9] to each task:

$$\forall c \in Channel : \mathrm{LTS}_c^2 = \mathrm{T7}(\mathrm{LTS}_c^1) \ \text{and} \ \forall t \in Task : \mathrm{LTS}_t^2 = \mathrm{T6}(\mathrm{T9}(\mathrm{T8}(\mathrm{LTS}_t^1)))$$

Notice that the transformation [T6] is reused at this level. Indeed, it consists in guaranteeing the exclusive access to resources.

Finally, the global LTS of the Level-3 (M^3) is obtained by the synchronized product of the models of channels and of tasks, plus the models of components introduced in the third step:

$$\mathrm{M}^3 = ((||_{t \in Task} \ \mathrm{LTS}_t^3) \ || \ (||_{c \in Channel} \ \mathrm{LTS}_c^2) \ || \ (||_{i \in Interface} \ \mathrm{LTS}_i) \ || \ (||_{a \in Arbiter} \ \mathrm{LTS}_a))$$

In the same way as previous steps, LTS_t^3 is obtained by applying transformations [T10] and [T11] to the models of tasks.

2.4.3 Proof of Property Preservation

Once the behavioral models have been generated, we prove complete and infinite trace inclusion between lower levels and higher levels by proving the existence of a simulation relation between two successive models (e.g., $\forall i : \mathrm{M}^{i+1} \sqsubseteq \mathrm{M}^i$); this

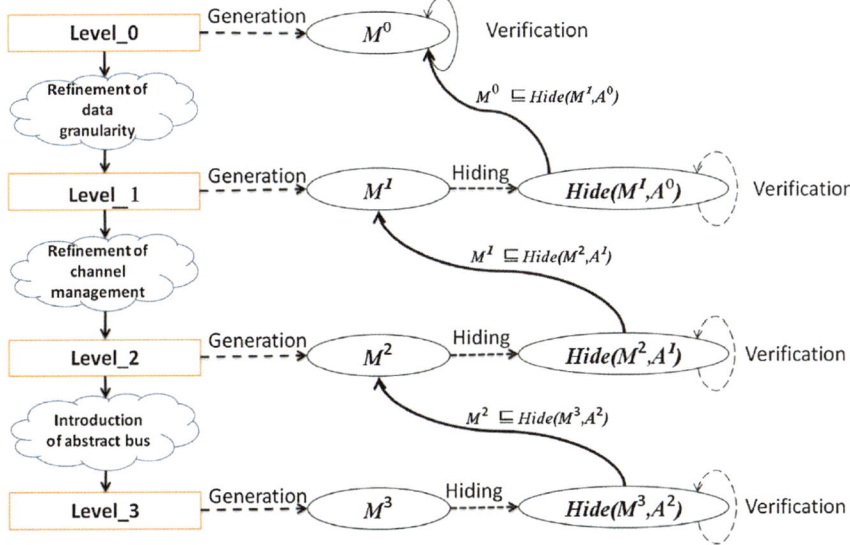

Fig. 2.7 Verification process with the first strategy of hiding operation

result ensures the preservation of stuttering linear-time properties from M^0 down to M^3. Actually, the refinement process introduces new actions, so that the set of actions of abstract level is included in the set of concrete level, $A^i \subseteq A^{i+1}$. Refinement checking between successive levels requires the hiding of the additional details about its behavior. To perform this, we used two strategies:

1. we kept the transitions of M^i (for $i > 0$) labelled over A^{i-1} but the new ones (from $A^i \backslash A^{i-1}$), introduced by the refinement, were considered as *non observable* τ actions. In terms of traces, we can find the trace of the path σ_{i-1} by removing the transitions labelled by new actions from the path σ_i (see Fig. 2.7).
2. We kept the transitions of M^i (for $i > 0$) labelled over A^0. So we can find the trace of the path σ_0 embedded into the trace of the transitions in σ_i (see Fig. 2.8).

The second-solution is more scalable. Indeed, the full system size obtained with the second strategy is much smaller than the one obtained with the first strategy since we hide more actions. However, the first solution appears more interesting. It allows us to prove a large set of properties: properties related to the different levels not only those related to application level. In our methodology, first, we try the experiment with the first strategy. If it fails, we apply the second one.

2.5 Case Study

This section illustrates the use of the proposed methodology for the design and functional verification of a digital camera initially presented in Vahid and Givargis [17]. The functional specification is partitioned into five modules, namely Charge-Coupled

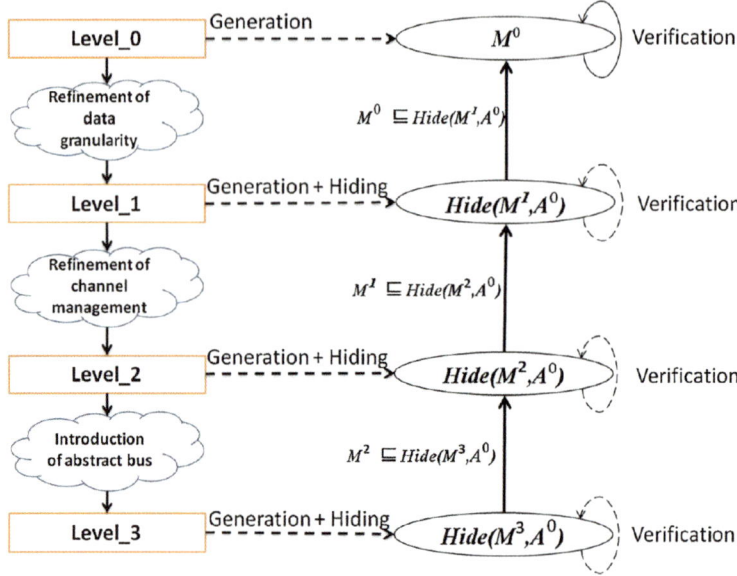

Fig. 2.8 Verification process with the second strategy of hiding operation

Device (CCD), CCD PreProcessing (CCDPP), Discrete Cosine Transformation + quantization (CODEC), transmitter (TRANS), and controller (CNTRL).

The digital camera captures, processes, and stores pictures into an internal memory. This task is initiated when the user presses the shutter button to take a picture. The CCD model simulates the capture of a picture and the transmission of pixels. The CCDPP module performs the luminosity adjustment on each pixel received from CCD module. The CODEC module applies the Discrete Cosine Transformation (DCT) algorithm to each bloc transmitted from CNTRL before being retransmitted into the CNTRL. The CNTRL module serves as the controller of the system. It also executes the quantization and Huffman compression algorithm after receiving the transformed bloc from CODEC. The camera is able to upload the stored picture to external permanent memory. The TRANS module takes care of this task, when it receives data from CNTRL.

Based on the SystemC code of the application given in Vahid and Givargis [17], we encoded it into TML language (the code is shown in Fig. 2.9), we provide a target architecture, which can support the application and a mapping relationship between them (Fig. 2.10). The architecture consists of five Processing Elements (PE1 to PE5) equipped with their own local memory. PE1 and PE2 communicate through a dedicated buffered line SE2; PE2 to PE5 as well as a shared memory SE1 are connected through a bus, which access is controlled by an arbiter. The allocation is represented with dashed lines from the task graph given on the upper part of Fig. 2.10; it associates one TML task per processing element; channel CI1 will be implemented on buffered line SE2 while all other channels are implemented into the shared mem-

```
CHANNEL  CI1  Image1 BRBW  1  CCD   CCDPP
CHANNEL  CI2  Image2 BRBW  1  CCDPP CNTRL
CHANNEL  CI3  Image2 BRBW  1  CNTRL TRANS
CHANNEL  CB1  Bloc   BRBW  1  CNTRL CODEC
CHANNEL  CB2  Bloc   BRBW  1  CODEC CNTRL

TASK CCDPP{       TASK CNTRL{        TASK CODEC
    read CI1          read CI2           {
    exec              Repeat N times     Repeat N times
    write CI2             write CB1          read CB1
    }                    read CB2           exec
TASK CCD {               exec               write CB2
    write CI1        Endrepeat          Endrepeat
    }               write CI3           }
TASK TRANS{         }
    read CI3
    }
```

Fig. 2.9 TML code of digital camera

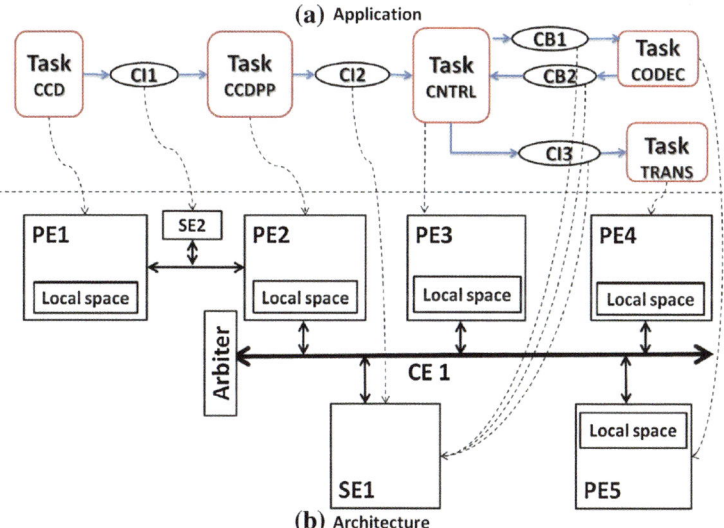

Fig. 2.10 Architecture and mapping of digital camera

ory SE1. We consider that a maximal capacity is associated to each memory space. We built the global models M^0, M^1, M^2, and M^3 of the platform corresponding to the different levels of refinement from Level-0 up-to Level-3 by following the generation scheme described in Sect. 2.4.2. In the current state of our research, we encode manually the models (LTSs) into Fiacre language [5].

We ran the use case with different architectural parameters summarized in Table 2.2: each PE has 1 core, each PE except PE2 has 1 interface, PE2 has 2 interfaces, and each local memory size equals to 2 UNITS. The size of shared memories

Table 2.2 Summary of architectural parameters with different sizes

	TML					PE1	PE2	PE3	PE4	PE5	SE2	SE1
	IMAGE1	IMAGE2	IMAGE3	BLOC1	BLOC2	MEMORY					SPACE	
Case 1	2×3	2×2	2×2	2	2	6	4	2	2	2	$size(CI1)$	$\sum_{c \in C} size(c)$
Case 2	16×18	16×16	16×16	64	64	16×18	18	16×16	64	16×16	$size(CI1)$	$\sum_{c \in C} size(c)$
Case 3	32×34	32×32	32×32	64	64	32×34	34	32×32	64	32×32	$size(CI1)$	$\sum_{c \in C} size(c)$

are defined by the required size for all the channels, which are mapped onto it. So, for SE2, the required size is defined as $\sum_{c \in C} size(c)$ such that $C = \{CB1, CB2, CI3\}$. We experimented the use case with different sets and different sizes of data, which are expressed by the number of UNITS.

2.5.1 Refinement Checking

At each refinement step, we built the corresponding models. By using the equivalence checker BISIMULATOR of the CADP toolbox [7], we compared the models at successive levels of abstraction. The refinement preorder relation, which we used, takes finite stuttering into account. It verifies the non-introduction of new traces. Moreover, for verifying the inclusion of complete and infinite traces, we also verify at each level the non-introduction of new blocking state (deadlock freedom) and the non-introduction of τ-cycles (livelock freedom).

   The full state generation fails with the two last test cases. We chose the second solution for the hiding operation to further reduce and to generate the state space of the system. Then, we verified the trace inclusion between successive models, so $M^0 \sqsubseteq M^1 \sqsubseteq M^2 \sqsubseteq M^3$. Table 2.3 summarizes quantitative results obtained from the experiment with the first and the last test case: they show the system sizes (states/transitions, after minimization) as well as the time consumption for the refinement verification.

2.5.2 Properties Verification

We also verified several properties that express various facets of the system correctness. They are expressed using the Model Checking Language (MCL) logic [12], which is an extension of the alternation-free regular μ-calculus and supported by CADP toolset. The MCL formulae are logical formulae built over regular expressions using boolean operators, modalities operators (necessity operator denoted by "[]" and the possibility operator denoted by "<>") and maximal fixed point operator (denoted by "mu"). Notice that atomic propositions are the actions labels of the Level-0, which should be preserved under the refinement process.

- Deadlock freedom: Absence of states without successors.

  P0: `[true*] <true> true`

- Initially, no reading action on channel **CB1** can be reached before the corresponding writing:

  P1: `[true*.(not write_CB1)*.read_CB1] false`

Table 2.3 Computation times of refinement and verification analysis

	Test case 1				Test case 3			
	Level-0	Level-1	Level-2	Level-3	Level-0	Level-1	Level-2	Level-3
♯ States	336	123088	437782	173546709	2542	141302	516326	2369408
♯ Transitions	872	431622	1646712	472413097	7212	419758	1038037	6593033
Verif time P0	0.8 s	13 s	9 min	26 min	3 s	3 min	11 min	40 min
Verif time P1	0.5 s	8 s	5 min	16 min	2 s	2 min	5 min	25 min
Verif time P2	0.7 s	9 s	5 min	15 min	2 s	2 min	6 min	25 min
Verif time P3	0.9 s	21 s	13 min	41 min	5 s	4 min	16 min	66 min
Verif time P4	0.8 s	14 s	11 min	39 min	5 s	3 min	16 min	64 min
Refine time	n/a	8 s	3 min	13 min	n/a	3 min	25 min	75 min

- No more than two actions of writing on channel **CB1** is possible before the corresponding reading:

  P2:`[true*.write_CB1.(not read_CB1)*.write_CB1.`
  `                    (not read_CB1)*.write_CB1] false`

- A writing action on channel **CI1** will be eventually reached:

  P3:`mu X.[true*] (<true> true and [not write_CI1] X)`

- After a writing action onto channel **CI1**, the corresponding reading is eventually reachable:

  P4:`[true*.write_CI1] mu X.(<true> true`
  `                    and [not (read_CI1)] X)`

The properties are verified at each level hence preserved from M^0 down-to M^3. Table 2.3 shows the time consumption for the properties verification at each level from the experiment (and for each test case). Furthermore, we compare the time required for the verification of these properties at each level of refinement with the time required by the refinement-based strategy. We observe that without using the strategy of checked refinement, the properties have to be verified at each level until Level-3. In the refinement-based strategy, once the validity of the properties has been established on M^0 and the refinement relation satisfied, the properties are guaranteed to be true in subsequent levels.

The Table 2.3 shows the benefits of the refinement-based strategy by comparing *verif time of any property at Level*-0 + *Refine time of any Level-i (for i > 0)* with *verif time of the property at Level-i*. When multiple properties have to be satisfied (which is generally the case), it is worth using our refinement strategies instead of a direct verification of properties on the Level-3 models. On small models (see left part of Table 2.3), the refinement verification is always much smaller than any single property verification time. For more complex models, the refinement time becomes similar to single property verification time, and this approach remains interesting when multiple behavioral properties have to be checked.

2.6 Conclusion

We presented a refinement-based methodology for design-space exploration of system on chip. Our approach provides guidelines to assist the designer in the refinement process, focusing on communication refinement. We established well-identified abstraction levels and transformation rules to derive a more concrete model from a more abstract one. Each abstraction level can be associated with a verification environment, in order to prove functional properties or refinement properties between different abstraction levels. This last point allows us to establish the validity of functional properties of concrete descriptions by testing the property on the most abstract level and proving the refinement, which is less costly than verifying the property on the concrete model directly; we exemplified this fact on a digital camera case study.

These encouraging results draws several perspectives. A first direction consists in proving that the transformation algorithm always produces a refinement, for *any* initial (deadlock-free) model; up to now, the refinement is established when the transformations are applied to a particular initial model. We saw in the experimental section that with more complex systems, this application-dependent refinement verification becomes as costly as the verification of a single property (and remains interesting in case of the verification of multiple properties). A generic proof, based on the formal definitions of our transformations is under study and would simplify the overall verification process. Another perspective concerns the extension of the approach to task computation refinement. For instance, we shall add computation details as computation scheduling.

References

1. Abdi S, Gajski D (2006) Verification of system level model transformations. Int J Parallel Prog 34:29–59. doi:10.1007/s10766-005-0001-y
2. Abrial JR (1996) The B-book: assigning programs to meanings. Cambridge University Press, New York
3. Apvrille L, Muhammad W, Ameur-Boulifa R, Coudert S, Pacalet R (2006) A UML-based environment for system design space exploration. In: Proceedings of the 13th IEEE international conference on electronics, circuits and systems (ICECS) 2006, pp 1272–1275. doi:10.1109/ICECS.2006.379694
4. Arnold A (1994) Finite transition systems—semantics of communicating systems. Prentice Hall, New Jersey
5. Berthomieu B, Bodeveix J, Farail P, Filali M, Garavel H, Gaufillet P, Lang F, Vernadat F (2008) Fiacre: an intermediate language for model verification in the TOPCASED environment. In: Proceedings of the embedded real time software and systems (ERTS2) 2008, Toulouse, France
6. Colley JL (2010) Guarded atomic actions and refinement in a system-on-chip development flow: Bridging the specification gap with Event-B. PhD thesis, University of Southampton
7. Garavel H, Lang F, Mateescu R, Serwe W (2013) CADP 2011: A toolbox for the construction and analysis of distributed processes. Int J Softw Tools Technol Transfer (STTT) 15:89–107. doi:10.1007/s10009-012-0244-z
8. Kahn G (1974) The semantics of a simple language for parallel programming. In: Rosenfeld JL (ed) Information Processing '74: Proceedings of the IFIP Congress, North-Holland

9. Kempf T, Doerper M, Leupers R, Ascheid G, Meyr H, Kogel T, Vanthournout B (2005) A modular simulation framework for spatial and temporal task mapping onto multi-processor SOC platforms. In: Proceedings of DATE'05. Munich, Germany, pp 876–881

10. Lieverse P, van der Wolf P, Deprettere E (2001) A trace transformation technique for communication refinement. In: CODES'01: Proceedings of the ninth international symposium on Hardware/software codesign, ACM

11. Marculescu R, Ümit Y (2006) Computation and communication refinement for multiprocessor SoC design: A system-level perspective. ACM Trans Des Autom Electron Syst 11:564–592

12. Mateescu R, Thivolle D (2008) A model checking language for concurrent value-passing systems. In: Proceedings of the 15th international symposium on formal methods (FM), Springer, Berlin, pp 148–164. doi:10.1007/978-3-540-68237-0_12

13. Mokrani H (2014) Assistance au raffinement dans la conception de systèmes embarqués. PhD thesis, LTCI/Telecom-ParisTech

14. Mokrani H, Ameur-Boulifa R, Coudert S, Encrenaz E (2011) Approche pour l'intégration du raffinement formel dans le processus de conception des SoCs. Journal Européen des Systèmes automatisés, MSR'11 pp 221–236

15. Pimentel A, Erbas C, Polstra S (2006) A systematic approach to exploring embedded system architectures at multiple abstraction levels. IEEE Trans Comput 55(2):99–112

16. Pratt VR (1984) The pomset model of parallel processes: Unifying the temporal and the spatial. In: Proceedings of the seminar on concurrency

17. Vahid F, Givargis T (2002) Embedded system design—a unified hardware/software introduction. Wiley, New York

18. Zivkovoc V, Deprettere E, van der Wolf P, de Kock E (2002) Design space exploration of streaming multiprocessor architectures. In: Proceedings of SIPS'02, San Diego, CA

Part II
Embedded Analog and Mixed-Signal System Verification

Chapter 3
A New Assertion Property Language for Analog/Mixed-Signal Circuits

Andrew N. Fisher, Dhanashree Kulkarni and Chris J. Myers

Abstract In automating the verification of *Analog/Mixed-Signal*(AMS) circuits, it is essential to have a specification language that can describe the behavior that needs to be checked. Although powerful and very expressive, many such languages have a steep learning curve for designers and are complicated to use. This chapter describes a simpler, more intuitive language called the *Language for Analog/Mixed-Signal Properties* (LAMP) that is incorporated into our LEMA verification tool, and demonstrates how this language can be used for AMS verification.

Keywords Formal verification · Model checking · Analog/Mixed-Signal (AMS) circuits · Property language · Assertions · Language for Analog/Mixed-Signal Properties (LAMP) · Temporal logic · Labeled petri nets · LPN Embedded Mixed-Signal Analyzer (LEMA) · Phase interpolator · Voltage-Controlled Oscillator (VCO)

3.1 Introduction

It is well-known that the process of verifying analog circuits is not nearly as automated as its digital cousin. The difficulty is exacerbated when these areas are combined to create AMS circuits. To address this, there has been significant recent interest in developing formal approaches for verifying AMS circuits [16]. In order to apply formal verification approaches, such as *model checking*, it is necessary to create

A.N. Fisher (✉) · C.J. Myers
University of Utah, Salt Lake City, UT 84112, USA
e-mail: andrew.n.fisher@utah.edu

C.J. Myers
e-mail: myers@ece.utah.edu

D. Kulkarni
Intel Corporation, Hillsboro, OR 97124, USA
e-mail: dhanashree.rk@gmail.com

© Springer International Publishing Switzerland 2015
M.-M. Louërat and T. Maehne (eds.), *Languages, Design Methods, and Tools for Electronic System Design*, Lecture Notes in Electrical Engineering 311, DOI 10.1007/978-3-319-06317-1_3

or extend formal languages to be able to describe the time-dependent properties of AMS designs. Several languages have been proposed, which for the most part fall into two categories. They are either inspired by temporal logics like Linear Temporal Logic (LTL) and Computational Tree Logic (CTL) or have a grammar closer to a programming language in the vein of SystemVerilog Assertions (SVA) [5].

A few examples of languages inspired by LTL/CTL are Metric Temporal Logic (MTL), Metric Interval Temporal Logic (MITL), and Signal Temporal Logic (STL). MTL augments LTL with timing [7], but unfortunately, it is not decidable in general [1]. MITL creates a balance between decidability and expressiveness by relaxing the continuous model of time [1]. In Maler and Ničković [10] and Maler et al. [12], the authors study the use of MITL in online monitoring while in Maler and Ničković [11], the authors extend MTL to create STL. These languages have been difficult to convince the analog and AMS community to use in practice since the formalism is so foreign to designers, and it is often difficult to determine which expression is needed to capture a desired property.

In addition to languages being built from LTL or CTL like formalisms, several languages have been proposed taking inspiration from assertion languages. A prominent example is the Property Specification Language (PSL), which can be used for specifying properties both in the digital and AMS domain. For example, Boulé and Zilic [3] uses PSL to express temporal properties of AMS designs and Jones et al. [6] uses PSL to describe the behavior of the DDR2 memory protocol in terms of assertions. Furthermore, Steinhorst and Hedrich [14] extends PSL to the Analog Specification Language (ASL) to better describe state space properties of AMS designs. As an alternative to PSL, Smith [13] uses SystemVerilog to describe the inherent asynchronous behavior in synchronous circuits, and Havlicek and Little [4] introduces Real-Time SystemVerilog Assertions (RT-SVA) as an extension of SVA adding more direct support for continuous assertions. Despite their generality and their aim to provide designers with a more program-like language, it is still difficult to craft a particular property of interest when using these languages.

This chapter introduces LAMP, the *Language for Analog/Mixed-Signal Properties*, to provide AMS designers with a more intuitive and easier to use property language. To demonstrate the utility of LAMP, this chapter describes how it can be used to specify verification properties for a Phase Interpolator (PI) and Voltage-Controlled Oscillator (VCO). In particular, the property for the PI that must be verified is that it changes to the appropriate phase for a given control signal and the property for the VCO is that the appropriate phase occurs after a suitable settling time. LAMP is incorporated into our AMS verification tool LEMA, which uses LPNs as its primary model for verification. Accordingly, LEMA includes a compiler for the language, which converts statements into a property LPN that can be combined with a model LPN for the AMS circuit, and then model checking techniques can be performed to check that the AMS circuit satisfies the property of interest.

This chapter is organized as follows. Section 3.2 provides more detail on LEMA, as well as, gives the necessary background for LPNs. Section 3.3 introduces the PI circuit, which is used as a motivating example. Section 3.4 describes LAMP and gives a sketch as to how the properties in LAMP are compiled by LEMA into LPNs.

Section 3.5 shows our results using LAMP for the verification of a PI and a VCO circuit, and finally, Sect. 3.6 gives our conclusions and ideas for future work.

3.2 Background

This section introduces our verification tool, LEMA, and the formal LPN model used by our tool. Although LEMA uses LPNs, most of the automatically generated LPNs can be thought of as state machines. In addition, the ideas presented in this chapter can just as easily be formulated with automata instead of LPNs, if desired.

3.2.1 LPN Embedded Mixed-Signal Analyzer (LEMA)

LEMA supports a Simulation-Aided Verification (SAV) approach that is depicted in Fig. 3.1. LEMA takes simulation traces generated by transistor-level SPICE simulations as input for a given circuit. These traces can be compiled by our model generator into an abstract SystemVerilog model for simulation-based verification. Alternately, our model generator can generate a formal model of the circuit in the form of an LPN [2, 8]. During the model generation process, LEMA identifies discrete variables by identifying variable traces that switch to different discrete levels within a certain time tolerance. Thus, LEMA can handle simulations with discrete variables as well as simulations where the discrete variables do not instantaneously change. To handle the continuous portion of the model, the generator relies on a set of thresholds that partition the traces into a set of *places*. Thus, it is the number of thresholds that influences the size of the resulting LPN model and not necessarily the size of any input file used in the generation of the model. This LPN can be combined with an LPN generated by the LAMP compiler introduced in this chapter. The AMS circuit can then be verified by passing the combined LPN for the circuit model and the property to one of LEMA's three model checkers: an exact Binary Decision Diagram (BDD) based model checker [15], a bounded Satisfiability Modulo Theory (SMT) based model checker [15], or a conservative *zone*-based model checker [9]. In like fashion, multiple properties can be checked simultaneously by compiling each into a corresponding property LPN and adding them to the model LPN. As before, the resulting combination can be checked by one of the model checkers.

3.2.2 Labeled Petri Net (LPN)

Labeled Petri Nets (LPNs) are the primary formalism used to model circuits when using LEMA for verification. This formalism is isolated from the user by providing the model generation procedure and verification property compilation as described

Fig. 3.1 LEMA
block diagram

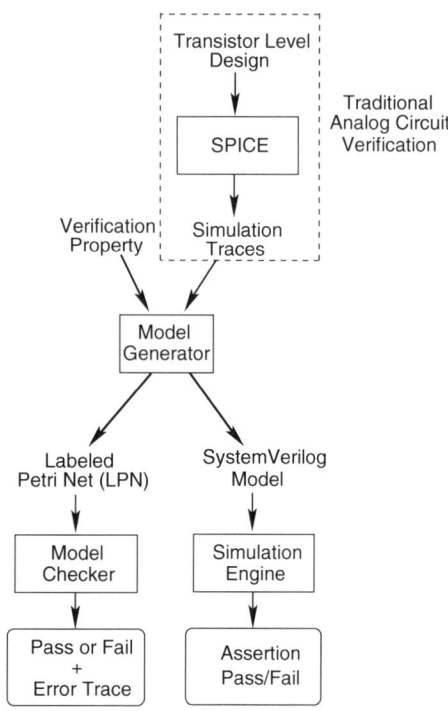

below; however, internally the property language is intimately tied to the LPN formalism. Thus, to provide context for this relationship, this section provides a formal definition. An LPN is a tuple $N = \langle P, T, T_f, V, F, M_0, Q_0, R_0, L \rangle$[1]:

- P: is a finite set of places;
- T: is a finite set of transitions;
- $T_f \subseteq T$: is a finite set of failure transitions;
- V: is a finite set of continuous variables;
- $F \subseteq (P \times T) \cup (T \times P)$ is the flow relation;
- $M_0 \subseteq P$ is the set of initially marked places;
- $Q_0: V \rightarrow (\mathbb{Q} \cup \{-\infty\}) \times (\mathbb{Q} \cup \{\infty\})$ is the initial range of values for each continuous variable;
- $R_0: V \rightarrow (\mathbb{Q} \cup \{-\infty\}) \times (\mathbb{Q} \cup \{\infty\})$ is the initial range of rates of change for each continuous variable;
- L: is a tuple of labels defined below.

An LPN consists of a finite set of *places*, P, and a finite set of *transitions*, T, together with a set of specially designated failure transitions, T_f. Failure transitions are used by LPNs to signal when a failure has occurred and are thus important to LPNs representing properties. The set V represents the set of continuous variables that are

---

[1] A simplified version of LPNs is used in this chapter that is sufficient for AMS circuit models.

used to represent the signal values in the AMS circuit. The *flow relation*, F, describes how places and transitions are connected. The sets M_0, Q_0, and R_0 represent the initial markings of the places, the initial values of the continuous variables, and the initial rates of change for the continuous variables, respectively.

The labels, L, for an LPN are defined by the tuple, $L = \langle En, D, VA, RA \rangle$:

- *En*: $T \rightarrow \mathcal{P}_\phi$ labels each transition $t \in T$ with an enabling condition.
- *D*: $T \rightarrow \mathcal{P}_\chi$ labels each transition $t \in T$ with a delay, for which t has to be enabled before it can fire.
- *VA*: $T \times V \rightarrow \mathcal{P}_\chi$ labels each transition $t \in T$ and continuous variable $v \in V$ with the continuous variable assignment that is made to v when t fires.
- *RA*: $T \times V \rightarrow \mathcal{P}_\chi$ labels each transition $t \in T$ and continuous rate variable $v \in V$ with the rate assignment that is made to v when t fires.

The enabling conditions are Boolean expressions, $\mathcal{P}_\phi$, with the following grammar:

$$\phi ::= \mathbf{true} \mid \neg\phi \mid \phi \wedge \phi \mid v_i \geq c_i$$

where $\neg$ is negation, $\wedge$ is conjunction, v_i is a continuous variable, and c_i is a rational constant. The assignments are numerical formulae, $\mathcal{P}_\chi$, with the following grammar:

$$\chi ::= c_i \mid \infty \mid v_i \mid (\chi) \mid -\chi \mid \chi + \chi \mid \chi * \chi \mid$$
$$\mathrm{INT}(\phi) \mid \mathrm{uniform}(c_i, c_j) \mid \mathrm{rate}(v_i)$$

where the function $\mathrm{INT}(e)$ converts a Boolean expression that evaluates to **true** or **false** to 1 or 0, respectively, the function uniform (l, u) gives a uniform random value in the interval (l, u), and the function rate (v_i) returns the current range of rates for the continuous variable v_i.

Although the above definition only refers to continuous variables, LPNs are able to handle discrete variables as well. Discrete variables can be modeled as simply continuous variables that evolve at a rate of zero.

3.3 Motivating Example

This chapter uses the PI circuit shown in Fig. 3.2 as an example. A PI circuit is a circuit that shifts the phase of an input clock, *phi*, according to the value of a control signal, *ctl*, to produce a shifted output clock, *omega*. Figure 3.3 gives an example of a LPN model generated by LEMA from simulation data for a PI circuit with four different phase shifts. In this diagram, the circles are the places and the intervening text between the places are the transitions. A transition is *enabled* when all the incoming places to the transition are marked, and the enabling condition (the expression in the curly braces) is true. A transition can *fire* after being enabled for the amount of time given by the delay assignment shown in the square brackets.

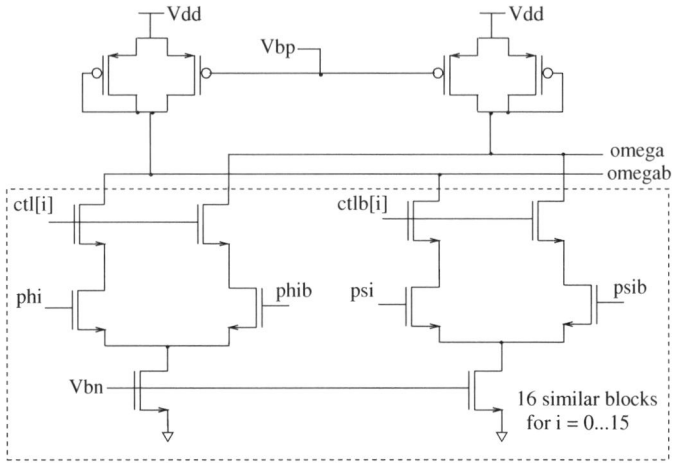

Fig. 3.2 Phase interpolator circuit implementation

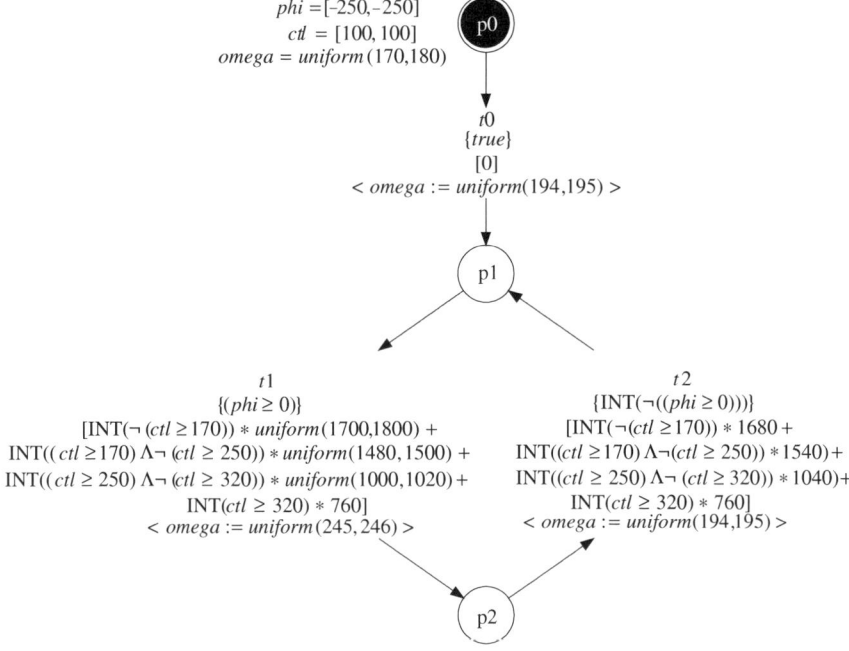

Fig. 3.3 Generated LPN model of a PI circuit

The firing of a transition removes the marking from all the incoming places of the transition, and adds a marking to each outgoing place leaving the transition. The firing of a transition also executes all of the variable and rate assignments found in the angle brackets. For the example LPN shown in Fig. 3.3, transition, *t*0, is enabled, and it fires immediately since its delay assignment is 0. This moves the marking from place *p*0 to *p*1 and changes *omega* to a value between 194 and 195. At this point, the circuit waits until *phi* goes high (i.e., above 0) then checks the value of the *ctl* signal to determine which *uniform* statement is evaluated. It then waits the specified amount of time before firing transition *t*1 to set *omega* to a value between 245 and 246, and moves the marking to place *p*2. Note that in Fig. 3.3 all the values are integers. During the model generation process, all of the continuous variables are scaled (with the same factor) to ensure that all the values are integers. Similarly, the time is scaled by a factor to ensure integer values as well. These two scaling factors are returned to the user to adjust their properties accordingly. For the LPN in Fig. 3.3, the time units are in picoseconds and the values of *phi*, *omega*, and *ctl* are in 10^{-2} V. While this scaling is not strictly necessary, it does simplify the implementation.

The verification property for the PI circuit can also be expressed as an LPN as shown in Fig. 3.4 [2]. This LPN checks that the phase shift of the output clock *omega* generated by the circuit matches the desired phase shift for the given control signal value. The LPN accomplishes this by first waiting for *phi* to go high which marks the places *pCheckMin* and *pCheckMax*. At this point, one of the *tMin* and one of the *tMax* transitions become enabled depending on the value of *ctl*. If the output clock, *omega*, goes high before the delay on the appropriate *tMin* transition passes, then the fail transition, *tFailMin*, fires indicating that the phase shift is too small. On the other hand, if the appropriate *tMin* fires first, then *pCheck* becomes marked, and the LPN is now waiting for *omega* to go high. If the delay on the appropriate *tMax* transition passes first, then the *tMax* fail transition fires indicating that the phase shift is too large. However, if *omega* goes high first, then *pReset* becomes marked, and the LPN waits for *phi* to go low and high again before checking the next phase shift. When this LPN is combined with the LPN for the circuit and an LPN describing the environment's behavior, reachability analysis can be performed to determine if a failure transition can fire [9].

As opposed to the model and environment LPNs, the property LPN must be constructed by hand, which is a very tedious and error-prone process. It is certainly

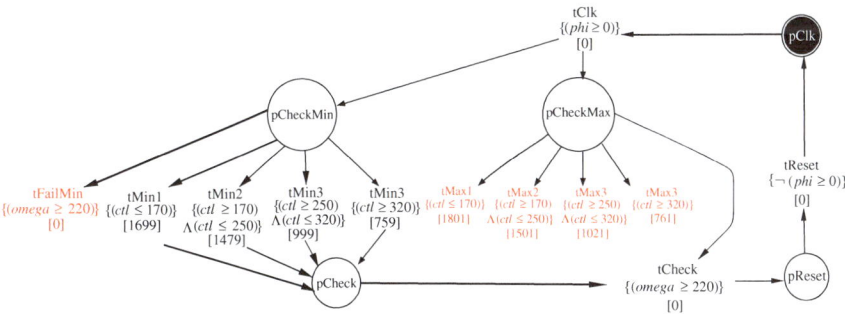

Fig. 3.4 An LPN for the phase interpolator verification property

not very reasonable to require designers to formulate their properties in this way. Therefore, a more intuitive property language is needed that can be readily compiled into property LPNs for verification purposes.

Before creating LAMP, we also considered using an existing AMS property language. For example, this same property can be written in STL as shown below:

$$\Box(((\uparrow (phi \geq 0) \wedge (ctl \leq 1.7)) \to$$
$$(\Box_{[0,1699]}omega < 2.2) \wedge (\Diamond_{[1699,1801]}omega > 2.2))$$
$$\vee((\uparrow (phi \geq 0) \wedge (ctl \geq 1.7) \wedge (ctl \leq 2.5)) \to$$
$$(\Box_{[0,1479]}omega < 2.2) \wedge (\Diamond_{[1479,1501]}omega > 2.2))$$
$$\vee((\uparrow (phi \geq 0) \wedge (ctl \geq 2.5) \wedge (ctl \leq 3.2)) \to$$
$$(\Box_{[0,999]}omega < 2.2) \wedge (\Diamond_{[999,1021]}omega > 2.2))$$
$$\vee((\uparrow (phi \geq 0) \wedge (ctl \geq 3.2)) \to$$
$$(\Box_{[0,759]}omega < 2.2) \wedge (\Diamond_{[759,761]}omega > 2.2)))$$

The $\Box$ notation requires that the property is always checked. The statement $\uparrow$ waits for the positive edge of the Boolean expression ($phi \geq 0$). Thus, collectively the first part of the statement checks that phi goes high and ctl is below 1.7. If this condition is satisfied, then the statement ($\Box_{[0,1699]}omega < 2.2) \wedge (\Diamond_{[1699,1801]}omega > 2.2)$) is checked. The interval subscript on $\Box$ indicates that the statement $omega < 2.2$ must remain true for 1699 time units. The next part of the statement ($\Diamond_{[1699,1801]}omega > 2.2$) requires that the statement $omega > 2.2$ become true between 1699 and 1801 time units. The rest of the statements are similar.

Writing a specification in a temporal logic is quite far away from the environment designers commonly use. Thus, it can be difficult for them. To address this issue somewhat, this property can be written using Signal Temporal Logic Property Specification Language (STL/PSL) as follows:

```
vprop PhaseInterpolator{
PI1 assert:
  always((rise(a:phi >= 0) and (a:ctl <= 1.7))
  → (always[0,1699](a:omega < 2.2)) and
  (eventually[1699, 1801](a:omega > 2.2)) or
  (rise(a:phi >= 0) and (a:ctl >= 1.7) and (a:ctl <= 2.5))
  → (always[0, 1479](a:omega <2.2)) and
  (eventually[1479, 1501](a:omega > 2.2)) or
  (rise(a:phi >= 0) and (a:ctl >= 2.5) and (a:ctl <= 3.2))
  → (eventually[0, 999](a:omega < 2.2)) and
  (eventually[999, 1021](a:omega > 2.2)) or
  (rise(a:phi >= 0) and (a:ctl >= 3.2))
  → (eventually[0, 759](a:omega < 2.2)) and
  (eventually[759, 761](a:omega > 2.2)));
}
```

This version is less intimidating, but it still requires designers to learn some temporal logic semantics in order to correctly use the always and eventually statements. Furthermore, it is difficult to determine how to convert this type of language into an LPN with failure transitions, which is needed for LEMA.

An alternative would be to write this property using RT-SVA [4] as shown below:

$$(phi \geq 0)[\sim> 1] \#\#0$$
$$(((ctl \leq 1.7)[\sim> 1] \#\#0$$
$$(!(omega > 2.2))[*1699 : 1801] \#\#1 \ (omega > 2.2))$$
$$\text{or}$$
$$(((ctl \geq 1.7)\&\&(ctl \leq 2.5))[\sim> 1] \#\#0$$
$$(!(omega > 2.2))[*1479 : 1501] \#\#1 \ (omega > 2.2])$$
$$\text{or}$$
$$(((ctl \geq 2.5)\&\&(ctl \leq 3.2))[\sim> 1] \#\#0$$
$$(!(omega > 2.2))[*999 : 1021] \#\#1 \ (omega > 2.2))$$
$$\text{or}$$
$$(((ctl \geq 3.2))[\sim> 1] \#\#0$$
$$(!(omega > 2.2))[*759 : 761] \#\#1 \ (omega > 2.2))) \#\#1$$
$$(phi < 0)[\sim> 1]$$

In RT-SVA, the expression $[\sim> 1]$ is an expression that waits for the preceding Boolean expression to become true. Thus, the first line waits for the input clock phi to become non-negative. The expression $A \#\#0 \ B$ is a concatenation operator that indicates the next expression B should become true at some time overlapping when A is true. Therefore, the next part checks the value of ctl when phi goes high. Finally, the $A[*l : u]$ statement specifies that A must remain true for between l time units and u time units. Therefore, the last part of the statement checks for a change in $omega$ from a low to a high value within a specified amount of time. Considering this example, it again appears to be tricky and somewhat tedious to read and write verification properties in RT-SVA. We also found it difficult to translate these properties into LPNs with failure transitions.

3.4 LAMP

To make properties easy to read and write, this chapter presents a more intuitive property language for AMS circuits called LAMP. LAMP is based on a set of functions that can be automatically compiled into a property LPN. Listing 3.1 is the format for a property in LAMP. The property starts with its name which is followed by the continuous variable declarations. The statements of the language are as follows:

1. **delay**(d)—wait for d time units. This statement is compiled into the LPN shown in Fig. 3.5a. The enabling condition is true, so the transition fires as soon as the delay d is met.

Listing 3.1 Format for a LAMP property

```
1   property <name> {
2     <declarations>
3     always {
4        <statements>
5     }
6   }
```

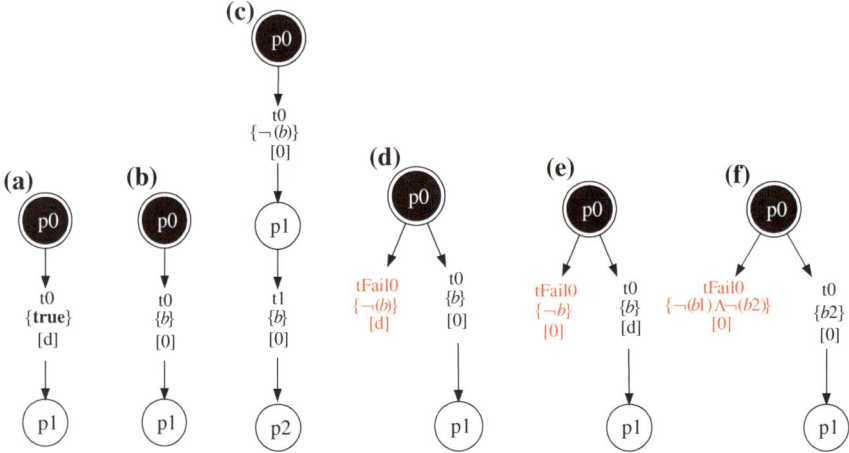

Fig. 3.5 LPNs for the basic LAMP statements: **a delay** (d), **b wait** (b), **c waitPosedge** (b), **d wait** (b, d), **e assert** (b, d), and **f assertUntil** (b1, b2)

2. **wait** (b) —wait until Boolean expression b becomes true. This statement is compiled into the LPN shown in Fig. 3.5b. In this LPN, transition *t0* fires when b becomes true. There is no time limit, which means that the firing of *t0* can wait as long as necessary for b to become true.

3. **waitPosedge** (b) —wait for a positive edge on the expression b (i.e., **wait** (b); **wait** (b)). The LPN for this statement is shown in Fig. 3.5c.

4. **wait** (b, d) —wait at most d time units for the expression b to become true. This statement is compiled into the LPN shown in Fig. 3.5d. If b is false initially, the failure transition, *tFail0*, is enabled, but it has a delay of d time units. If during this time interval b goes true, *t0* is fired immediately, since it has 0 delay. If, however, b remains false for d time units, *tFail0* fires and a failure is recorded.

5. **assert** (b, d) —ensures that b remains true continuously for d time units. This statement is compiled into the LPN shown in Fig. 3.5e. If b is true initially, the transition, *t0*, is enabled, but it has a delay of d time units. If during this time interval b goes false, the failure transition, *tFail0*, fires immediately indicating a failure. If, however, b remains true for d time units, *t0* fires.

6. **assertUntil** (b1, b2) —ensures that expression b1 remains true until b2 becomes true. This statement is compiled into the LPN shown in Fig. 3.5f.

(a)   **(b)**

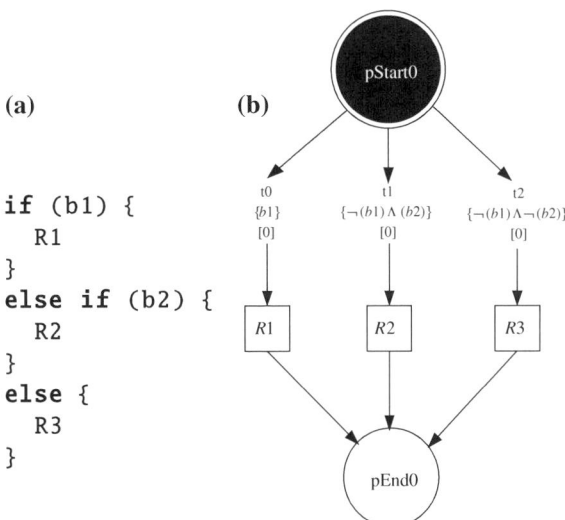

```
if (b1) {
    R1
}
else if (b2) {
    R2
}
else {
    R3
}
```

Fig. 3.6 LAMP syntax and LPN for an *if-else* statement

In this LPN, the failure transition, *tFail0*, fires if b1 and b2 are false before b2 becomes true.

7. The language also provides an **if-else** statement as shown in Fig. 3.6.

8. **always**(conditionsList) {statements} – continue to execute the statements until one of the signals in the list of variables conditionsList changes, then break out. The generated LPN is shown in Fig. 3.7, assuming a list containing at least the variables a and b. First, transition $t0$ fires and stores the current values of the variables in the list conditionsList in a set of new variables $\_a$, $\_b$, ... Then, the statements inside the always block continue to execute as long as the condition $alcond = (a = \_a) \wedge (b = \_b) \wedge \dots$ remains true. If *alcond* becomes false, an exiting transition fires leaving the loop. In particular, every transition in the always block has *alcond* added to the enabling condition while every place has an exit transition with $\neg alcond$. If the conditionsList is empty, then *alcond* is taken to be true and all the exiting transitions are removed.

Note that the formal semantics of each of these statements is defined by the corresponding LPN given in Figs. 3.5, 3.6 and 3.7.

A property compiler for LAMP is incorporated into the LEMA verification tool. This compiler generates a property LPN from a property written in this language as follows:

1. Create an LPN with the name of the property.
2. For each variable declaration, create a continuous variable in the LPN.

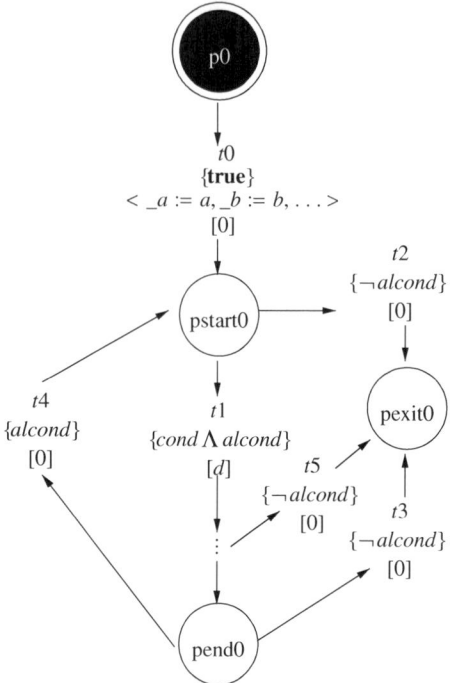

Fig. 3.7 LPN for an always statement with sensitivity list $\{a, b, \ldots\}$. The LPN associated with the statements in the always block go between the *pstart0* and *pend0* places. Each transition belonging to the statements has the condition *alcond* combined with the original condition. The expression *alcond* is $(a = \_a) \wedge (b = \_b) \cdots$

3. For each statement, construct an LPN using the templates described above making the last place for each statement the same as the first place for the following statement.
4. When an always block is encountered, create a new variable $\_a$ for each variable a in the condition list. Add the transition that stores the variables and add the starting place for the always block. Construct all the interior statements according to step 3 while adding the *alcond* to each transition constructed. A transition is added from the last place to the starting place with the *alcond* as its enabling condition. Finally, add the exit place and the exit transitions with enabling condition ¬*alcond*.

Using LAMP, the property for our PI circuit can be expressed as shown in Listing 3.2, which is clearly more intuitive then the approaches described earlier. Note that the values have been scaled according to the factor provided by the model generator to give integer values. After this property is compiled by LEMA, the property LPN shown in Fig. 3.8 is obtained.

Listing 3.2 PI circuit property using LAMP

```
1   property PhaseInterpolator {
2     real ctl;
3     real omega;
4     real phi;
5     always {
6       wait(phi >= 0);
7       if (~(ctl >= 170)) {
8         assert(omega < 220, 1699);
9         wait(omega >= 220, 102);
10      }
11      else if ((ctl >= 170) & ~(ctl >= 250)) {
12        assert(omega < 220, 1479);
13        wait(omega >= 220, 102);
14      }
15      else if ((ctl >= 250) & ~(ctl>320)) {
16        assert(omega < 220, 999);
17        wait(omega >= 220, 102);
18      }
19      else if (ctl >= 320) {
20        assert(omega < 220, 759);
21        wait((omega >= 220, 102);
22      }
23      wait(~(phi >= 0));
24    }
25  }
```

3.5 Results

We designed and simulated a circuit implementation for PI circuits with 4, 8, and 16 different phase shifts. Using simulation data and the model generator in LEMA, we created models for these three different PI circuits. Figure 3.3 shows the model generated for the circuit with four possible amounts of phase shift. Using LAMP, we constructed verification properties, such as the one shown in Listing 3.2, for each of these PI circuits to check that the phase shift generated by the circuit is correct. The LPN generated for this property is combined with the LPN for the model and one for the environment. Then, LEMA's zone-based model checker is used to verify whether the PI circuit satisfies the property (i.e., no fail transitions fired). The verification results of PI circuits for 4, 8, and 16 phases are given as the first three entries of Table 3.1. As can be seen from this table, LEMA is able to successfully verify that the output phase shift is correct.

We also considered a couple of variations that cause verification errors. First, to simulate an output clock that goes high too soon, we changed the property for the PI with four phases, so that the assert statement for the control value 40 asserts that *omega* is less than 220 for 680 time units instead of 759. As seen in the fourth entry of Table 3.1, the property correctly signals a failure. Each of these checks is done

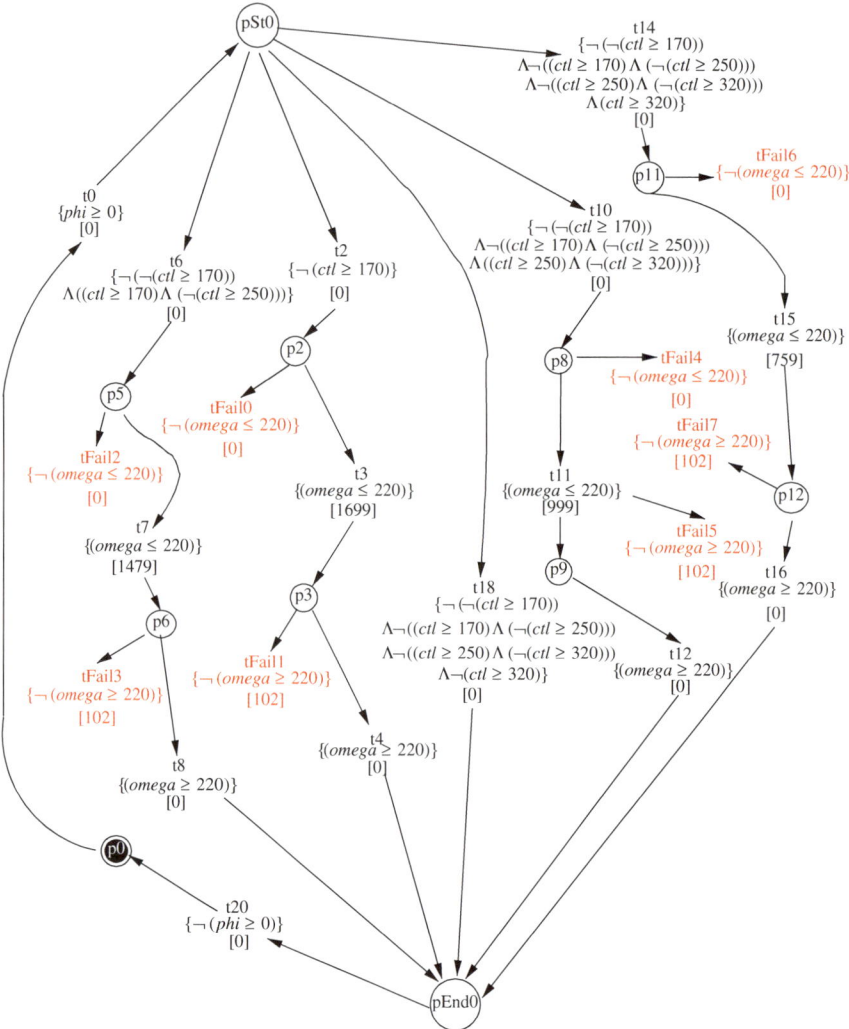

Fig. 3.8 The LPN generated for the property expressed in LAMP in Listing 3.2

with an environment that can non-deterministically change the control signal shortly before the next time the input clock goes high. If this restriction is removed and the control signal is allowed to change at any time, LEMA finds a failure as indicated by the fifth result of Table 3.1. This failure occurs because after the property LPN begins checking the output clock phase for one control signal, the environment can change the control signal to a different value, resulting in a different phase. The property then continues to check for the behavior for the previous control value, which indicates a failure. This failure can be fixed by adding a second always block around the whole

Table 3.1 Verification results for a PI circuit

Property	Time (s)	States	Verifies?
PI with 4 control signals	0.135	126	Yes
PI with 8 control signals	0.277	300	Yes
PI with 16 control signals	1.362	769	Yes
PI with short delay	0.083	14	No
PI with changing controls	0.779	2,407	No

These results are generated using LEMA, a Java-based verification tool, on a 64-bit machine running an Intel Core i5 CPU M 480 @ 2.67 GHz with 4 processors and 4 GiB of memory

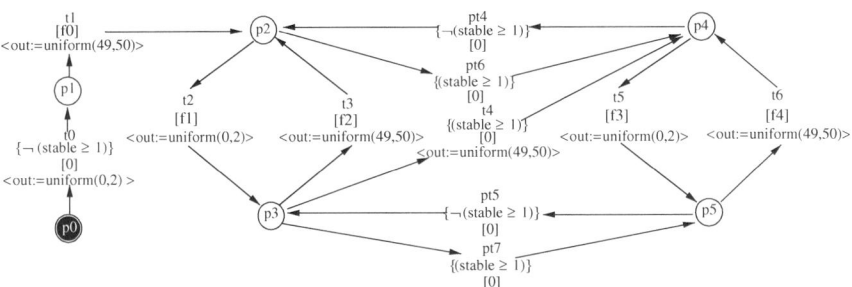

Fig. 3.9 Model of a VCO with a stable and unstable phase

property and adding the control signal *ctl* to the condition list. Dealing with such transient behavior is further illustrated in the next example, a VCO circuit.

A VCO is a circuit that outputs a clock signal, *out*, which frequency changes according to the voltage level of a control signal, *ctl*. A model for a VCO is shown in Figs. 3.9 and 3.10. This model is generated using LEMA's model generator on simulation data for three control voltages 2, 3, and 4 together with interpolation between these values as described in Sect. 4 of Kulkarni [8]. The model consists of two phases: an *unstable* phase signified by *stable* = 0 and a *stable* phase signified by *stable* = 1. The unstable state is modeled by the *p2*, *t2*, *p3*, and *t3* loop while the stable state is modeled by the *p4*, *t5*, *p5*, and *t6* loop. When the control signal changes, it takes the system some amount of time before the system settles into the expected phase. This transient behavior is modeled by setting the *stable* signal to 0 when the control changes and then the *stable* signal is set to 1 after some delay. The setting of the *stable* signal is handled by the model in Figure 3.10. The delays in Fig. 3.9 are:

f0 $((ctl \geq 2) \wedge \neg(ctl \geq 3)) * 19 + ((ctl \geq 3) \wedge \neg(ctl \geq 4)) * 19 + (ctl \geq 4) * 19$
f1 $(\neg(stable \geq 1) \wedge (ctl \geq 2) \wedge \neg(ctl \geq 3)) * uniform((ctl * (-2) + 17), (ctl * (-2) + 113)) + (\neg(stable \geq 1) \wedge (ctl \geq 3) \wedge \neg(ctl \geq 4)) * uniform((ctl * (0) + 11), (ctl * (0) + 107)) + (\neg(stable \geq 1) \wedge (ctl \geq 4)) * uniform((ctl * (0) + 11), (ctl * (0) + 107))$

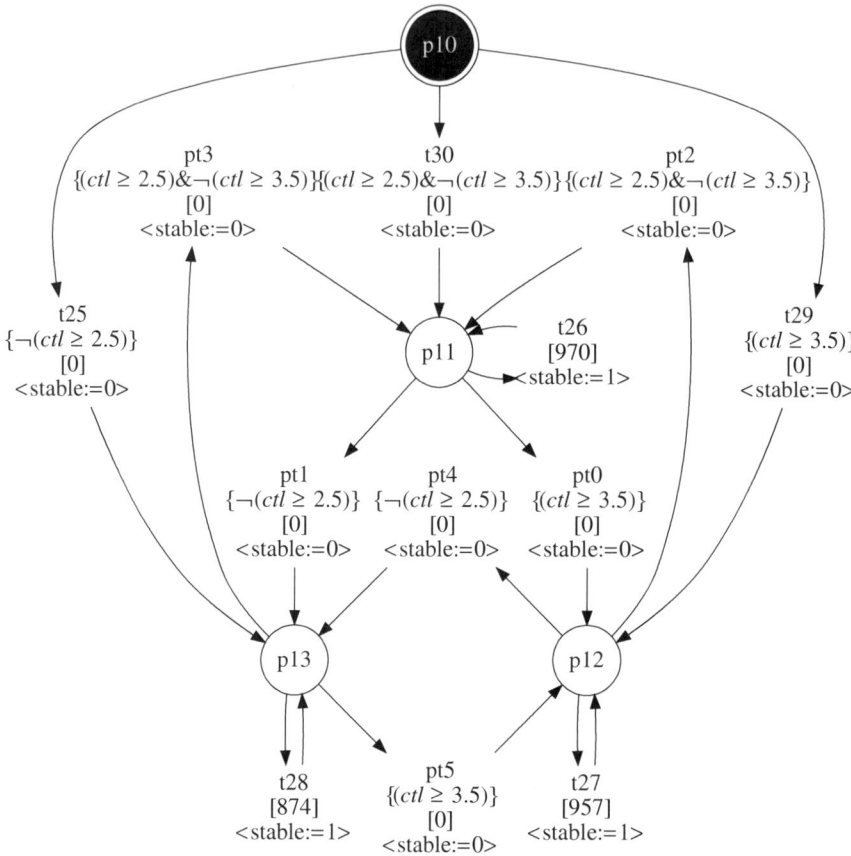

Fig. 3.10 Model for the process that changes the stable and unstable phases for Fig. 3.9

f2 $(\neg(stable \geq 1) \wedge (ctl \geq 2) \wedge \neg(ctl \geq 3)) * uniform((ctl * (-2) + 21), (ctl * (-3) + 26)) + (\neg(stable \geq 1) \wedge (ctl \geq 3) \wedge \neg(ctl \geq 4)) * uniform((ctl * (-2) + 21), (ctl * (-2) + 23)) + (\neg(stable \geq 1) \wedge (ctl \geq 4)) * uniform((ctl * (-2) + 21), (ctl * (-2) + 23))$

f3 $(\neg(stable \geq 1) \wedge (ctl \geq 2) \wedge \neg(ctl \geq 3)) * (ctl * (-2) + 19) + (\neg(stable \geq 1) \wedge (ctl \geq 3) \wedge \neg(ctl \geq 4)) * (ctl * (-1) + 16) + (\neg(stable \geq 1) \wedge (ctl \geq 4)) * (ctl * (-1) + 16)$

f4 $(\neg(stable \geq 1) \wedge (ctl \geq 2) \wedge \neg(ctl \geq 3)) * (ctl * (-5) + 3) + (\neg(stable \geq 1) \wedge (ctl \geq 3) \wedge \neg(ctl \geq 4)) * (ctl * (-2) + 21) + (\neg(stable \geq 1) \wedge (ctl \geq 4)) * (ctl * (-2) + 21)$

To verify that the VCO has the correct delay after a suitable time has passed for the unstable period, one could use the property in Listing 3.3 and compile it into the LPN shown in Fig. 3.11. This property declares a control system, *ctl*, and the clock signal, *out*. The first **always** block indicates that the following property

Listing 3.3 VCO circuit property using LAMP

```
1   property VCO {
2      real ctl;
3      real out;
4      always {
5         delay(1000);
6         waitPosedge(out >= 40);
7         always (ctl) {
8            assert(out >= 40, f3);
9            wait(out <= 30, 3);
10           assert(out <= 30, f4);
11           wait(out >= 40, 5);
12        }
13     }
14  }
```

should be repeatedly checked. Next, there is a **delay** to wait for the clock to stabilize followed by a **waitPosedge** to ensure that the frequency check starts when the clock, *out*, first goes high. The second **always** block includes *ctl* in its condition list, thus the next statements are repeatedly checked unless *ctl* changes. On the event of *ctl* changing, the second always block exits and the outer **always** block starts again. The statements inside the inner **always** check that the output clock remains high (indicated by *out* being at least 40 units) for the appropriate delay it f3. Then, the clock must go low (*out* less than 30) within 3 time units. The clock must remain low for a delay of *f4* and finally go high again with 5 time units.

The results of applying the property in Fig. 3.11 to the VCO model in Figs. 3.9 and 3.10 are contained in Table 3.2 with the label *Limited Phase Checker*. The first three lines show the results when the control is set to a single control value of 2, 3, or 4 and in each case the system is verified. Next, the environment is modified to non-deterministically change to one of the three values 2, 3, or 4 every 3000 time units and again the system is verified. Finally, the environment is modified to allow the voltage level to change to 2, 3, or 4 at any time. In this case, the property fails. The reason is due to the placement of the **delay** statement outside the second **always** block. If the control changes just prior to entering the second **always** block, then the model is in the unstable state while the property LPN checks for the stable frequency. Once the **delay** and **waitPosedge** statements are placed inside the second **always** block, the property is verified in all cases of the environment as shown in the last five lines of Table 3.2.

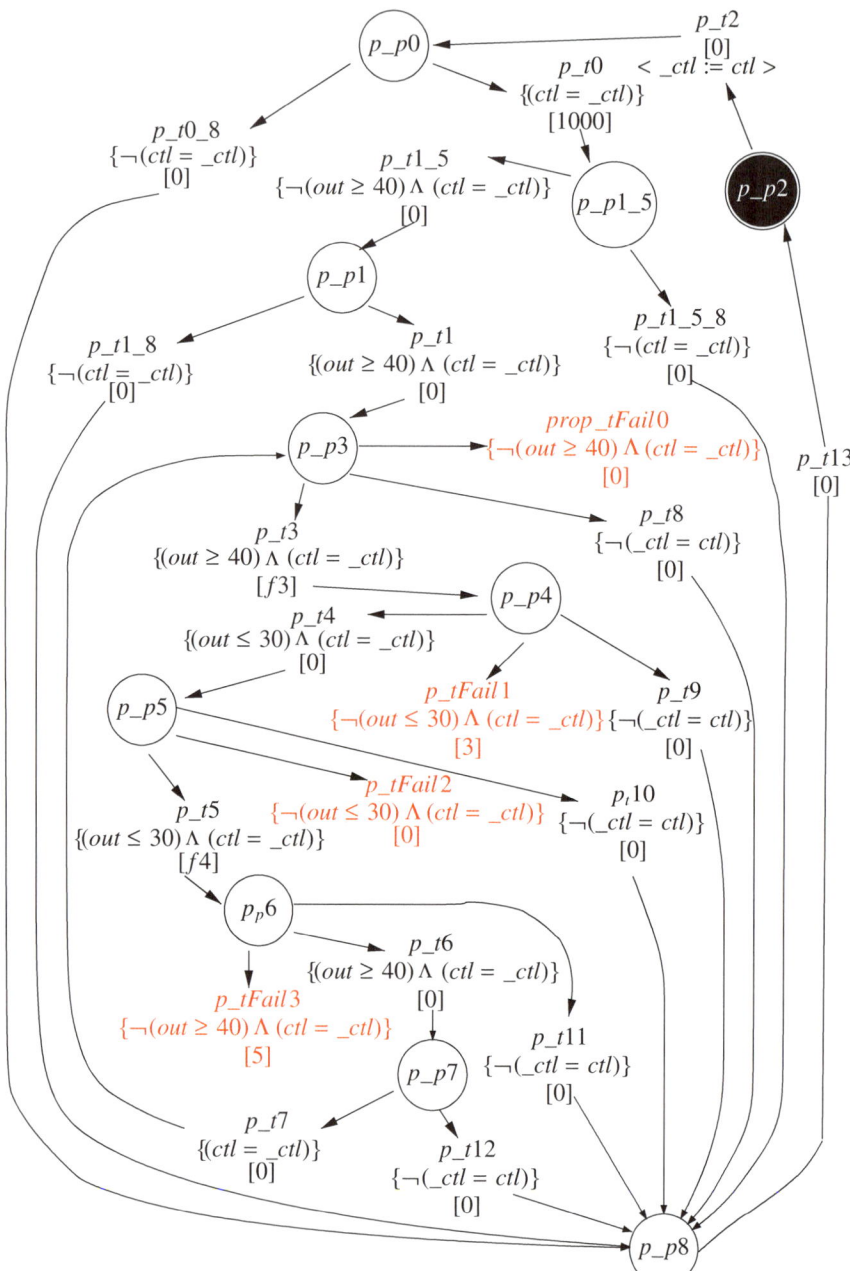

Fig. 3.11 The LPN generated for the property expressed in LAMP in Listing 3.3

Table 3.2 Verification results for a VCO circuit

Property	Control signals	Time (s)	States	Verifies?
Limited phase checker	2	0.144	22	Yes
Limited phase phecker	3	0.177	22	Yes
Limited phase checker	4	0.177	136	Yes
Limited phase checker	2, 3, 4 reg. int.	0.223	185	Yes
Limited phase checker	2, 3, 4 random	0.419	322	No
General phase checker	2	0.158	18	Yes
General phase checker	3	0.161	18	Yes
General phase checker	4	0.161	24	Yes
General phase checker	2, 3, 4 reg. int.	0.195	24	Yes
General phase checker	2, 3, 4 random	1.411	336	Yes

These results are generated using LEMA, a Java-based verification tool, on a 64-bit machine running an Intel Core i5 CPU M 480 @ 2.67 GHz with 4 processors and 4 GiB of memory

3.6 Conclusion

In order to verify whether an AMS circuit is correct given a model of the behavior, one needs to start with a property to verify. Several options have been proposed that are primarily inspired by LTL/CTL-like formalisms or by programming-like languages such as PSL and SystemVerilog. Although these methods are powerful and quite general, these languages often are difficult to convince designers to use since they have a steep learning curve.

This chapter presents LAMP, a more intuitive language for AMS property specification. This chapter demonstrates the utility of LAMP by showing how it can be used to express a desired property of a PI and VCO circuit. For the PI, the property is a precise phase shift should be produced by this circuit under the control of its input signal. This property is shown to be simple to express in LAMP while it is more opaque in formalisms such as STL and RT-SVA. Furthermore, this chapter demonstrates the use of LAMP in a verification setting by verifying that the output phase for various control signals is correct for a PI circuit with 4, 8, or 16 different phase shifts. This chapter further demonstrates that our tool can detect failures in the phase shift for a PI circuit. For the VCO example, the property is to verify the appropriate frequency is produced according to what the voltage signal is, after a suitable delay for the frequency to stabilize. This property is checked for three control voltages and it is shown that the property can be written to verify the circuit in an environment that randomly changes the control voltage.

For future work, one direction is to extend the language to support more types of constructs. For example, as is seen in Listing 3.2, the property has to be scaled according to how the original model is scaled. To aid in such circumstances, the language could be augmented to more gracefully handle scaling factors. In addition, it may be helpful to have the ability for the property language to hold state in more cases then in the condition list for an always block. As a second direction, it is useful to study the exact class of properties that can be expressed and those that cannot. In addition to being useful in knowing when the language is applicable, this also aids in comparisons between this language and the myriad of other possibilities. Finally, one notable drawback to LAMP is that it is unable to verify liveness properties beyond the level of bounded liveness. Thus, it would be useful to investigate how to incorporate such a check into the LEMA framework.

Acknowledgments This material is based upon work supported by the National Science Foundation under Grant No. CCF-1117515. Any opinions, findings, and conclusions or recommendations expressed in this material are those of the author(s) and do not necessarily reflect the views of the National Science Foundation. The work was also supported by SRC contract 2008-TJ-1851 and Intel Corporation.

References

1. Alur R, Feder T, Henzinger TA (1996) The benefits of relaxing punctuality. J ACM 43(1):116–146
2. Batchu S (2010) Automatic extraction of behavioral models from simulations of analog/mixed-signal (AMS) circuits. Master's thesis, University of Utah, Salt Lake City
3. Boulé M, Zilic Z (2008) Automata-based assertion-checker synthesis of PSL properties. ACM Trans Des Autom Electron Syst (TODAES) 13:4:1–4:21. doi: 10.1145/1297666.1297670
4. Havlicek J, Little S (2011) Realtime regular expressions for analog and mixed-signal assertions. In: Proceedings of the international conference on formal methods in computer-aided design (FMCAD) 2011. FMCAD Inc, Austin, pp 155–162
5. Std IEEE, 1800–2009 (2009) IEEE standard for system Verilog-unified hardware design, specification, and verification language, IEEE. doi:10.1109/IEEESTD.2009.5354441
6. Jones KD, Konrad V, Ničković D (2010) Analog property checkers: a DDR2 case study. Formal Methods Syst Des 36:114–130. doi:10.1007/s10703-009-0085-x
7. Koymans R (1990) Specifying real-time properties with metric temporal logic. Real-Time Syst 2:255–299. doi:10.1007/BF01995674
8. Kulkarni D (2013) Formal verification of digitally-intensive analog/mixed signal circuits. Master's thesis, University of Utah, Salt Lake City
9. Little S, Seegmiller N, Walter D, Myers C, Yoneda T (2011) Verification of analog/mixed-signal circuits using labeled hybrid petri nets. IEEE Trans Comput Aided Des Integr Circ Syst 30:617–630. doi:10.1109/TCAD.2010.2097450
10. Maler O, Ničković D (2004) Monitoring temporal properties of continuous signals. Proceedings of FORMATS-FTRTFT, Springer, LNCS 3253:152–166
11. Maler O, Ničković D (2013) Monitoring properties of analog and mixed-signal circuits. Int J Softw Tools Technol Transfer 15(3):247–268. doi:10.1007/s10009-012-0247-9
12. Maler O, Ničković D, Pnueli A (2008) Checking temporal properties of discrete, timed and continuous behaviors. In: Avron A, Dershowitz N, Rabinovich A (eds) Pillars of computer science. Springer, Berlin, pp 475–505
13. Smith D (2010) Asynchronous behaviors meet their match with System Verilog assertions. In: Proceedings of the design and verification conference (DVCON) 2010
14. Steinhorst S, Hedrich L (2008) Model checking of analog systems using an analog specification language. In: DATE '08: Proceedings of the conference on design, automation and test in Europe, ACM, New York, pp 324–329. doi: 10.1145/1403375.1403453

15. Walter D, Little S, Myers C, Seegmiller N, Yoneda T (2008) Verification of analog/mixed-signal circuits using symbolic methods. IEEE Trans Comput Aided Des Integr Circ Syst 27(12):2223–2235. doi:10.1109/TCAD.2008.2006159
16. Zaki MH, Tahar S, Bois G (2008) Formal verification of analog and mixed signal designs: A survey. Microelectron J 39(12):1395–1404. doi:10.1016/j.mejo.2008.05.013

Chapter 4
Integrating Circuit Analyses for Assertion-Based Verification of Programmable AMS Circuits

Dogan Ulus, Alper Sen and Faik Baskaya

Abstract Digitally-programmable analog circuits provide reconfigurability and flexibility for next-generation electronic systems and modern electronic systems need such circuits more than ever. For verification of these circuits, the change in analog characteristics according to digital inputs should be monitored and checked to determine whether measured analog characteristics satisfy desired conditions in a unified Analog/Mixed-Signal (AMS) verification environment. Therefore, we integrate common analog circuit analyses into an assertion-based verification flow, and we verify time-varying analog characteristics of digitally-programmable AMS circuits. We use results of DC, AC, and Fourier transform based analyses in our AMS assertion language, and monitor violations caused by any change in digital inputs. We show an application of our approach on a programmable low-pass filter circuit where cut-off frequency can be digitally controlled.

Keywords Assertions · Assertion checkers · Assertion languages · Temporal logic · Formal methods · Verification · Monitors · Simulation · Performance analysis · Circuit analysis · Analog/Mixed-Signal (AMS) integrated circuits · Frequency-domain analysis · Time-domain analysis

D. Ulus (✉)
Verimag, Centre Equation, 2 Av. de Vignate, 38610 Gieres, France
e-mail: dogan.ulus@imag.fr

A. Sen
Department of Computer Engineering, Bogazici University, 34342 Istanbul, Turkey
e-mail: alper.sen@boun.edu.tr

F. Baskaya
Department of Electrical and Electronics Engineering, Bogazici University,
34342 Istanbul, Turkey
e-mail: faik.baskaya@boun.edu.tr

© Springer International Publishing Switzerland 2015
M.-M. Louërat and T. Maehne (eds.), *Languages, Design Methods,
and Tools for Electronic System Design*, Lecture Notes in Electrical Engineering 311,
DOI 10.1007/978-3-319-06317-1_4

4.1 Introduction

There are growing demands for mixed-signal circuits especially for system-on-a-chip applications. However, a desire for more functionality in a more compact device has increased the level of complexity of these applications and made their verification harder than ever. Especially bugs that can occur at the boundaries of analog and digital blocks may remain hidden until the very end because analog and digital blocks are verified separately. To prevent such bugs and verify analog and digital blocks together, some proven techniques from digital domain have been extended toward analog domain in recent years.

Among these digital techniques, Assertion-Based Verification (ABV) has gained popularity for its practicality and scalability. Formal description of AMS properties and automatic evaluation of simulation runs are two main benefits of ABV flow for mixed-signal designs. Assertion-based methodology reduces manual effort, increases reusability and leads to a productivity increase in AMS verification.

These digital techniques are built on top of the analog verification techniques such as circuit analyses. Assertions can check properties from time-domain simulation (transient analysis) results but these assertions are mostly safe operating area checks, that observe voltages and currents in design. For performance related properties, analog designers perform other types of analog analyses for their designs besides transient analysis. Therefore, assertions for other analog analyses and their integration into mixed-signal time-domain verification still remains a challenge.

In this chapter, we integrate circuit analyses into assertion-based verification flow and we verify a specific class of mixed-signal circuits, that is digitally-programmable analog circuits, where analog circuit characteristics can be controlled by switching digital inputs. In Fig. 4.1, we show our unified assertion-based verification flow that includes circuit analyses such as DC, AC, transient, Fourier transform based analyses as well as digital verification. In our approach, assertion-based flow can check and monitor results of circuit analyses whenever a digital event changes the state of the analog circuit.

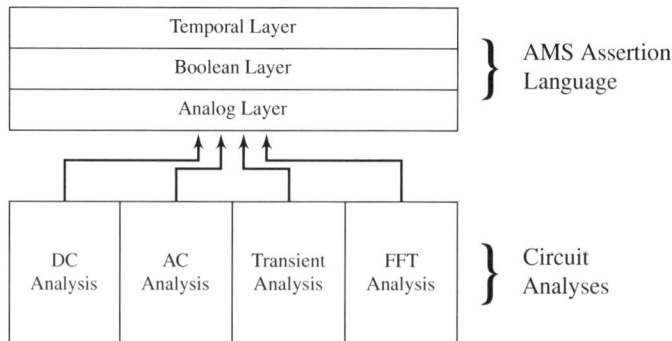

Fig. 4.1 Structure of an assertion-based AMS verification environment including circuit analyses

In the next section, we briefly give a summary of assertion-based AMS verification from the literature. In Sect. 4.3, we present the grammar of our assertion language integrating circuit analyses for assertion based verification flow. In Sect. 4.4, we explain DC, AC, and Fourier transform based analyses and their integration into mixed-signal verification and we validate our approach on a programmable low-pass filter circuit in Sect. 4.5.

4.2 Related Work

Several formal or semi-formal approaches for AMS circuit verification are presented in [3, 16]. Formal approaches like [1, 13] use state-space discretization and symbolic computations for AMS verification. Alternatively, works proposing semi-formal methods use simulation traces to verify AMS properties. Expressive monitors and checkers are crucial for simulation-based approach, therefore several specification languages have been proposed to describe assertions addressing different aspects of AMS verification.

In most monitoring techniques, predicates over real-valued signals convert real-valued signals into Boolean signals and temporal properties are monitored and checked in simulation traces. Among them, *Signal Temporal Logic* (STL) [9], which extends the *Metric Interval Temporal Logic* (MITL) [2], is presented to monitor time-domain properties of continuous signals. The underlying idea and logic of STL is extended with many useful constructs such as auxiliary state machines in [10], frequency-domain operators in [5], haloes in [14] and measurement operators in [15]. Similarly, specification languages in [8, 11] use assertions to verify AMS designs but they use a discrete-time notion in synchronization between analog and digital domains unlike a continuous time notion as in STL.

Expansion towards frequency domain is a natural and required step for more feature-rich assertion-based verification. Therefore, the authors of [8] include an FFT operator in their analog layer but it is limited to amplitude measurement only. In [5], the relation between time and frequency domain is discussed from AMS assertion perspective, and Time-Frequency Logic (TFL) is introduced. However, important tools for analog verification such as DC, AC, and other FFT-based analyses are missing in current assertion languages. In this chapter, we investigate the benefits of integration between circuit analyses and current assertion languages.

4.3 A New AMS Assertion Language

AMS assertion languages consist of several abstraction layers and are influenced by *Property Specification Language* (PSL) [6]. Alongside *Boolean* and *Temporal* layers as in PSL, AMS assertion languages introduce an *Analog* layer to capture analog properties. In early AMS assertion languages, analog layer is only capable of

Table 4.1 Grammar of our AMS Assertion Language

Temporal layer	TempExpr	::=	⊚BoolExpr
		|	⊚TempExpr
		|	TempExpr • TempExpr
		|	¬TempExpr
Boolean layer	BoolExpr	::=	SignalExpr ⊠ SignalExpr
		|	BoolExpr • BoolExpr
		|	¬BoolExpr
Analog layer	SignalExpr	::=	RawSignal ⊙ RawSignal
		|	RawSignal
		|	Const
	RawSignal	::=	dcExpr
		|	acExpr
		|	tranExpr
		|	fftExpr
		|	dataExpr
	dcExpr	::=	$\mathcal{D}(Node)@DC(\,EventsExpr\,)$
	acExpr	::=	$\mathcal{A}(Node)@AC(\,EventsExpr\,)$
	tranExpr	::=	Node
	fftExpr	::=	$\mathcal{F}(SignalExpr)@FFT(\,EventsExpr\,)$
	dataExpr	::=	datatowf (data)
	EventsExpr	::=	$\mathcal{E}^a(SignalExpr)$
		|	$\mathcal{E}^d(BoolExpr)$

Operators			
⊚	Temporal operators	$\mathcal{D}$	DC property operators
•	Binary boolean operators	$\mathcal{A}$	AC property operators
¬	Boolean negation operator	$\mathcal{F}$	FFT Property operators
⊙	Arithmetic operators	$\mathcal{E}^a$	Analog event operators
⊠	Comparison operators	$\mathcal{E}^d$	Digital event operators

processing transient analysis results. However, other circuit analyses used in traditional analog verification is crucial to verify all aspects of analog design. Therefore, we extend our AMS assertion language to handle other types of circuit analyses such DC, AC, and FFT analysis.

We show the grammar of our AMS assertion language in Table 4.1. The temporal and Boolean operators have their usual semantics. *DC*, *AC*, and *FFT* operators denote corresponding operators. Note that *FFT* operator needs parameters such as the number of points and the window function, which we do not detail in Table 4.1. The operator *datatowf* converts custom time-data points into analog signals, which is useful to compare analysis result with desired values. DC property operators, shown as $\mathcal{D}$, extracts DC values from a given DC analysis. AC property operators, shown as $\mathcal{A}$, calculate AC properties like the bandwidth *BW* from a given AC analysis.

Similarly, FFT property operators, shown as $\mathcal{F}$, calculate FFT properties like Total Harmonic Distortion (THD) from a given FFT analysis. Events are used to determine analysis points in time. Event detection is done by digital event operators $\mathcal{E}^d$ such as *rise* on digital signals and by analog event operators $\mathcal{E}^a$ such as *crossing* detect events on analog signals. An example property using our grammar is as follows:

$$Always(10\mathrm{e}6 < BW(out)@AC(\ rise(ctrl)) < 12\mathrm{e}6\ )$$

where *out* denotes an output node of an analog filter. For this assertion, the bandwidth operator (BW) in the analog layer calculates the bandwidth from AC analysis results for each rise event on the Boolean signal ctrl. Then, the bandwidth value is checked to see whether it is between 10 and 12 MHz in Boolean layer, and the *Always* operator in the Temporal layer checks whether it is true for all simulation times.

4.4   Circuit Analyses

Circuit analyses are well-defined ways of collecting information (metrics) about analog designs. Because circuit analyses provide valuable metrics for verification, we see a great benefit to integrate them into assertion-based mixed-signal verification.

Among circuit analyses, DC operating point analysis determines the quiescent state of circuits. AC analysis extracts small-signal linear response of the circuit for a single frequency input around the DC operating point. Transient analysis computes time-domain response of the circuit by iteratively solving the algebraic differential system of equations. Fast Fourier Transform (FFT) analysis, returns power spectrum of time-domain signal so that we can see how much power resides in frequency components of that signal. FFT analysis allows us to analyze noise and linearity characteristics of analog circuits from time-domain simulation. In AMS design, designers need to combine these analyses with digital properties for a unified verification environment.

Therefore, we focus on DC, AC, and FFT specifications of AMS circuits, and express properties for their time-varying characteristics using our AMS assertion language. We propose to perform circuit analyses at the specific time instants specified by events during transient analysis, and monitor time-varying characteristics. In following sections, we explain DC, AC, and FFT analysis in detail and events to integrate these analyses into transient analysis (time-domain simulation).

4.4.1   Circuit Analyses at Events

Events in our context correspond to a change on either an analog or a digital signal. Events can be classified as analog events (such as a signal crossing a certain voltage value) or digital events (a transition in corresponding digital signal). We use events

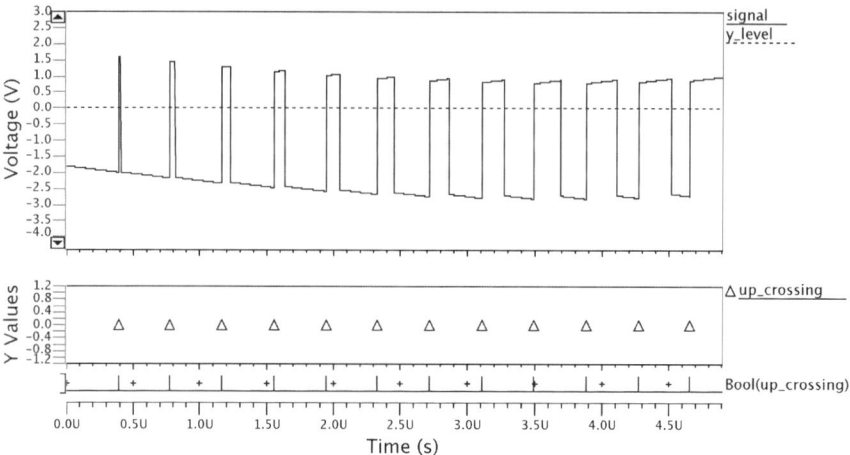

Fig. 4.2 Crossing detection for threshold $y\_level = 0.0$ and upward direction on an analog signal

to determine the time instants to perform circuit analyses in the middle of transient analysis. It means that we start a transient analysis and we pause transient analysis whenever a specified event occurs and we check the corresponding property for that time instant.

Event operators observe either an analog or a digital signal and return a set of events denoted by $\mathcal{E}$. For example, the event operator *crossing* on the analog signal returns a set of events such as $\mathcal{E} = \{e_0, e_1, \ldots, e_n\}$ where each event e_n indicates that the signal crosses specified threshold in specified direction. In Fig. 4.2, we illustrate a crossing detection for the threshold $y\_level = 0.0$ and the upward direction on an analog signal. Then, we are able to perform circuit analyses for these events, and check DC, AC, and FFT properties whenever a control input changed in AMS circuit. This way, we can monitor these properties dynamically without specifying external or pre-determined time points for analyses.

4.4.2 DC Operating Point Analysis and DC Assertions

DC operating point analysis determines the quiescent state or stable initial condition of the circuit. The quiescent state is computed by the simulator with capacitances opened and inductances short-circuited. It provides operating point information before a small-signal (AC) analysis or a transient analysis so it is the starting point in the analog design flow.

In traditional analog design flow, designers make assumptions and choices for DC values inside their analog blocks. Assertions capture these assumptions and design choices and they report if there is an undesired situation. DC characteristics of analog circuits include DC voltage levels of circuit nodes, DC current values for circuit branches as well as operating modes and conditions for circuit devices.

Related analog circuit specifications that can be extracted via DC analysis include offset voltage values, offset voltage drifts, bias current values, current drifts and operation modes of transistor devices. For example, if intended DC voltage level for the output node is between 0.85 and 0.95 V for an amplifier design, we can capture this specification in the following formula:

$$0.85 < V(out)@DC(\mathcal{E}) < 0.95$$

For time-invariant analog circuit blocks, DC analysis is performed at the start of simulation and it is considered as valid for all simulation. Therefore, single DC analysis is sufficient to verify DC characteristics for time-invariant systems. However, in the case of circuits having time-varying DC characteristics such as adaptive or digitally controlled analog circuits, it is beneficial to integrate DC analysis checks into time-domain simulation (transient analysis). This way, the interaction between digital and analog domains can be investigated in a single simulation environment. To ensure desired DC characteristics are achieved during such a mixed-signal simulation, we integrate DC analysis checks into AMS assertions. For example, we want to check DC voltage level for the output node of an amplifier with digitally-adjustable transconductance. Because transconductance adjustment by digital events can shift DC voltage levels in the analog circuit, output DC voltage level should be monitored and checked during time-domain simulation via the following assertion:

$$Always(0.85 < V(out)@DC(\{e_1, e_2, e_3\}) < 0.95)$$

This assertion monitors the DC value of the output node and checks if it satisfies the desired condition when events e_1, e_2, and e_3 occur during all simulation times.

4.4.3 AC Analysis and AC Assertions

Traditionally, designers analyze analog circuits by exciting them with a single-frequency input signal, and they measure the output signal to see how the magnitude and phase of the input signal is changed by the circuit. In general, only linear analog circuits can be analyzed using this technique; therefore, a nonlinear circuit should be linearized by assuming input signals are small enough. This technique, called as small-signal analysis or AC analysis, is a common and useful procedure when analyzing the frequency response of analog circuits.

Related analog circuit specifications that can be extracted via AC analysis include DC gain, bandwidth, phase margin, gain margin and roll-off slopes. In Fig. 4.3, we show an example AC analysis plot, where we can see AC metrics for the output node of a low-pass filter circuit. We can write assertions to check whether the measured metrics satisfy the specifications in assertion-based verification. For example, if the

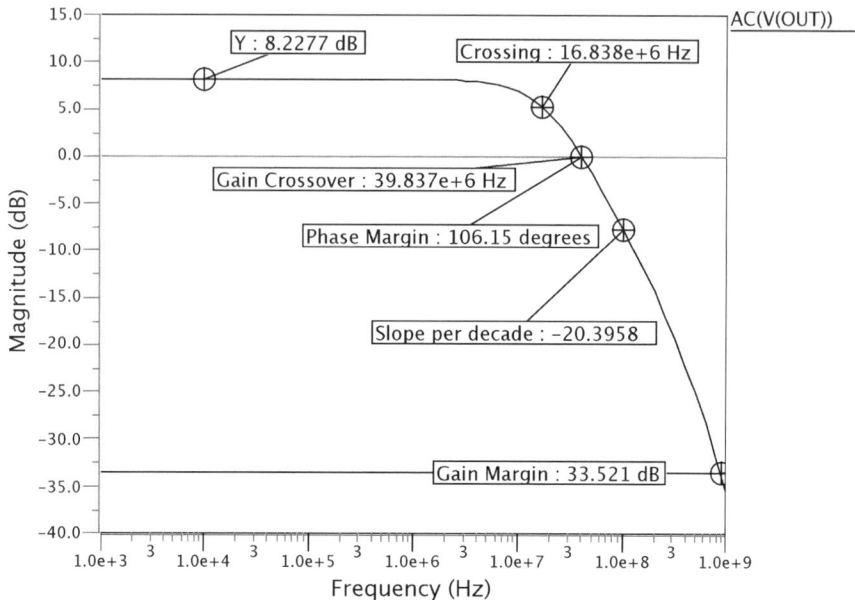

Fig. 4.3 An example AC analysis plot and measured AC specifications

desired value for DC gain is between 0 and -0.5 dB for a low-pass filter design, this specification is captured as:

$$-0.5 < dcgain(out)@AC(\mathcal{E}) < 0.0$$

In Linear Time-Invariant (LTI) analog blocks, AC analysis is performed at the start of simulation after DC analysis and it is considered as valid for all simulation times. Therefore, single AC analysis is sufficient to verify AC characteristics for LTI systems. However, in case of circuits having time-varying AC characteristics such as adaptive or digitally controlled analog circuits, it is beneficial to integrate AC analysis checks into time-domain simulation (transient analysis) as we did for DC assertions.

For example, if we want to check the DC gain value for a low-pass filter with digitally-adjustable cut-off frequency, then we write the following assertion to capture DC gain property:

$$Always(-0.5 < dcgain(out)@(\{e_1, e_2, e_3\}) < 0.0)$$

where e_1, e_2, and e_3 denote discrete events such as changes in digital control inputs. In Fig. 4.4, we illustrate how AC analysis is performed in the middle of a time-domain simulation. During time-domain simulation, AC analysis is performed for time instants when events e_1, e_2, e_3 occur. Then, our assertion monitors DC gain at

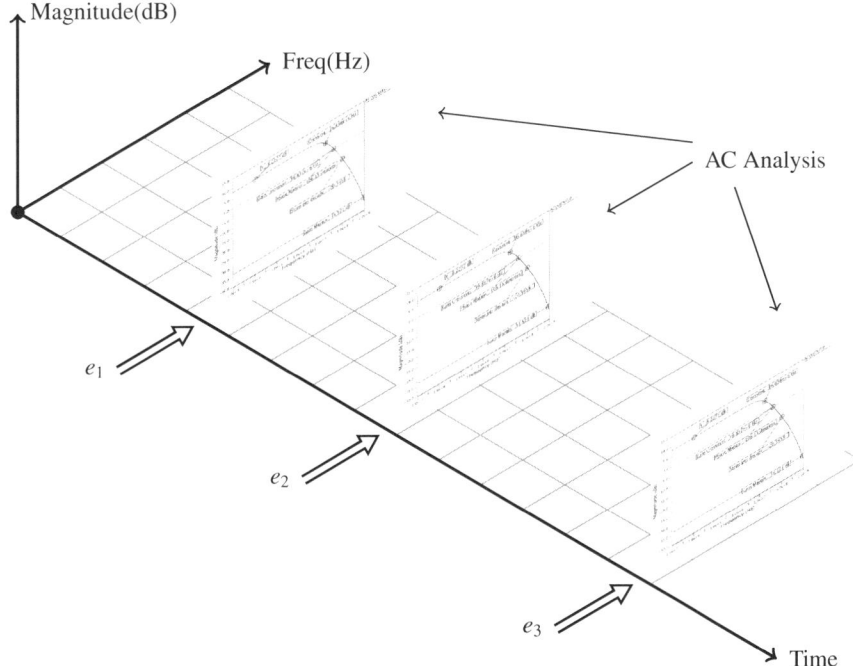

Fig. 4.4 Monitoring time-varying AC characteristics in the middle of time-domain circuit simulation. A new AC analysis is performed when an event changes the state of the circuit

the output node from AC analysis results and checks if it always satisfies the desired condition when these events occur.

An important limitation of our work is as follows. Although current simulators can perform AC analysis in the middle of transient analysis, they use current node voltages for AC analysis instead of performing true DC analysis. This limitation makes each AC analysis that is performed at the exact time of the event wrong. Therefore, we solve this problem practically by delaying the corresponding event times (until DC voltages are stabilized) for AC analyses. With simulators allowing event-driven DC and AC analyses, analysis and verification of time-varying analog circuits are easier and more comfortable.

4.4.4 FFT Analysis for Noise and Linearity

Noise and linearity characteristics of analog circuits determine the range of useful signals that the circuit processes as intended for. The smallest value of signals that can occur is limited by the noise floor of the circuit whereas the largest value of signals is limited by the nonlinearity of the circuit. Therefore, we use Fast Fourier

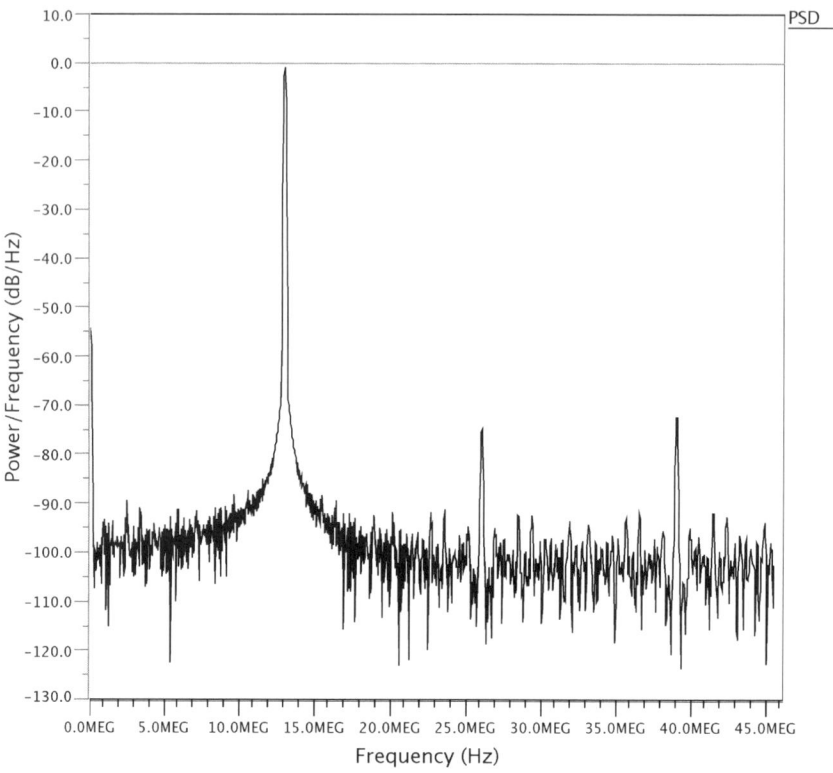

Fig. 4.5 Power spectrum example

Transform (FFT) based analyses to verify the dynamic range of analog circuits in practice.

Dynamic range metrics include Total Harmonic Distortion (THD), Signal-to-Noise and Distortion Ratio (SNDR), and Spurious Free Dynamic Range (SFDR). We can extract these metrics by performing an FFT to get the power spectrum of time-domain simulation results if the circuit is excited with a single-frequency input. On the other hand, we should consider all factors to compute a healthy FFT such as the number of points, windowing choice, and sampling frequency during these analyses.

In Fig. 4.5, we illustrate an example power spectrum plot, which shows the power of each frequency component along frequency axis. The biggest peak indicates the fundamental frequency (13 MHz in this case), where most of the power is concentrated, and we can see the noise and distortion components in other frequencies.

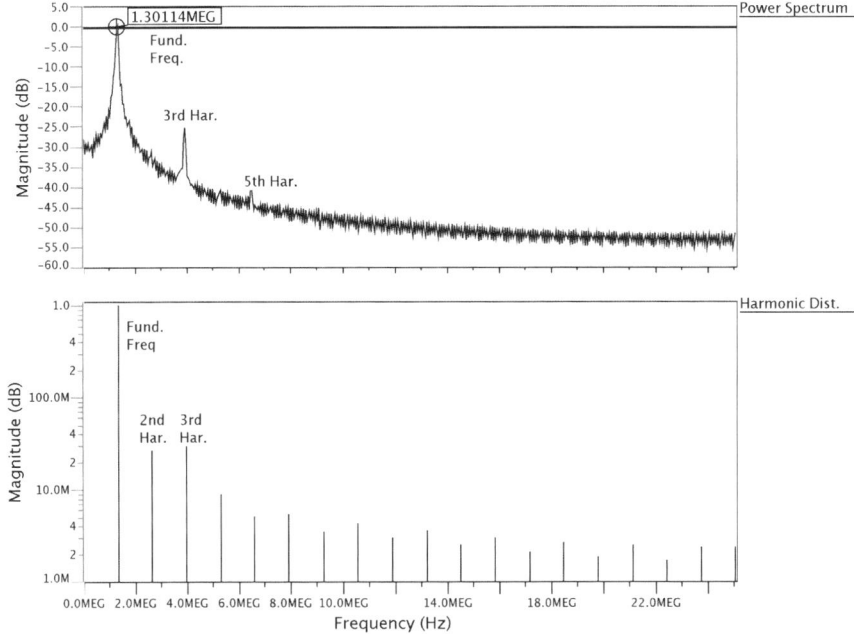

Fig. 4.6 Total Harmonic Distortion (THD) example

4.4.4.1 Total Harmonic Distortion

The Total Harmonic Distortion (THD) of a signal is the ratio of the total power of all second and higher harmonic components to the power of the fundamental harmonic (first harmonic) for that signal. Because a nonlinear system produces second, third and higher-order distortion components at the harmonics of the input (fundamental) frequency, when excited with a sinusoidal source, it is used as a measurement for the nonlinearity of a system. The THD metric is calculated by the formula below and is expressed in gain (dB) or percentage:

$$THD = 10 \log \frac{H_{D2}^2 + H_{D3}^2 + \cdots + H_{Dn}^2}{H_{D1}^2}$$

where H_{D1}^2, H_{D2}^2, H_{D2}^2, and H_{Dn}^2 represent the power value of first-, second-, third- and nth-order harmonics, respectively.

In Fig. 4.6, we illustrate an example THD analysis of an analog circuit, which is excited by an input frequency of 1.3 MHz. In the normalized power spectrum of output signal, we can see the biggest peak at 1.3 MHz, which is the fundamental frequency, and a few smaller peaks at the integer-multiples of fundamental frequency, which are called harmonics. Then, the THD operator calculates the ratio of the power of fundamental frequency and the power of all other harmonics. We write an assertion to monitor time-varying THD characteristics of any weakly nonlinear analog circuit as follows:

$$Always(0.0 < THD(out)@FFT(\{e_1, e_2, e_3\}) < 2.5)$$

This assertion checks if calculated the THD value always satisfies the desired specification at the time instants when events e_1, e_2, and e_3 occur.

4.4.4.2 Signal-to-Noise and Distortion Ratio

Another widely-used metric for noise and linearity specification is Signal-to-Noise and Distortion Ratio (SNDR). It is the ratio of signal power to power sum of all other spectral components, and expressed in dB. SNDR is a good indicator about system's performance because it takes both distortion and noise components into account. SNDR is calculated by using the formula below:

$$SNDR = 10 \log \frac{P_{signal}}{P_{noise} + P_{distortion}}$$

where P_{signal}, P_{noise} and $P_{distortion}$ denote the value of power of signal, noise, and distortion components. We write an assertion to monitor time-varying SNDR characteristics of any weakly nonlinear analog circuit as follows:

$$Always(SNDR(out)@FFT(\{e_1, e_2, e_3\}) > 30)$$

This assertion checks if calculated SNDR value always satisfies the desired specification at the time instants when events e_1, e_2, and e_3 occur.

4.4.4.3 Spurious Free Dynamic Range

Spurious Free Dynamic Range (SFDR) is the ratio of the input signal to the peak spurious component. Spurs can occur at the harmonics of fundamental frequency because of nonlinearity or at other frequency values because of mismatches in the circuit. In Fig. 4.7, we illustrate an example SFDR analysis of an analog circuit. In the normalized power spectrum, we see that the ratio of the input signal power and the power of biggest peak component (third harmonic) is 23 dB.

We write an assertion to monitor time-varying SFDR characteristics of any weakly nonlinear analog circuit as follows:

$$Always(SFDR(out)@FFT(\{e_1, e_2, e_3\}) > 20)$$

This assertion checks if the calculated SFDR value always satisfies the desired specification at the time instants when events e_1, e_2, and e_3 occur.

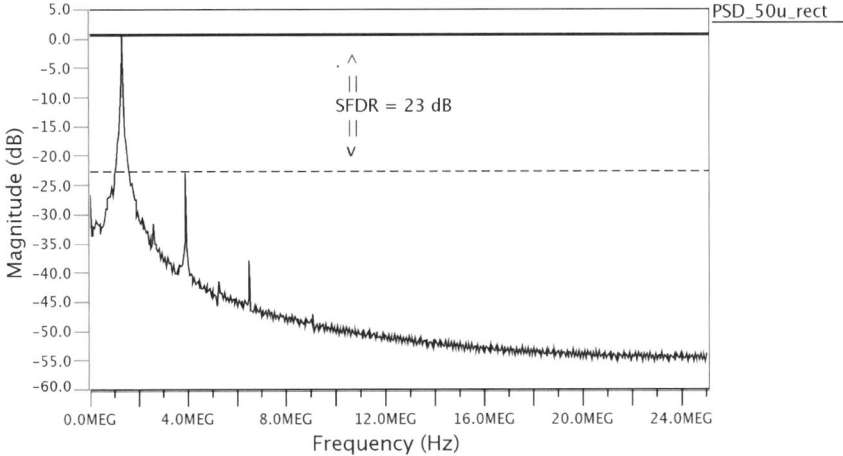

Fig. 4.7 Spurious Free Dynamic Range (SFDR) example

4.5 Case Study

We investigate a digitally-programmable continuous-time Gm-C low-pass filter design to integrate circuit analyses in our assertion-based verification flow. In a typical receiver system used in communication applications, three basic operations are performed: amplification, filtering and data conversion. Programmable Gain Amplifierss (PGAs), Low-Pass Filters (LPFs), Analog-to-Digital Converters (ADCs), and Digital-to-Analog Converters (DACs) are designed to implement these operations in actual circuits. In next-generation applications, these circuits should be very flexible and capable of adapting their performance to different standard requirements with reduced power consumption [4]. To achieve this, we need to design and verify analog circuits with digital control of biasing, gain, circuit-level, and stage-level reconfiguration, block power down/up.

Designers can implement filter designs in both digital and analog domains. Domain selection includes many design trade-offs and challenges. Although filtering at digital domain is preferred over analog domain, because of digital domain's robustness and scalability, the overwhelming requirements for the following data conversion step makes programmable analog filters attractive for new generation mixed-signal applications.

To demonstrate our approach, we have designed a digitally-programmable low-pass Gm-C filter by using architectures and ideas presented in [7, 12] for $0.18\,\mu\mathrm{m}$ technology and we simulated the circuit with the Eldo SPICE simulator.

Cut-off frequency for this circuit is determined using the following equation:

$$f_c = k \times \frac{gm_0}{2\pi\sqrt{C_1 \cdot C_2}} \times 0.6412 \tag{4.1}$$

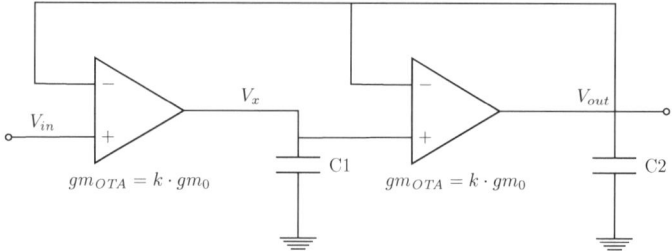

Fig. 4.8 The programmable low-pass filter circuit

where f_c denotes the cut-off frequency of the filter, g_m denotes the transconductance value of amplifiers, C_1, C_2 denote capacitor values in the circuit and the k value provides programmability for this filter. We can change cut-off frequency of the low-pass filter during operation by changing k (Fig. 4.8).

As in [12], we achieve such programmability in actual circuit by connecting three Operational Transconductance Amplifier (OTA) in parallel, where the OTAs have binary weighted transconductance values, gm_0, $2gm_0$, and $4gm_0$, respectively. Each OTA has a dedicated digital control bit to drive transistor switches inside the OTA circuit and this way we can turn it on or off. Because transconductance values of parallel OTAs are added up if OTAs are turned-on, we can control overall transconductance by adjusting three control bits of OTAs. The k value represents a binary-coded decimal 3-bit digital input and it is connected to the control bits of the OTAs for transconductance adjustment.

Note that $k = 0$, thus binary "000", is an invalid value for the low-pass filter circuit according to Eq. (4.1). Therefore, before starting to check circuit characteristics, we write an assertion to ensure that k is never equal to zero. We captured this property as follows:

$$Always(\ k[0] \vee k[1] \vee k[2])$$

where $k[n]$ denotes the Boolean value of the nth bit binary-coded decimal k. This assertion warns if all bits of k are logic zero, thus $k = 0$.

In verification of programmable analog filters, we should check DC operating points of different states. The simulator performs a DC analysis and computes DC levels of all nodes. Then, we check if these values satisfy desired conditions for all time instants. However, changing 3-bit digital k value can disrupt DC levels in the circuit and cause a shift in DC level of the output node. On the other hand, ensuring that DC levels remain in a specified range is important for robustness of a system time-varying characteristics. Therefore, the simulator performs a DC analysis whenever a change in k occurs and we check the DC levels for each change during simulation. In Fig. 4.9, we illustrate the change in DC level of the output node based on time and varying k. If we want to keep the value of output DC level of the filter between 880 and 920 mV, we write this property:

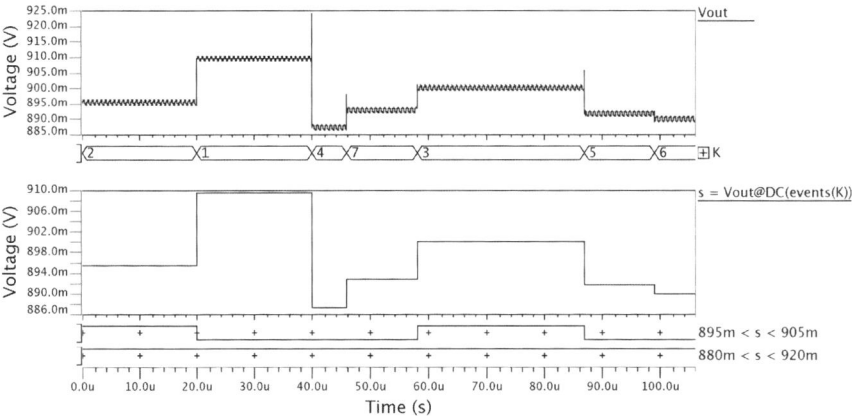

Fig. 4.9 Monitoring time-varying DC level of the filter output according to k

$$Always(880e - 6 < V(out)@DC(\mathcal{E}_k) < 920e - 6)$$

where $\mathcal{E}_k$ denotes an event set for changes in k. In Fig. 4.9, we can see that varying DC level for output node is always between specified values so the assertion is satisfied. However, a stricter requirement for output DC level such as [895, 905 m] would fail for this circuit. We annotate the results of these checks in the form of boolean signals in Fig. 4.9.

As the next step, we verify time-varying AC characteristics of the our low-pass filter. Design specification states that bandwidth value of the low-pass filter should be changed linearly depending on k. According to Eq. (4.1), the bandwidth value of the filter should be approximately equal to $k \cdot 2.1$ MHz (e.g., 2.1 MHz if $k = 1$, 8.4 MHz if $k = 4$) in the ideal case if we select circuit parameters, gm_0, C_1 and C_2 as 25 μA/V, 1 and 4 pF, respectively. To verify linearity of bandwidth adjustment, one approach we can take is to extract *bandwidth per k* metric and check whether it always stays in an acceptable region. This way, we ensure that the error deviated from expected value stays in specified limits and we capture this property as follows:

$$Always(2.0e6 < BW(out)@AC(\mathcal{E}_k)/k < 2.2e6)$$

To monitor this property, an AC analysis is performed and a frequency plot is returned for each change in k. Bandwidth operator (BW) computes bandwidth value from each AC analysis, and we plot bandwidth of the filter versus time in Fig. 4.10. We can see that this assertion is evaluated true in Fig. 4.10 because the *bandwidth per k* metric always remains inside the [2, 2.2 M] region.

According to Eq. (4.1), incorrect transconductance values of Operational Transconductance Amplifiers (OTAs) can lead to erroneous behavior. Therefore, we write the following assertions to check whether transconductance values of both (OTAs) in the filter circuit satisfy desired values:

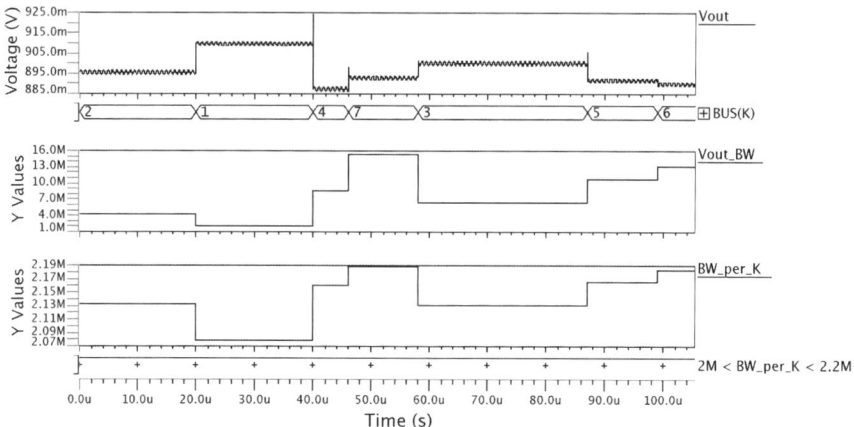

Fig. 4.10 Monitoring time-varying *bandwidth per k* value of the filter according to k factor.

$$Always(-5e6 < GM(x, in, out)@AC(\mathcal{E}_k) - datatowf(gm(k)) < 5e6)$$
$$Always(-5e6 < GM(out, x, out)@AC(\mathcal{E}_k) - datatowf(gm(k)) < 5e6)$$

where the *GM* operator calculates the transconductance value using AC analysis results of input and output nodes of OTAs, and *datatowf*(*gm*(*k*)) denotes a custom waveform for theoretical values of transconductance. This way we can trace the error up to the smallest circuit block and we can say which block is not working properly. Note that tracing error is an important benefit of our verification flow. Although writing assertions in hierarchical manner still requires design expertise, the checking process is now automatic, standardized, and less error-prone.

Ultimately, a filter is designed to implement a linear operation however actual implementations of filter circuits (or any other circuit that implements a linear operation) are slightly nonlinear because of nonlinearity of transistor devices. FFT-based THD and SNDR metrics are used to measure the amount of nonlinearity of designs from transient simulation results. We captured the THD and SNDR specifications as follows:

$$Always(0 < THD(out)@FFT\{\mathcal{E}_k\} < 3)$$
$$Always(SNDR(out)@FFT\{\mathcal{E}_k\} > 30)$$

In Fig. 4.11, we plot the output node of the filter and monitor the THD and SNDR properties for our filter design according to the change in k and we annotate calculated THD and SNDR values on the plot. We see that the THD and SNDR values are not always in the specified range so the THD and SNDR specifications are not satisfied for all k. Unsatisfied linearity specifications usually do not have easy fixes and designers may require a change in circuit topology to achieve desired linearity. However, the same assertions can be used for several topologies; therefore, verification effort across different topologies is reduced compared to manual verification.

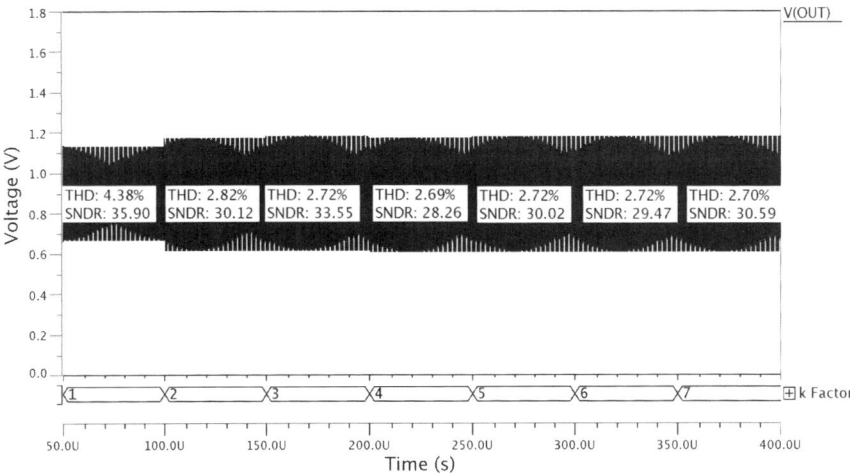

Fig. 4.11 Monitoring time-varying THD and SNDR values of the filter according to k

4.6 Conclusion

Any simulation-based AMS verification methodology would be incomplete without an integrated support for circuit analyses. Therefore, we integrated circuit analyses such as DC, AC, and FFT analyses into an assertion-based verification flow, and proposed an AMS verification language including support for these analyses as well as transient analysis. This is a required step for a more unified and expressive environment for assertion-based AMS verification. In current setup, we used traditional SPICE simulators to perform circuit analyses where you have limited control over simulation. However, if we model our circuits at higher abstraction levels, we are able to use simulators like SystemC-AMS, where we have much better control over simulation. We performed our experiments on a specific class of analog circuits, digitally-programmable analog circuits whose analog characteristics can change in time. Extending our work to support the SystemC-AMS simulator is a natural next step in the future. At the end, we aim to use the same assertions for both high-level SystemC-AMS and low-level SPICE models. Also, we will analyze the applicability of our approach for other cases that include time-varying characteristics such as analog circuits with aging transistors.

Acknowledgments This work was supported in part by the Semiconductor Research Corporation under task 2082.001, Marie Curie European Reintegration Grant within the 7th European Community Framework Programme, the Turkish Academy of Sciences, and Bogazici University Research Fund 7223.

References

1. Al-Sammane G, Zaki M, Tahar S (2007) A symbolic methodology for the verification of analog and mixed signal designs. In: Proceedings of the conference on design, automation and test in Europe (DATE), pp 249–254
2. Alur R, Feder T, Henzinger T (1996) The benefits of relaxing punctuality. JACM 43(1):116–146
3. Barke E, Grabowski D, Graeb H, Hedrich L, Heinen S, Popp R, Steinhorst S, Wang Y (2009) Formal approaches to analog circuit verification. In: Proceedings of the conference on design, automation and test in Europe (DATE), pp 724–729
4. de la Rosa JM, Castro-Lopez R, Morgado A, Becerra-Alvarez EC, del Rio R, Fernandez FV, Perez-Verdu B (2009) Adaptive CMOS analog circuits for 4G mobile terminals—review and state-of-the-art survey. Microelectron J 40(1):156–176
5. Donze A, Maler O, Bartocci E, Nickovic D, Grosu R, Smolka S (2012) On temporal logic and signal processing. In: Proceedings of the symposium on automated technology for verification and analysis (ATVA), pp 92–106
6. Foster H, Marschner E, Wolfsthal Y (2005) IEEE 1850 PSL: The next generation. In: Proceedings of design and verification conference and exhibition (DVCON)
7. Geiger RL, Sanchez-Sinencio E (1985) Active filter design using operational transconductance amplifiers: a tutorial. IEEE Circuits Devices Mag 1(2):20–32
8. Lämmermann S, Ruf J, Kropf T, Rosenstiel W, Viehl A, Jesser A, Hedrich L (2010) Towards assertion-based verification of heterogeneous system designs. In: Proceedings of the conference on design, automation and test in Europe (DATE), pp 1171–1176
9. Maler O, Ničković D (2012) Monitoring properties of analog and mixed-signal circuits. Int J Softw Tools Technol Transfer 15(3):247–268
10. Mukherjee S, Dasgupta P, Mukhopadhyay S (2011) Auxiliary specifications for context-sensitive monitoring of AMS assertions. IEEE Trans Comput Aided Des Integr Circuits Syst 30(10):1446–1457
11. Mukhopadhyay R, Panda SK, Dasgupta P, Gough J (2009) Instrumenting AMS assertion verification on commercial platforms. ACM Trans Des Autom Electron Syst (TODAES) 14(2), Article No. 21
12. Pavan S, Tsividis YP, Nagaraj K (2000) Widely programmable high-frequency continuous-time filters in digital CMOS technology. IEEE J Solid-State Circuits 35(4):503–511
13. Steinhorst S, Hedrich L (2012) Analog assertion-based verification on partial state space representations using ASL. In: Proceedings of the forum on specification and design languages (FDL), pp 98–104
14. Ulus D, Sen A (2012) Using haloes in mixed-signal assertion based verification. In: Proceedings of the workshop on high level design validation and test (HLDVT), pp 49–55
15. Ulus D, Sen A, Baskaya F (2013) Analog layer extensions for analog/mixed-signal assertion languages. In: Proceedings of the conference on very large scale integration (VLSI-SoC), pp. 66–71
16. Zaki MH, Tahar S, Bois G (2008) Review: formal verification of analog and mixed signal designs: a survey. Microelectron J 39:1395–1404

Part III
Embedded Analog and Mixed-Signal System Design

Chapter 5
Hybrid Dynamical Systems for Memristor Modelling

An Approach to Avoid the Terminal-State Problem

Joachim Haase and André Lange

Abstract Leon O. Chua introduced the memristor as the fourth circuit element to complete the set of fundamental passive two-terminal elements in 1971. For a long time it seemed as if memristors were just toys in the sandbox of network theorists. The situation abruptly changed in 2008 when scientists from HP reported on a nanoelectronic device, which showed a memristive behaviour. Main hopes for new opportunities and circuit concepts in the transition to increasingly smaller integrated circuits are going to be related to this discovery. For an examination of these possibilities by means of simulation, a large number of memristor models has been developed in recent years. A special property of the behavioural models of memristive nanoelectronic devices is the restricted range of internal state variables. A number of tricky solutions has been developed up to now to handle this problem. In this section we present a straightforward solution for this problem within the framework of hybrid dynamical systems.

Keywords Memristor · Hybrid dynamical systems · Nanoelectronic devices · Memristive systems · Terminal state-problem · Linear drift · Nonlinear drift · Titanium dioxyde memristive device · Analog/Mixed-Signal (AMS) · VHDL-AMS

5.1 Introduction

The classical passive two-terminal network elements are resistor, inductor, and capacitor. The resistor is defined by a relationship between voltage and current of the branch that connects the terminals. A relationship between the integral of the voltage (i.e., the flux-linkage) and the current defines an inductor. Similarly, we can define a capacitor by a relationship between the integral of the current (i.e., the charge) and the voltage. Chua [5] noted that for reasons of completeness a forth ele-

J. Haase (✉) · A. Lange
Fraunhofer Institute for Integrated Circuits IIS, Design Automation Division EAS,
Zeunerstraße 38, 01069 Dresden, Germany
e-mail: joachim.haase@eas.iis.fraunhofer.de

A. Lange
e-mail: andre.lange@eas.iis.fraunhofer.de

© Springer International Publishing Switzerland 2015
M.-M. Louërat and T. Maehne (eds.), *Languages, Design Methods,
and Tools for Electronic System Design*, Lecture Notes in Electrical Engineering 311,
DOI 10.1007/978-3-319-06317-1_5

ment was missing that is described by a relationship between flux-linkage and charge. He called the missing circuit element memristor because "it behaves somewhat like a nonlinear resistor with memory". Thus, memristor is an abbreviation for memory and resistor. The current value of its resistance—called memresistance—depends on terminal voltages and currents in the past. The value of the memristance is held even if the power is taken away. This makes the memristor an attractive candidate for low-power computation and memory elements.

In the conclusions of his famous 1971 paper Chua remarked: "It is perhaps not unreasonable to suppose that such a device might already have been fabricated as a laboratory curiosity but was improperly identified!" However, it took more than 35 years that Stanley Williams and his colleagues Dimiriy Strukov, Gregory Snider, and Duncan Stewart from the Hewlett Packard Labs reported on the long-sought realization of the missing element [23]. They presented a nanoelectronic device with memristive behavior based on ionic drift in solid-state TiO_2 thin films in 2008 and used the generalized memristor concept of Chua and Kang [7] to discuss the behaviour of the device. A state space equation describes the dynamic behaviour of the drift whereas the memristance depends on the continuous state.

The article of Wiliams et al. sparked numerous other activities. The interest in this field has been increasing enormously ever since. On the one hand, this work has aimed at a better understanding of memristive effects in nanoelectronic devices and issues of technological realization of these components. On the other hand, work on new circuit concepts and applications of memristive devices has been carried out. Chua and Williams gave an insight into the current investigations of possibilities of future electronic system design in a DATE tutorial in 2012. They addressed the potential of memristors for nonvolatile memory devices to replace current Flash, DRAM, and SRAM architectures as well as their potential for brain-like machines using nano-scale neuromorphic chips [24]. An overview on other applications is summarized in Tetzlaff and Schmidt [25]. Further aspects of memristor-based Resistive RAM (ReRAM) can be found in Niu et al. [20].

A large number of studies in this field is based on simulation experiments. These experiments require accurate and robust models of memristors. Thus, a lot of work has been done in this field over the last years (see for instance Batas and Fiedler [3], Biolek et al. [4], Corinto and Ascoli [8], Corinto et al. [9], Johlekar and Wolf [14], Kvatinsky et al. [16], Lehtonen and Laiho [18], Pickett et al. [21], Prodromakis et al. [22]). To improve the accuracy, an appropriate description of the nonlinear behaviour, especially the nonlinear ionic drift in memristive devices is required and has been included into the models. Regarding the robustness of the models, we have to mention an essential difference between an "ideal" memristor and its nanoelectronic realisations. While the "ideal" memristor allows an unlimited integration of voltage and current, the range of continuous state variables of models of nanoelectronic devices is limited. If the state variable hits the boundaries then no further enlargement or reduction is possible. Special precautions are necessary to ensure that the limits can be left. This behaviour is called the "terminal state-problem" [22]. More or less sophisticated solutions have been proposed to solve this memristor modeling problem (see for instance Biolek et al. [4], Corinto and Ascoli [8], Corinto et al. [9], Johlekar and Wolf [14], Kvatinsky et al. [16], Prodromakis et al. [22]).

However, we know a similar question handling the integrator windup [2] if saturation at the integrator's output occurs. It is obvious that we can try to handle the terminal state-problem similar to modeling a limited integrator block. We will show how this can be done in a straightforward way using hybrid dynamical systems (see for instance Abate et al. [1], Goebel et al. [12], Lee and Seshia [17]) as modeling approach for nanoelectronic memristive devices. This concept can be simply applied using behavioural modeling languages for analog mixed-signal systems as VHDL-AMS and Verilog-AMS. At the present time, besides SPICE models [3, 4], Verilog-A(MS) models for memristors are proposed [10, 15]. However, existing models do not make use of mixed-signal languages' capabilities. In the subsequent sections, we will figure out the basics of the modeling approach and describe its implementation using VHDL-AMS [13] examples.

5.2 Basic Concepts

5.2.1 Hybrid Dynamical Systems

We start with a short account on some basics of how to describe hybrid dynamical systems. Discrete modes and time-continuous state variables characterize the behaviour of hybrid dynamical systems. We recap some fundamental ideas regarding the modelling requirements based on Abate et al. [1], Goebel et al. [12].

A hybrid dynamic system is described by a finite set Q of discrete modes (also referred to as discrete states). We can assign a domain $D_i \subseteq \mathbb{R}^{n_i}$ to each mode $i \in Q$. Regarding the current status of mixed-signal behavioural modelling languages, we will only consider the case that all n_i are equal. The time-continuous behaviour of a mode is given by a differential inclusion F (flow map) with $x'(t) \in F_i(x(t), inp(t), t)$ with $x(t) \in D_i$ and input $inp(t)$. Such an inclusion can be described as an explicit or implicit system of differential and algebraic equations. The output $y(t) \in Y \subseteq \mathbb{R}^{n_Y}$ of the hybrid dynamical system depends on the input u of the system and the state x of the active mode.

Possible transitions from one mode to another can be represented by a graph whose nodes form the set Q of the modes. The transitions are directed edges in this graph. The start mode of an edge e is referred as $s(e)$. The target mode is $z(e)$. A transition defined by an edge e is activated if a guard condition $G_{s(e),z(e)}(x, u, t)$ is fulfilled and $s(e)$ is the current mode. Furthermore, we have to describe how after the activation of an edge at time $t = t_a$ the new state $x^+(t_a)$ in $D_{z(e)}$ can be determined based on the value $x(t_a)$ in $D_{s(e)}$. This procedure is called set action.

In the initialization phase at time $t = 0$ and during a transition from one mode to another at time t_a, we need to assure that the values $x(0)$ and $x^+(t_a)$, respectively, are consistent initial values for the corresponding flows. If the flow is given by an explicit

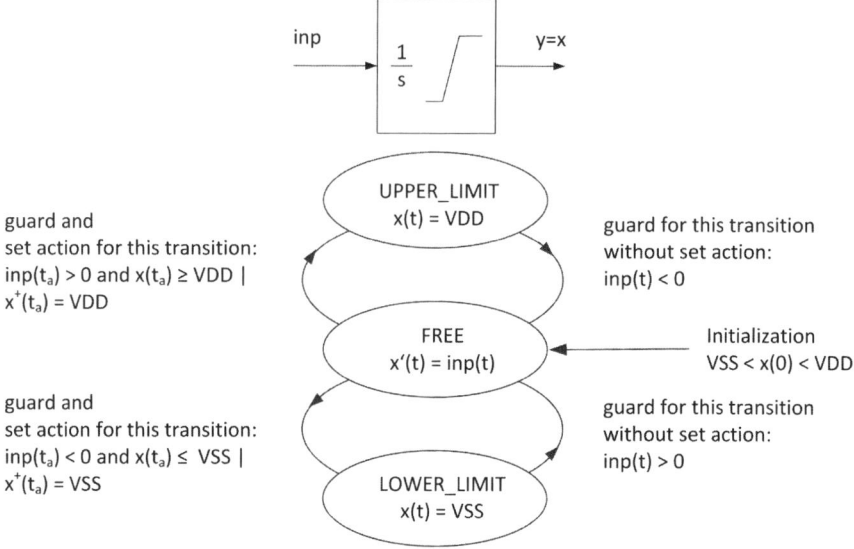

Fig. 5.1 Description of a limited integrator by a hybrid dynamical system

system of differential equations, it is in general sufficient that the value belongs to the domain of the associated mode.

Example

Figure 5.1 illustrates the concept on behalf of a limited integrator, where x is the integral of the input *inp*. However, it is limited to the closed interval [*VSS, VDD*]. After the limits *VSS* or *VDD* have been reached, no further integration is carried out until the input *inp* will result in increasing or decreasing the value of the integral, respectively. The output y equals to the continuous state x.

5.2.2 Memristor Modelling Aspects

Chua and Kang [7] extended the memristor concept to memristive systems. Chua [6] remarked that it is a nonlinear circuit foundation for nanodevices. Further extension to memcapacitors and meminductors were figured out in Di Ventra et al. [11].

We restrict the subsequent representation to time-invariant first-order memristive one-ports. A first-order time-invariant current-controlled memristive one-port is defined by:

$$x'(t) = f(x(t), i(t)) \tag{5.1}$$
$$v(t) = R(x(t)) \cdot i(t) \tag{5.2}$$

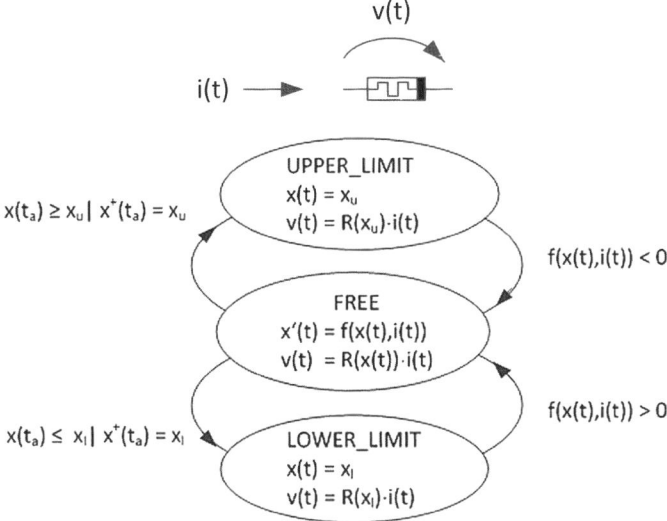

Fig. 5.2 Current-controlled memristive device with $x(t) \in [x_l, x_u]$

$v(t)$ and $i(t)$ denote the voltage and current across the device at time t, respectively. x represents one internal state variable. f is a map $f : \mathbb{R} \times \mathbb{R} \to \mathbb{R}$. R is called memristance. The device is passive if and only if $R(x(t)) \geq 0$ for all t. The current i controls the state x that condenses the past. The memristor remembers the past even without power if the state x is always stable when no current flows through the device, that means $\bigvee x(t)(f(x(t), 0) = 0)$.

In a corresponding manner, we can define a first-order time-invariant voltage-controlled memristive one-port

$$x'(t) = f(x(t), v(t)) \tag{5.3}$$
$$i(t) = G(x(t)) \cdot v(t) \tag{5.4}$$

where G is called the memductance.

For the current-controlled memristive device, we are going to discuss the consequences of a limitation of the state variable x to a closed interval $I = [x_l, x_u]$. Thus, if we note $x(t) \leq x_l$, we must stop the integration given by the Eq. (5.1) and set $x(t) = x_l$. We can only continue with the integration when $f(x(t), i(t)) > 0$. Analogously, we have to stop if $x(t) \geq x_u$ and set $x(t) = x_u$. We continue with the integration when $f(x(t), i(t)) < 0$. Figure 5.2 shows this behaviour.

The voltage-controlled memristive device can be handled in an analogous manner.

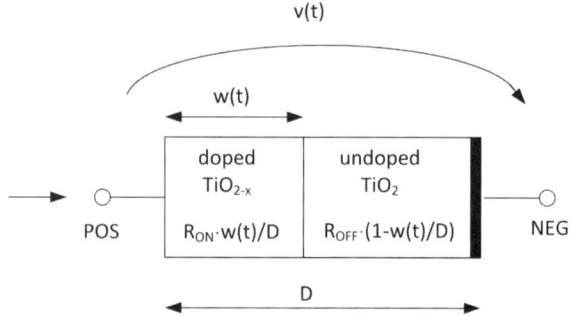

Fig. 5.3 Principle arrangement of the TiO₂ memristive device [16, 23]

5.3 Memristor Models

5.3.1 Model with Linear Drift

The TiO$_2$ device (Fig. 5.3) of the HP Labs is based on changing the boundary between a lower resistance part and a high resistance part [23].

The position w of the boundary is influenced by the current i through the device. Using $x = \frac{w}{D}$ as state variable, the behaviour of the device is characterized by [23]:

$$x'(t) = \mu_V \cdot \frac{R_{ON}}{D^2} \cdot i(t) \tag{5.5}$$

$$v(t) = (R_{ON} \cdot x(t) + (1 - x(t)) \cdot R_{OFF}) \cdot i(t) \tag{5.6}$$

μ_V is the ion mobility for the linear ion-drift model that is given by Eq. (5.5). D is the physical width of the device. The width w can change between 0 and D. Thus the state variable x is in the interval [0, 1]. We can distinguish three modes. The memristance is R_{ON} if the width w is at the upper limit D. This is mode ONE (value of state x is 1). If the width w is at the lower limit 0, the memristance is R_{OFF}. We call this case mode ZERO (value of state x is 0). If the width w is between 0 and D, the memristance is $R_{ON} \cdot x + (1 - x) \cdot R_{OFF}$ and the mode is FREE. To avoid "chattering" of the model, we require $i(t) > \varepsilon_i > 0$ and $i(t) < -\varepsilon_i < 0$ to switch from mode ZERO to FREE and from mode ONE to FREE, respectively, where ε_i is a sufficient small value.

Figure 5.4 shows the corresponding state diagram. Based on the diagram, a model can easily be implemented using a mixed-signal behavioural modelling language. A VHDL-AMS implementation that illustrates the principal procedure is given in Listings 5.1 and 5.2. The entity declaration in Listing 5.1 describes the interface as shown in Fig. 5.3. The architecture in Listing 5.2 represents the behaviour from Fig. 5.4.

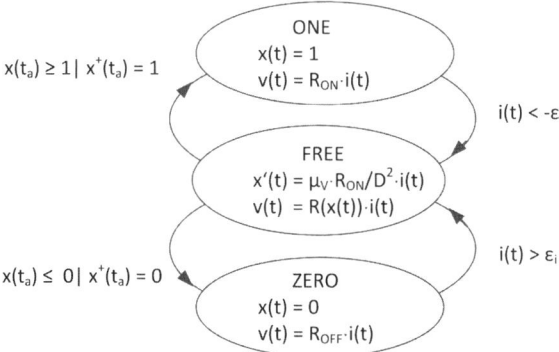

Fig. 5.4 Memristor model with linear drift with $x(t) \in [0, 1]$

Listing 5.1 VHDL-AMS entity declaration of the memristor model with linear drift

```
1   library IEEE;
2   use IEEE.ELECTRICAL_SYSTEMS.all;
3   use IEEE.MATH_REAL.all;
4
5   entity MEMRISTOR is
6     generic (-- Initial state (between 0 and 1)
7               X0          : REAL := 0.1;
8               R_OFF       : REAL := 16.0E3;
9               R_ON        : REAL := 100.0;
10              D           : REAL := 10.0E-9;
11              UV          : REAL := 1.0E-14);
12     port    (terminal POS  : ELECTRICAL;
13              terminal NEG  : ELECTRICAL);
14     constant EPS_I : REAL := 1.0E-9;
15   begin
16     assert (X0 > 0.0) and (X0 < 1.0)
17       report "ERROR: X0 in (0,1) required."
18       severity ERROR;
19   end entity MEMRISTOR;
```

The model code is largely self-explanatory. The simultaneous statements (with == equal sign) represent the Eqs. (5.5) and (5.6). The **process** statement describes the update of the mode. The **wait on** statement checks whether one of the guard conditions is activated. Afterwards, the reaction on the active guard is carried out. The model must be initialized with a value X0 greater than 0 and less than 1 (mode FREE). This is only done to simplify the representation. With some small extensions for the operating point analysis (DOMAIN equals QUIESCENT_DOMAIN), the initial values 0 and 1 are also possible. An implementation using Verilog-A/Verilog-AMS can be carried out in the same manner. It is of course possible to reproduce the results from Strukov et al. [23] that are used to discuss properties of memristive devices, which are based on the linear drift

Listing 5.2 VHDL-AMS architecture definition of the memristor model with linear drift

```
1    architecture LINEAR of MEMRISTOR is
2        type    MODE_TYPE is (FREE, ZERO, ONE);
3        signal MODE      : MODE_TYPE := FREE;
4        quantity V across I through POS to NEG;
5        quantity X        : REAL;
6    begin
7        if      DOMAIN = QUIESCENT_DOMAIN use
8                X   == X0;
9        elsif   MODE    = ZERO                    use
10               X   == 0.0;
11       elsif   MODE    = ONE                     use
12               X   == 1.0;
13       else
14               X'DOT == (UV/(D**2))*R_ON*I;
15       end use;
16
17       break on MODE;
18       V    == (R_ON*X + (1.0-X)*R_OFF)*I;
19
20   P1: process is
21       begin
22       wait on DOMAIN;
23       while TRUE loop
24        wait on X'ABOVE(0.0),    X'ABOVE(1.0),
25                I'ABOVE(EPS_I), I'ABOVE(-EPS_I);
26        if     MODE = FREE then
27                if    X <= 0.0 then
28                        MODE  <= ZERO;
29                elsif X >= 1.0 then
30                        MODE  <= ONE;
31                end if;
32        elsif MODE = ZERO then
33                if    I > EPS_I then
34                        MODE <= FREE;
35                end if;
36        elsif MODE = ONE   then
37                if    I < -EPS_I then
38                        MODE <= FREE;
39                end if;
40        end if;
41        end loop;
42        end process P1;
43
44   end architecture LINEAR;
```

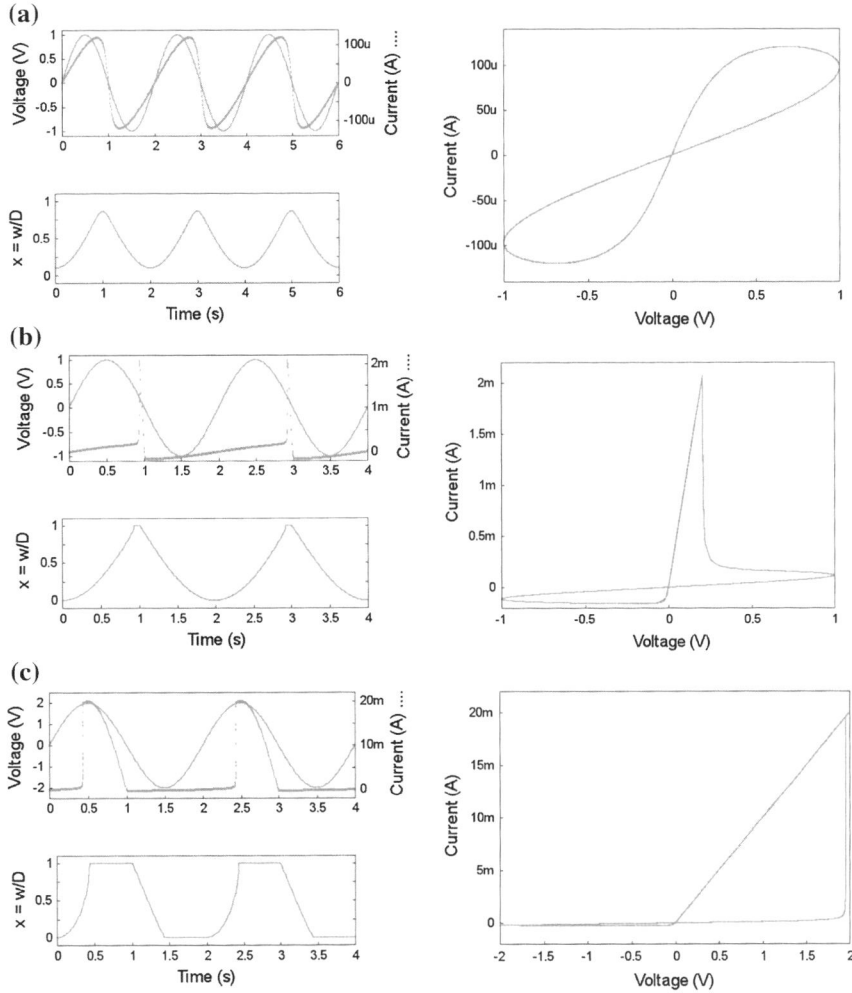

Fig. 5.5 Simulation results for the linear drift model (parameters from Strukov et al. [23])

model ($\mu_V = 10^{-14} \text{m}^2/\text{Vs}$, $D = 10nm$). In the upper left diagrams of Fig. 5.5, i and v depending on time t are shown. The lower left diagrams show x. The right diagrams represent $i - v$ plots, where i depending on v is shown.

5.3.2 Models with Nonlinear Drift

Strukov et al. [23] already mentioned that the linear drift model is not in accordance with the observation in thin film devices. To describe the qualitative behaviour correctly, they multiplied Eq. (5.5) by a window function that is equivalent to

Table 5.1 Window functions

Name	Description of the window function	$g(0)$	$g(1)$	Ref.
Strukov	$g(x) = x(1-x)$	0	0	[23]
Joglekar	$g(x) = 1 - (2x-1)^{2p}$ with $p \in \mathbb{Z}^+ = \mathbb{N} \setminus \{0\}$	0	0	[14]
Biolek	$g(x) = 1 - (x-1)^{2p}$ for current $i \leq 0$ $g(x) = 1 - x^{2p}$ for current $i > 0$ with the same $p \in \mathbb{Z}$ in both cases	0 1	1 0	[4]
Prodromakis	$g(x) = j \cdot ((x-0.5)^2 + 0.75)^p$ where $p \in \mathbb{R}^+$ and the window function can be scaled to lower or higher values with j	0	0	[22]

$g(x) = x \cdot (1-x)$. This function is sometimes called Strukov function after the first name in the list of authors of Strukov et al. [23].

Using this window function, there is no further drift when the state reaches the boundaries ($x(t) = 0$ or $x(t) = 1$). This is in accordance with the physics of the device. For state values between 0 and 1, the drift velocity $x'(t)$ is not constant. It depends for a constant current i on x. That means, nonlinear drift is described by a model with $g(x) \neq \text{const}$ in the range of x:

$$x(t) = \mu_V \cdot \frac{R_{ON}}{D^2} \cdot g(x(t)) \cdot i(t) \tag{5.7}$$

$$v(t) = (R_{ON} \cdot x(t) + (1 - x(t)) \cdot R_{OFF}) \cdot i(t) \tag{5.8}$$

Other window functions $g : \mathbb{R} \to \mathbb{R}$ have been proposed and applied over the last years. Table 5.1 gives an overview over them.

Using these window functions and values $x_0 \in (0, 1)$ to initialize a memristor model, the x values can move asymptotically to 0 or 1. For numerical reasons, they can also reach 0 and 1. Looking closer at Fig. 5.2, we detect the difficulties to move back from the lower limit mode (LOWER_LIMIT or ZERO) or the upper limit mode (UPPER_LIMIT or ONE) to FREE. For window functions g with values $g(0) = 0$ or $g(1) = 0$, we cannot leave mode LOWER_LIMIT (ZERO) and UPPER_LIMIT (ONE), respectively, because $\bigvee i(t)(f(0, i(t)) = \mu_V \cdot \frac{R_{ON}}{D^2} \cdot g(0) \cdot i(t) = 0)$ and $\bigvee i(t)(f(1, i(t)) = \mu_V \cdot \frac{R_{ON}}{D^2} \cdot g(1) \cdot i(t) = 0)$, respectively. This means that the required guard condition is never fulfilled. This situation describes the problem that is usually called the terminal-state problem [22].

Biolek window function avoids this problem by switching between two function descriptions depending on the direction of the current i. This might be one reason for using Biolek window function as a popular candidate for model implementations [4, 10].

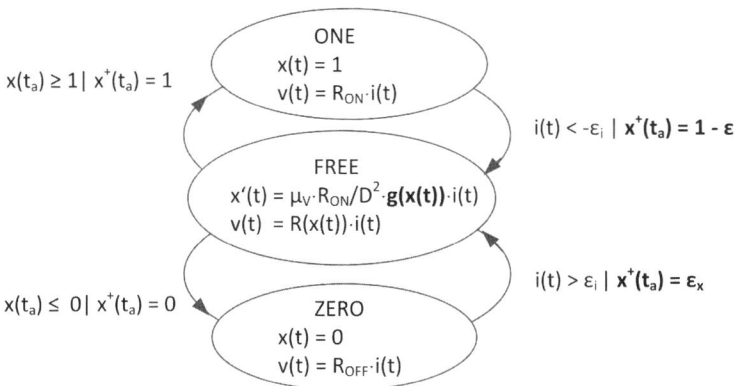

Fig. 5.6 Memristor model with nonlinear drift (based on window functions g) with $x(t) \in [0, 1]$

We sketch a more general solution for the family of models described by Eqs. (5.7) and (5.8) in Fig. 5.6. The current i is guarded in modes ZERO and ONE. If a guard condition is fulfilled, a set action assigns a new value to x for the continuation in mode FREE. We assign $x^+(t) = \epsilon_x > 0$ or $x^+(t) = 1 - \epsilon_x < 1$ using a small, but sufficiently large enough, $\epsilon_x > 0$. This approach assures that we avoid the terminal-state problem in a systemic manner.

The VHDL-AMS implementation requires only some small modifications of the code. For instance, the statement:

```
X'DOT  ==  (UV/(D**2))*R_ON*I;
```

has to be replaced by:

```
X'DOT  ==  (UV/(D**2))*R_ON*G(X)*I;
```

to consider a window function G. The set action can be carried out in the process statement using the **break** statement. The affected part can be modified to:

```
   . . .
   elsif MODE = ZERO then
           if I > EPS_I then
               MODE <= FREE;
               break X => EPS_X;
 -- inserted
           end if;
   elsif MODE = ONE  then
           if I < -EPS_I then
               MODE <= FREE;
               break X => (1.0 - EPS_X);
 -- inserted
           end if;
   . . .
```

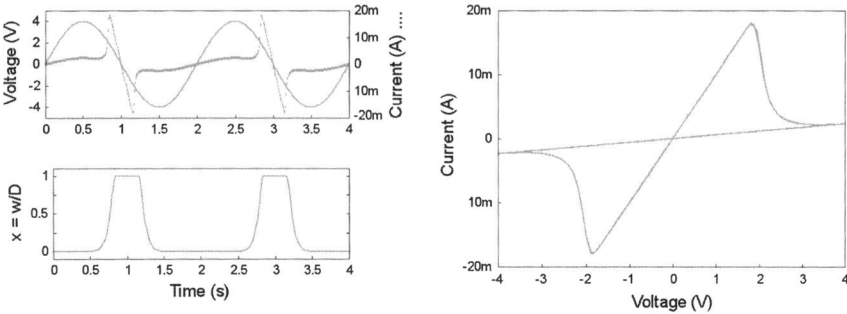

Fig. 5.7 Simulation results for the nonlinear drift model with Strukov window ($v(t) = 4\,\text{V} \cdot \sin\left(\frac{2\pi}{2\text{s}} \cdot t\right)$, $R_{\text{ON}} = 100\,\Omega$, $R_{\text{OFF}} = 1.7\,\text{k}\Omega$ (instead of $\frac{R_{\text{OFF}}}{R_{\text{ON}}} = 50$ for Fig. 5.3c in Strukov et al. [23]))

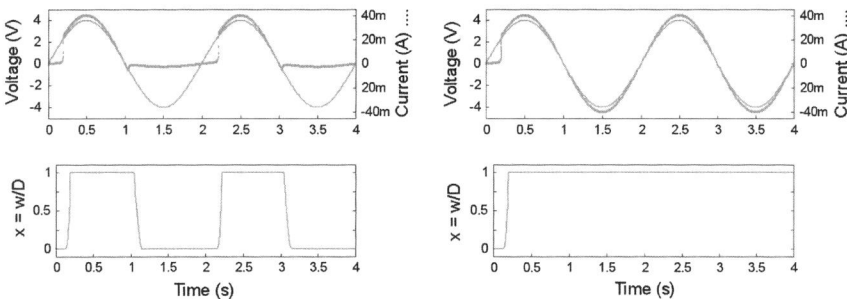

Fig. 5.8 Simulation results for the nonlinear drift model with Joglekar window (left with state handling as shown in Fig. 5.6, right without – x is fixed w). $v(t) = 4\,\text{V} \cdot \sin\left(\frac{2\pi}{2\text{s}} \cdot t\right)$, $R_{\text{ON}} = 100\,\Omega$, $R_{\text{OFF}} = 1.7\,\text{k}\Omega$ (Joglekar window function with $p = 3$)

Figure 5.7 shows simulation results with the Strukov window function. Figure 5.8 shows simulation results with a Joglekar window function. The left diagrams show voltage v and the corresponding state x using a model that applies the state handling as figured out in this section. The right diagrams show results of a simulation with a model without state handling. We observe that the state x is fixed after x hits once the upper limit 1. The terminal-state problem is not handled correctly.

5.3.3 Further Models

Over the past years, further models have been investigated to describe the behaviour of nanoelectronic memristive devices that use Strukov et al. [23] as starting point.

Lehtonen and Laiho [18] presented a nonlinear voltage controlled memristor model. The drift velocity of the state x is given by $x'(t) = a \cdot g(x(t)) \cdot v(t)^q$, where a is a real constant, g is a window function, and p is an odd natural number. They required to restrict x in $[s_1, s_2]$, where $\bigvee x(t) \in [s_1, s_2] \, (g(x(t) > 0)$. Kvatinsky et al. [16], Pickett et al. [21] developed a further model where a resistor is in series with an electron tunnel barrier also known as Simmons tunnel barrier model. The velocity of the tunnel barrier width is described by a complex analytical relation that depends on the direction of the current. The width movement is restricted to a limited range. Kvatinsky et al. [15, 16] introduced simplified assumptions to make the Simmons tunnel diode model computational more efficient. They proposed equations for their ThrEshold Adaptive Memrister Model (TEAM) that constrain the state variable to $x(t) \in [x_{ON}, x_{OFF}]$. Restrictions also occur in the Boundary Condition-based Model (BCM) [8, 9]. The approach presented in this paper can successfully be applied in all these cases.

5.4 Conclusion

We demonstrated that hybrid dynamical systems are a promising approach to overcome the terminal-state problem in current models of nanoelectronic memristive devices. We can easily implement such models using modelling languages such as VHDL-AMS that support mixed-signal modelling features. Verilog-AMS/ Verilog-A are also good candidates for doing this. Development of memristive devices continues. Investigators study fundamental issues and problems in the realisation of memristors [19] as well as they apply new principles. For instance, Valov et al. [26] report on an arrangement that is no longer passive and describe it using a model with more than one state. However, the terminal-state problem remains. Thus, we expect that the presented approach can also help with solving future modelling requirements.

Acknowledgments The work reported in this chapter was done within the project CoolEDesign that is part of the Leading-Edge Cluster "Cool Silicon" which is sponsored by the German Federal Ministry of Education and Research (BMBF) within the scope of the Leading-Edge Cluster Competition.

References

1. Abate A, D'Innocenzo A, Di Benedetto MD, Sastry S (2009) Understanding deadlock and live-lock behaviors in hybrid control systems. Nonlinear Anal: Hybrid Syst 3(2):150–162. doi:10.1016/j.nahs.2008.12.005
2. Åström KJ, Rundqwist L (1989) Integrator windup and how to avoid it. In: Proceedings of the American control conference 1989, pp 1693–1698
3. Batas D, Fiedler H (2011) A memristor SPICE implementation and a new approach for magnetic flux-controlled memristor modeling. IEEE Trans Nanotechnol 10(2):250–255. doi:10.1109/TNANO.2009.2038051
4. Biolek Z, Biolek D, Biolova V (2009) SPICE model of memristor with nonlinear dopant drift. Radioengineering 18(2):210–214. http://radioeng.cz/fulltexts/2009/09_02_210_214.pdf

5. Chua LO (1971) Memristor—the missing circuit element. IEEE Trans Circuit Theory 18:507–519. doi:10.1109/TCT.1971.1083337
6. Chua LO (2003) Nonlinear circuit foundations for nanodevices. Part I: the four-element torus. Proc IEEE 91(11):1830–1859. doi:10.1109/JPROC.2003.818319
7. Chua LO, Kang SM (1976) Memristive devices and systems. Proc IEEE 64(2):209–223. doi:10.1109/PROC.1976.10092
8. Corinto F, Ascoli A (2012) A boundary condition-based approach to the modeling of memristor nanostructures. IEEE Trans Circuits Syst I: Regul Pap 59:2713–2726. doi:10.1109/TCSI.2012.2190563
9. Corinto F, Ascoli A, Gilli M (2012) Mathematical models and circuit implementations of memristive systems. In: Proceedings of the international workshop on cellular nanoscale networks and their applications (CNNA) 2012. doi:10.1109/CNNA.2012.6331458
10. da Costa HJB, de Assis Brito Filho F, de Araujo do Nascimento PI (2012) Memristor behavioural modeling and simulations using Verilog-AMS. In: Proceedings of the IEEE Latin American symposium on circuits and systems (LASCAS) 2012, pp 1–4. doi:10.1109/LASCAS.2012.6180334
11. Di Ventra M, Pershin YV, Chua LO (2009) Circuit elements with memory: memristors, memcapacitors, and meminductors. Proc IEEE 97(10):1717–1724. doi:10.1109/JPROC.2009.2021077
12. Goebel R, Sanfelice RG, Tee AR (2012) Hybrid dynamical systems: modeling, stability, and robustness. Princeton University Press, Princeton
13. IEC 61691–6/IEEE Std 1076.1-2009 (2009) Behavioural languages—Part 6: VHDL analog and mixed-signal extensions, 1st edn. doi:10.1109/IEEESTD.2009.5464492
14. Joglekar YN, Wolf SJ (2009) The elusive memristor: properties of basic electrical circuits. Eur J Phys 30(4):661. http://stacks.iop.org/0143-0807/30/i=4/a=001
15. Kvatinsky S, Talisveyberg K, Fliter D, Friedman EG, Kolodny A, Weiser UC (2011) Verilog-A for memristor models. Technical Report 801, Center for Communication and Information Technlogies (CCIT). http://webee.technion.ac.il/people/skva/Memristor
16. Kvatinsky S, Friedman EG, Kolodny A, Weiser UC (2013) TEAM: ThrEshold adaptive memristor model. IEEE Trans Circuits Syst I: Regul Pap 60:211–221. doi:10.1109/TCSI.2012.2215714
17. Lee EA, Seshia SA (2011) Introduction to embedded systems, A cyber-physical systems approach. Lulu.com. http://LeeSeshia.org/
18. Lehtonen E, Laiho M (2010) CNN using memristors for neighborhood connections. In: Proceedings of the international workshop on cellular nanoscale networks and their applications (CNNA) 2010. doi:10.1109/CNNA.2010.5430304
19. Meuffels P, Soni R (2012) Fundamental issues and problems in the realization of memristors. Preprint: arXiv:1207.7319v1 [cond-mat.mes-hall]. http://arxiv.org/pdf/1207.7319v1
20. Niu D, Xiao Y, Xie Y (2012) Low power memristor-based ReRAM design with error correcting code. In: Proceedings of the 17th Asia and South Pacific design automation conference (ASP-DAC) 2012, pp 79–84. doi:10.1109/ASPDAC.2012.6165062
21. Pickett MD, Strukov DB, Borghetti JL, Yang JJ, Snider GS, Stewart DR, Williams RS (2009) Switching dynamics in titanium dioxide memristive devices. J Appl Phys 106(7):074508–074508-6. doi:10.1063/1.3236506
22. Prodromakis T, Peh BP, Papavassiliou C, Toumazou C (2011) A versatile memristor model with nonlinear dopant kinetics. IEEE Trans Electron Devices 58(9):3099–3105. doi:10.1109/TED.2011.2158004
23. Strukov DB, Snider GS, Stewart DR, Williams RS (2008) The missing memristor found. Nature 453:80–83. doi:10.1038/nature06932
24. Tetzlaff R, Bruening A (2012) Memristor technology in future electronic system design. In: Proceedings of the design, automation test in Europe conference and exhibition (DATE) 2012, p 592. doi:10.1109/DATE.2012.6176541

25. Tetzlaff R, Schmidt T (2012) Memristors and memristive circuits—an overview. In: Proceedings of the IEEE international symposium on circuits and systems (ISCAS) 2012, pp 1590–1595. doi:10.1109/ISCAS.2012.6271557
26. Valov I, Linn E, Tappertzhofen S, Schmelzer S, van den Hurk J, Lentz F, Waser R (2013) Nanobatteries in redox-based resistive switches require extension of memristor theory. Nat Commun 4: http://dx.doi.org/10.1038/ncomms2784

Chapter 6
Code Generation Alternatives to Reduce Heterogeneous Embedded Systems to Homogeneity

Franco Fummi, Michele Lora, Francesco Stefanni and Sara Vinco

Abstract The high level of heterogeneity of modern embedded systems forces designers to use different computational models and formalisms, thus making reuse and integration very difficult tasks. Reducing such an heterogeneity to a homogeneous implementation is a key solution to allow both simulation and validation of the system. Furthermore, the implementation may be executed on highly optimized architectures or used as a starting point for redesign flows. This paper proposes two novel flows to gain a homogeneous implementation of a starting heterogeneous system, thus showing how heterogeneity can be reconciled to a single language, still preserving correctness. The target languages are SystemC-AMS, that enhances support for continuous behaviors and allows complete validation, and C++, an executable implementation that can be the starting point of redesign flows. The approaches are compared with respect to state-of-the-art techniques in terms of performance and accuracy, also through the application to a complex case study.

Keywords Heterogeneous embedded systems · Homogeneous code generation · SystemC code generation · SystemC-AMS Linear Signal Flow (LSF) · C++ code generation · Heterogeneous components integration · Heterogeneous system simulation · Component-based design · Model of computation · Automata management

F. Fummi (✉) · F. Stefanni
EDALab s.r.l., Verona, Italy
e-mail: franco.fummi@edalab.it

F. Stefanni
e-mail: francesco.stefanni@edalab.it

F. Fummi · M. Lora · S. Vinco
University of Verona, Verona, Italy
e-mail: franco.fummi@univr.it

M. Lora
e-mail: michele.lora@univr.it

S. Vinco
e-mail: sara.vinco@univr.it

© Springer International Publishing Switzerland 2015
M.-M. Louërat and T. Maehne (eds.), *Languages, Design Methods, and Tools for Electronic System Design*, Lecture Notes in Electrical Engineering 311, DOI 10.1007/978-3-319-06317-1_6

6.1 Introduction

Since their introduction, embedded systems have become more and more heterogeneous [9]. On the one hand, analog and digital HW coexists with the SW managing the system, thus introducing integration issues focused on HW-SW communication. On the other hand, embedded systems are more embedded in the physical environment, that must be included as a design constraint to preserve a correct design flow.

This heterogeneity makes integration and reuse of already existing components a very difficult task [7]. Designers are forced to use formalisms, languages, and Models of Computation (MoCs) that fit better with each specific domain. Thus, interaction across different domains becomes very challenging and time demanding. Integrating such existing components, through co-simulation [2] or component-based flows [4], is error prone as there is no guarantee on the correctness of the integrated system.

A key solution is to support integration and reuse by generating of a *homogeneous implementation* of the starting heterogeneous system. Indeed, this implementation can simulate the system behavior and validate both interaction and functionality of the result of integration. On the other hand, the implementation is as an executable specification of the functionality, and it thus can be run on optimized architectures providing a SW version of the system. Finally, the generated code can be the starting point of redesign flows, thus enhancing design space exploration.

In literature, many approaches tried to tackle embedded system heterogeneity. On one hand, *Model-Based Design (MBD) approaches* propose top-down flows, where a model of the system is gradually refined, providing as a result a code implementation of the whole functionality in the chosen MoCs [1, 6]. Unfortunately, adopting strict top-down methodologies makes reuse of already existing components very difficult, as integrating different MoCs is far from trivial [9]. On the contrary, bottom-up methodologies and component reuse are supported by several *co-simulation frameworks* [2], where each component is simulated in its own native simulator. As a result, co-simulation assembles components without providing a rigorous formal support and preserving the degree of heterogeneity of the starting system. This makes integration and validation very hard tasks. Furthermore, homogeneous code implementation of the system functionality is not supported.

This posed a urge for a computational model able to cover the heterogeneity typical of embedded systems, but supported by automatic tools to allow and ease reuse [9]. To cover this lack, Di Guglielmo et al. [7] proposed UNIVERCM, an automaton-based formalism that allows to model both analog and discrete behaviors, together with typical SW behaviors. They also presented the mapping from heterogeneous components to UNIVERCM. Unfortunately, so far the effectiveness of UNIVERCM is limited, as code generation from the formal homogeneous representation has been targeted only weakly.

This chapter exploits UNIVERCM to reduce heterogeneous components to a homogeneous formalism (top of Fig. 6.1). Indeed, UNIVERCM covers a wide range of heterogeneous domains [7] and it allows to reduce the heterogeneity to a single formalism. From this starting point, the paper targets a comparison between various

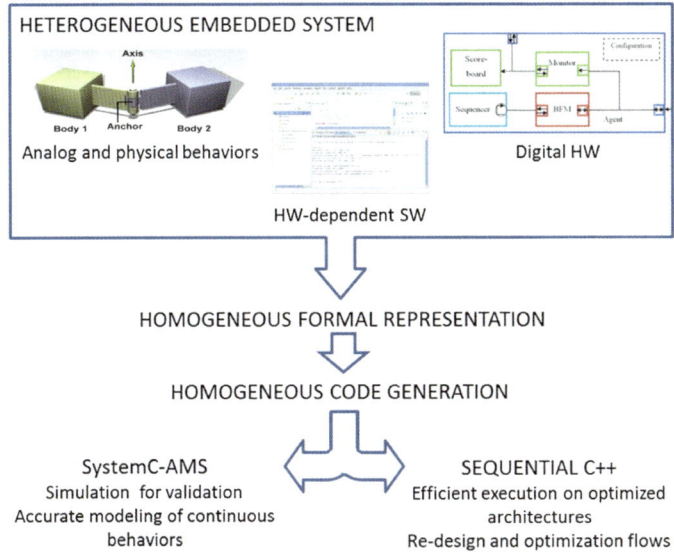

Fig. 6.1 Proposed methodology: the heterogeneous system is reconciled to a homogeneous formalism, to achieve then code generation for simulation, execution in SW, and to apply redesign

flavors of C++ code generation, targeting the implementation of a homogeneous representation of the starting heterogeneous embedded system. In detail, the chapter focuses on two flavors: sequential C++ and SystemC-AMS [11] (bottom of Fig. 6.1). The generated code can be used to validate the integrated system by checking the result of integration. To this extent, SystemC-AMS is adopted, as it fully supports analog evolution. The generated code may also be used to apply redesign flows and to execute the system functionality on optimized and parallel processors. In this case, C++ code generation provides a flexible starting point.

As a result, the focus of the chapter will be on two methodologies that start from a heterogeneous system to automatically generate 1. an efficient C++ implementation and 2. a SystemC-AMS implementation. The target is to obtain a simulatable description of a system by using the MoC, that allows to fully validate the system reproducing also its continuous behaviors. The proposed flows are compared also with state of the art approaches, to highlight both effectiveness and portability.

6.2 The UNIVERCM Computational Model

The UNIVERCM formalism is an automaton-based representation that unifies the modeling of both the analog (i.e., continuous) and the digital (i.e., discrete) domains, as well as hardware-dependent SW. A formal and complete definition is available in Di Guglielmo et al. [7]. UNIVERCM has been chosen as starting point of the

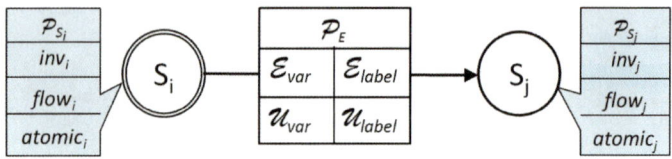

Fig. 6.2 Graphical representation of a UNIVERCM automaton

proposed methodology for two different reasons. The paper also proposed a deep analysis on the applicability of UNIVERCM to heterogeneous domains, thus showing that the computational model allows to cover the heterogeneity that characterizes embedded systems. Also, the paper showed that it is possible to provide formal rules and automatic tools to convert the heterogeneity to UNIVERCM and to produce a homogeneous simulatable implementation of the generated UNIVERCM system. Thus, reuse and bottom-up design are enhanced.

The rest of this section highlights the main characteristics of UNIVERCM. In each UNIVERCM automaton, states model the continuous dynamics of the system, whereas edges between states model its discrete dynamics. A UNIVERCM automaton can be depicted as shown in Fig. 6.2.

States are characterized with three predicates. The flow predicate (*flow*) constrains the evolution of continuous variables into the state. The invariant predicate (*inv*) specifies whether it is possible to remain into the state or not, depending on a set of conditions on variables. Finally, the atomicity predicate (*atomic*) allows to specify sections of the UNIVERCM automaton that are traversed as one single transition when executing more automata in parallel.

The activation of an *edge* between two states is constrained by a guard ($\mathcal{E}_{var}$) and a set of incoming labels ($\mathcal{E}_{label}$), i.e., the edge can be traversed only if the guard is satisfied and the incoming synchronization labels are received. When the edge is traversed, the values of continuous and discrete variables are updated as specified by an update function associated to the edge ($\mathcal{U}_{var}$). Furthermore, a set of synchronization labels listed into the set of outgoing labels of the edge are activated ($\mathcal{U}_{label}$), to allow synchronization with other automata.

Both edges and states are associated with a *priority*, that allows to partially order states and edges and to model non-deterministic as well as deterministic behaviors.

6.3 Code Generation from UNIVERCM

When generating code from a UNIVERCM description, the UNIVERCM semantics must be mapped to the semantics of the target language and platform. This step requires to provide support for three main aspects of UNIVERCM:

- Management of the *evolution*, as activation of automata and update of variables and synchronization labels by respecting the UNIVERCM semantics;

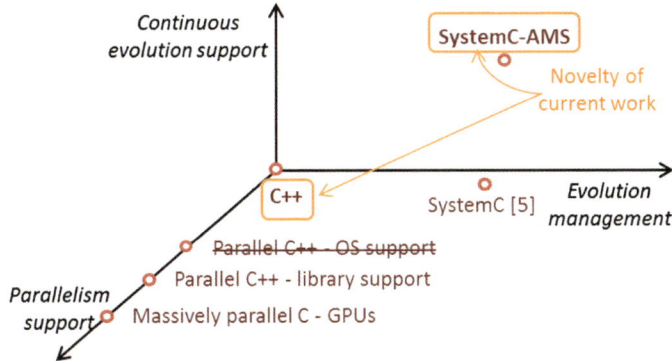

Fig. 6.3 Taxonomy of code generation alternatives. This chapter focuses on the generation of SystemC-AMS code and of sequential C++ code

- *Parallelism*, to reproduce the composition semantics of UNIVERCM, realized as automata that evolve in parallel and share variables and synchronization events;
- Implementation of *continuous behaviors*, with support libraries or through discretization steps.

These aspects constitute the three dimensions of the taxonomy of target languages proposed in Fig. 6.3.

The barest support is provided by *sequential C/C++*, as C++ does not natively support any of the listed aspects. As such, code generation must implement also a scheduling capability to reproduce the UNIVERCM semantics, while continuous evolution must be discretized with mathematical approaches. Parallelism support can be provided either by the operating system (e.g., with the `pthread` library) or with ad-hoc libraries (e.g., OpenMP and MPI). Unfortunately, the OS support implies a heavy overhead that makes this alternative non viable for efficient parallelism [8]. On the other hand, ad-hoc libraries require to take into account restrictions of the target architecture, such as constraints on resource usage or communication overhead. This part will be developed as future work.

Support for evolution management is provided by *SystemC*, that has been first explored for code generation in Di Guglielmo et al. [5]. SystemC supports execution with a complex event-driven semantics. The scheduler, in charge of activating computation and of managing data and event update in the system, is exploited to map the UNIVERCM semantics. Parallelism is emulated, as all parallel entities behave as if they were executed in parallel, despite of their execution being sequential. The main drawback of SystemC is that continuous evolution is not supported. To cover this lack, the Accellera consortium developed the *SystemC-AMS* standard as an extension of SystemC. It provides the possibility of describing continuous dynamics through a variety of styles such as Laplace domain descriptions or by connecting continuous base constructs.

Given this wide range of alternatives, this chapter will focus on sequential C++ code generation (for efficiently executing the system functionality in SW or to apply redesign flows) and SystemC-AMS code generation (for validating the system with the support of analog behaviors and non-determinism).

6.3.1 Generation of C++ Code

6.3.1.1 Automata Management

Each UNIVERCM automaton is mapped to a C++ function representing the whole automaton evolution. A state variable is used to store the current state of the automaton. The state variable is defined as an enumeration over the set of automaton states and its default value is associated to the initial state s_0 of the automaton.

The function generated from an automaton is built as a **switch** statement, where each case label represents one of the automaton states. Each state case lists then the implementation of all the outgoing edges and of the delay transition provided for the state, as shown in Fig. 6.4 on the right side.

Each $edge \in EDG$ is implemented as an **if** or **else if** statement, which guard is a logic conjunction of the enabling condition of the edge $\mathcal{E}_{var}$ and of the activation condition on labels $\mathcal{E}_{label}$. The body, which is executed when the guard is satisfied, includes the update of variables $\mathcal{U}_{var}$ and the activation of labels $\mathcal{U}_{label}$. Furthermore, the state variable is updated to the destination state of the edge (e.g., lines 3–7 and 12–16 of the non-deterministic code of Fig. 6.4).

The default case of the **switch** statement is an *error handling* branch, which is managing cases when the automaton can neither perform continuous evolution nor traverse outgoing edges. An error is risen to notify this exception and execution is interrupted (e.g., with an **assert** checked at runtime) (lines 21–22 of the non-deterministic code of Fig. 6.4).

Continuous evolution is implemented as an **if** or **else if** statement, which guard is the invariant condition *inv*. The body, which is executed when the guard is true, describes the continuous evolution as a discretized *flow* function (lines 8–11 of the non-deterministic code of Fig. 6.4). The approach adopted in this chapter exploits the *Euler numerical integration algorithm* taking as input a time discretization step chosen by the designer [3]. The Euler method is a first-order method, i.e., the error at any given time is proportional to the step size. As such, it is possible to replace the Euler classic algorithm with one of the many available algorithms for the approximation of solutions of ordinary differential equations [3], such as its *backward* or *exponential* variants or algorithms from the family of Runge-Kutta methods.

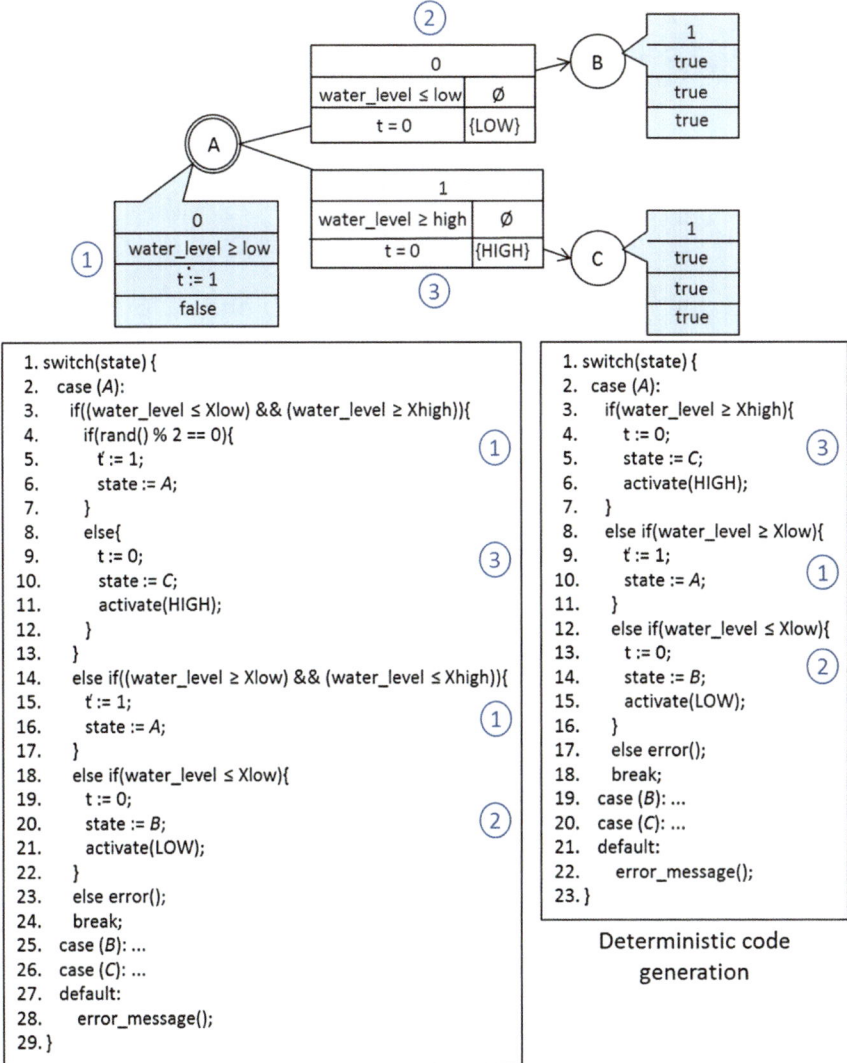

Fig. 6.4 UNIVERCM automaton (*top*) as well as the corresponding generated code by preserving non determinism (*left*) and by choosing an arbitrary order between transitions with the same priority (*right*)

6.3.1.2 Priority Management

In UNIVERCM, both edges and states are annotated with a priority, to determine which one prevails when one or more conditions ($\mathcal{E}_{label}$ and $\mathcal{E}_{var}$ for edges, *inv* for states) are satisfied. To respect the relationship between such priorities in the code generation process, transitions are listed in the **if-else** enumeration by respecting the priority, i.e., higher priority transitions come first so that they are evaluated and, if conditions are satisfied, activated before checking other possibly active transitions.

Whenever two transitions have the same *priority*, it is possible to preserve the non-determinism of UNIVERCM by using a pseudo-random generator, so that the choice of the transition to take is non-deterministic. As an example, in Fig. 6.4, two transitions have the same priority when the current state of the automaton is A: the delay transition (①) and the edge with destination state C (③). Figure 6.4 on the left shows the generated code. Whenever both the enabling conditions are satisfied, the code non deterministically chooses the path to follow (lines 3–13). This approach is effective in a validation phase, as it allows to activate all possible paths in the automaton and it respects the non-determinism intrinsic to the UNIVERCM formalism. However, it introduces a lot of overhead for managing randomness, that may not be necessary in a final C++ implementation.

As a result, the C++ generation algorithm may also list the transitions by choosing an arbitrary order between transitions with the same priority. Figure 6.4 on the right shows the implementation of the automaton that avoids non determinism. The transitions with the same priority are listed in an arbitrary order, i.e., first the edge to state C (③, lines 3–7) and then the delay transition (①, lines 8–11). Transitions with lower priority are still listed after higher priority transitions (②, lines 12–16) and the error statement closes the current case for state A (line 17). This approach is adopted when the goal is to see one of the possible implementations of the UNIVERCM system, rather than a simulation of UNIVERCM automata. Finally, it is worth noting in Fig. 6.4, how the automata and the code generated from it resemble each other, thus enhancing the readability of the code and easing the tracking of the MoC constructs inside the generated implementation.

6.3.1.3 Variable and Label Management

In UNIVERCM, variables can be written by multiple automata. When multiple automata write on the same variable at the same time, then the value is updated non-deterministically by choosing among the values assigned by each automaton.

To reproduce this behavior in the C++ implementation, each variable is assigned a *read value*, i.e., the current value that is read by all automata, and a set of *written values*, i.e., one per automaton writing on the variable. Such values are C++ variables, which type matches as much as possible the value range for the UNIVERCM variable. Each UNIVERCM variable is also provided with one boolean *flag* per automaton, which is stating whether the automaton tried to change the variable value. This flag is necessary to determine whether each automaton updated the variable called **written_value**: in each step, and thus whether such value must be considered in the choice of the updated the variable **read_value**. Variable management will rely on a management function, which is used to support execution and interaction between all automata (as explained in Sect. 6.3.1.5).

This approach preserves the UNIVERCM semantics, but it is complex and it introduces a lot of overhead. Thus, two complementary approaches are possible. On one hand, it is possible to deterministically choose that one automaton always prevails on others, i.e., the value written by such automaton is always chosen as next read value for the variable. In this way, there is no need for flags and the algorithm for choosing the next read value is straightforward. On the other hand, one can adopt only one written value per variable, thus implying that all automata can freely write on it. As a result, one of the values will non-deterministically prevail at any execution step. Once again, choosing one of the latter approaches implies that the generated code distances itself from the UNIVERCM semantics, thus gaining performance and more efficient execution.

Finally, *labels* $\mathcal{L}$ are represented with two boolean values. A current value flag states whether the label is active (i.e., **true**) in the current step. Complementary, a future value is set to activate a label in the next step. At the end of each simulation step, the management function sets the current value to the future one, and puts the future value flag to **false**.

6.3.1.4 Preservation of Atomicity

The *atomic* construct was added to UNIVERCM to preserve atomicity conditions of the starting description. Figure 6.5 represents this behavior on a UNIVERCM automaton where all the transitions between state A and state D belong to the atomic region.

To support this behavior in the final C++ code, the atomic region (Fig. 6.5a) must be collapsed into a single transition (Fig. 6.5b). The enabling condition of the transition invokes a function (*atomic_foo()* in the Fig. 6.5) emulating the evolution of the atomic region on local variables. If all edges of the atomic transition can be traversed, the return value is *true* (in Fig. 6.5b, execution reaches the edge between states S_2 and S_0). As a result, the atomic region is traversed correctly and the update function of the transition propagates the new variable values.

An error may occur if a necessary synchronization event is not available inside of the atomic region or if one of the guards is false, thus not allowing the activation of an internal edge (in Fig. 6.5b, execution reaches state S_E). In this case, the function returns *false*. As a result, the transition can not be activated and the automaton can evolve by following a different transition.

6.3.1.5 Role of the Management Function

The management function is a C++ function generated once for each system of UNIVERCM automata converted to C++. The role of this function is similar to a scheduling routine as it manages the status of the overall system and parallel composition of automata.

The Algorithm 6.1 presents a pseudo code of the management function. For each simulation step, the management function activates all automata by invoking the

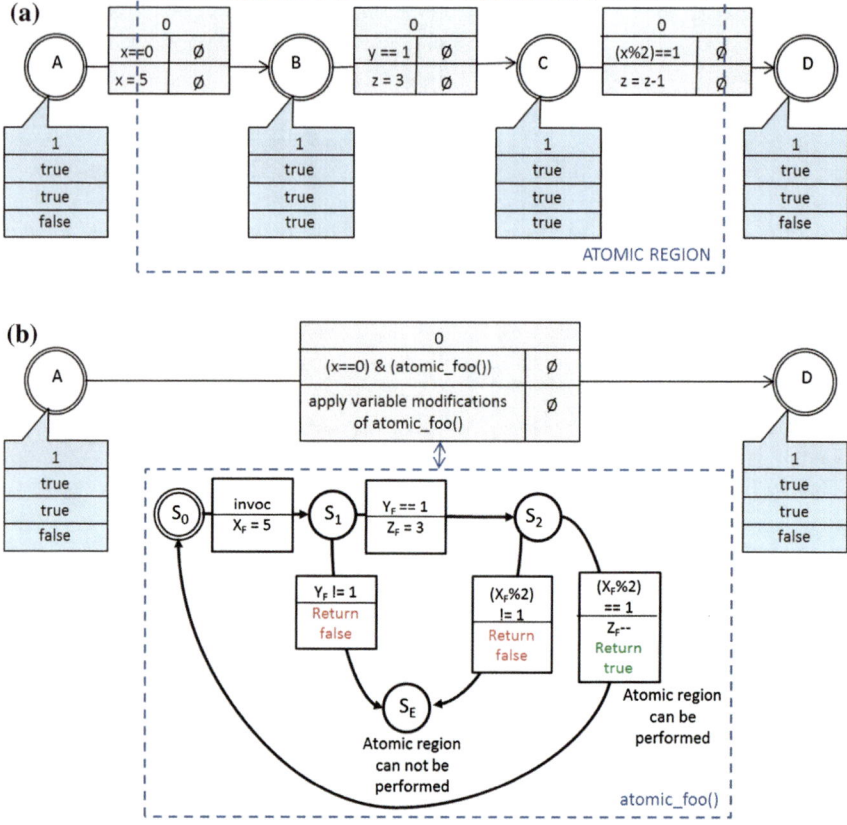

Fig. 6.5 Example of preservation of the *atomic* construct. The starting atomic region in the UNIVERCM automaton (**a**) is converted into a support function *atomic_foo()* (**b**)

Algorithm 6.1: Outline of the management function code

```
1 forall the simulation steps do
2 │  forall the automata A do
3 │  │  execution_step(A);
4 │  forall the variables v do
5 │  │  v.current_value = choose_next_value(v);
6 │  forall the synchronization labels l ∈ L do
7 │  │  l.current_value = l.future_value;
8 │  │  l.future_value = false;
```

---

corresponding generated C++ function (lines 2–3). Once that all of them advanced their execution, the function updates the system status by managing variables and events (lines 4–8), as explained in Sect. 6.3.1.3.

6.3.2 Generation of SystemC-AMS Code

SystemC code generation from UNIVERCM descriptions was first described in Di Guglielmo et al. [7], but such a work provided a very limited support for continuous evolution. SystemC-AMS is an extension of SystemC built for supporting the design of Analog/Mixed-Signal (AMS) systems. SystemC-AMS introduces new language constructs for the design of analog behaviors, defining a new execution semantics and different modeling formalisms. Thus, the adoption of SystemC-AMS allows a complete mapping of the UNIVERCM semantics.

6.3.2.1 SystemC-AMS and the Linear Signal Flow

SystemC-AMS extends standard discrete-event-based SystemC by providing more than one way for describing continuous behaviors. This chapter adopts the *Linear Signal Flow (LSF)* formalism, as it reflects many of the UNIVERCM characteristics. The LSF formalism produces a component-based structural description of a SystemC-AMS module by using basic analog components such as integrators, adders, gain operators, or delays. Other than continuous base elements, LSF descriptions gives the possibility to insert discrete components such as multiplexer or demultiplexer that can be controlled by standard, discrete-event-based SystemC processes. A major advantage of LSF is that time is considered as a continuous value (thus reflecting the semantics of UNIVERCM and of hybrid systems).

On the contrary, in Timed Data Flow (TDF) semantics, time is considered as a set of discrete points. Furthermore, TDF does not support continuous evolution (e.g., in the form of differential equations) and thus it can not be straightforwardly adopted to model continuous evolution. Consequently, exploiting the LSF formalism permits to fully express the UNIVERCM semantics.

It is important to note that LSF is non-conservative. Thus, if the starting code is conservative, all implicit relations (e.g., Kirchhoff's laws) must be explicitly modeled. Straightforwardly supporting conservative components will be part of future extensions. Figure 6.6c shows an example of the usage of LSF to represent the differential equation $w'(t) = 0.6v(t) - 0.03w(t)$, adopted in the following as reference example.

6.3.2.2 Continuous Time Description with SystemC-AMS

In the UNIVERCM formalism, each state of an automaton may have a different continuous evolution modeled with the *flow* predicate. Thus, each *flow* predicate is expressed by using LSF constructs. The differential equation modeled by *flow* is analysed to identify time-dependent variables. Then, the equation is decomposed into primitive operations, that are mapped to LSF basic constructs to build an interconnection of basic analog constructs.

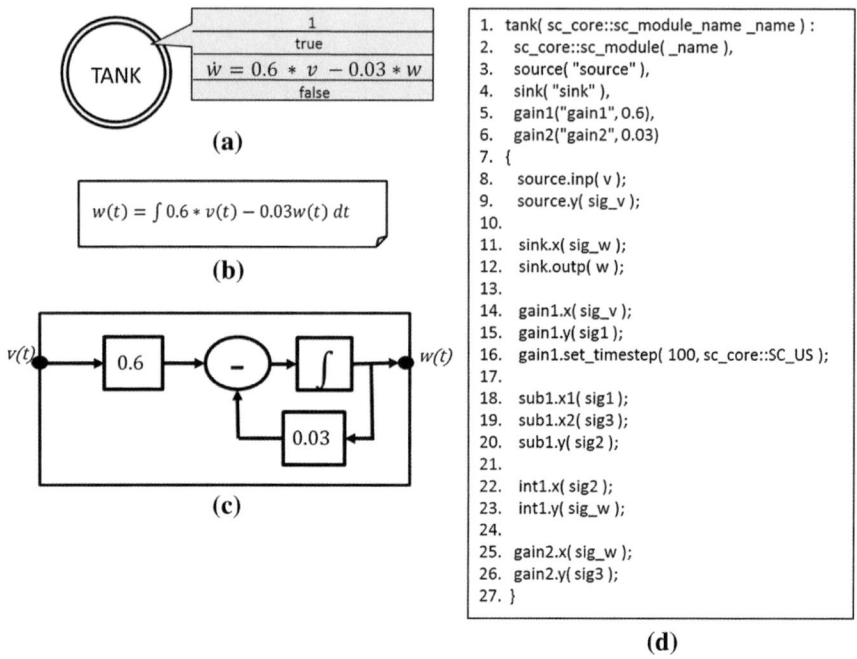

(a)

(b)

(c)

```
1.  tank( sc_core::sc_module_name _name ) :
2.    sc_core::sc_module( _name ),
3.    source( "source" ),
4.    sink( "sink" ),
5.    gain1("gain1", 0.6),
6.    gain2("gain2", 0.03)
7.  {
8.    source.inp( v );
9.    source.y( sig_v );
10.
11.   sink.x( sig_w );
12.   sink.outp( w );
13.
14.   gain1.x( sig_v );
15.   gain1.y( sig1 );
16.   gain1.set_timestep( 100, sc_core::SC_US );
17.
18.   sub1.x1( sig1 );
19.   sub1.x2( sig3 );
20.   sub1.y( sig2 );
21.
22.   int1.x( sig2 );
23.   int1.y( sig_w );
24.
25.   gain2.x( sig_w );
26.   gain2.y( sig3 );
27. }
```

(d)

Fig. 6.6 Continuous-time representation synthesis flow using the LSF formalism of SystemC-AMS applied to the UNIVERCM automaton representing the Tank component of the Water Tank example used in Sect. 6.4. **a** depicts the original UNIVERCM state, **b** represents the equation extracted by the state represented in (**a**), **c** depicts the block diagram derived from the equation in (**b**), **d** shows the block diagram implemented using SystemC-AMS LSF blocks

Figure 6.6 shows this approach. The *flow* predicate of the depicted state (Fig. 6.6a) is analyzed to reconstruct the differential equation (Fig. 6.6b) and then translated into a block diagram (Fig. 6.6c). Finally, the block diagram is realized as an interconnection of LSF basic blocks (Fig. 6.6d).

Given the LSF implementation of all *flow* predicates of an automaton, it is then necessary to build a mechanism to activate, at any execution time, only the portion of the LSF diagram implementing the *flow* predicate of the current state. To solve this problem, the LSF structural description of the interconnection is enriched with demultiplexers and multiplexers, inserted near the sources and the sinks of the interconnection respectively. This allows to preserve the correct evolution that is imposed by the current discrete state. The introduced multiplexers and demultiplexers are controlled by the process modeling the discrete state evolution of the automaton.

6.3.2.3 Discrete Behavior Description with SystemC-AMS

The formal semantics of UNIVERCM state that automata evolve in parallel. Thus, each automaton is mapped to a SystemC module composed by a process, implementing the state transition system, and some support functions. Furthermore, the SystemC module contains the LSF interconnection modeling the continuous behaviors.

The support for discrete behaviors reflects the approach proposed for the C++ code generation (Sect. 6.3.1). However, some differences are necessary to meet some limitations imposed by the LSF formalism. In the C++ mapping, UNIVERCM labels are represented by using couples of boolean variables. Instead, in SystemC-AMS, each label is expressed by one **sc_event** (declared in the global scope) and a flag for every automaton, where **true** states that the label is active. Using **sc_event**s permits to exploit the characteristics of the SystemC discrete-event scheduler, thus preserving the UNIVERCM semantics while reducing the overhead for synchronization. Therefore, the management function only handles non-determinism and concurrent write operations on shared variables. The management function is therefore encapsulated in a separate SystemC module.

6.4 Experimental Results

The effectiveness of the proposed methodologies is proven on a heterogeneous case study: a water tank system (Sect. 6.4.1) and a boiler system (Sect. 6.4.2). Experiments were run on an Intel Core i7 CPU working at 3.40 GHz, running kernel version 3.2.0 of Ubuntu Linux 12.04. Execution times are calculated as the average of execution time of 100 iterations of each functionality.

6.4.1 The Water Tank System

The water tank system is made of five components: 1. a *tank*, which is characterized by an uncontrolled outbound water flow; 2. an *evaluator*, which checks the level of water in the tank and compares it with the upper and lower bounds (if the water level is too low or too high, warnings are notified to the controller); 3. a *valve*, which aperture affects the incoming flow of water; 4. a *controller*, which acts on the aperture of the valve in order to keep the water level in a safe interval; 5. a software *driver*, which sets the legal upper and lower bounds accepted for the water level and the maximum number of warnings accepted before the system is halted.

6.4.1.1 Application of Code Generation Methodologies

In the starting heterogeneous system, analog components are implemented as CIF hybrid automata, the controller is implemented in VHDL while the driver is written in C.

First of all, the components have been converted to UNIVERCM automata [5], as shown in Table 6.1. Column *Language* shows the starting language of the reused component. Columns *States (#)* and *Transitions (#)* (actions or delays) outline the main characteristics of each automaton. Column *Conversion time* (ms) shows the time spent in the conversion of each component by means of automatic tools [5].

The application of the *C++ generation methodology* (Sect. 6.3.1) leads to the generation of a C++ function for a each automaton. Table 6.1 (Column *C++*) shows the main characteristics of the generated code. Column *Switches/branches cases* shows the number of cases contained by the **switch** construct of each C++ function (left) and the maximum number of branches per each case (right). After generating the C++ function corresponding to each automata, it is necessary to implement the *management function*, which occupies 102 lines of code. Column *Implementation* shows the number of generated lines of code, both for the deterministic and non-deterministic versions. As the water tank system does not contain a high level of *non-determinism*, the two versions have few differences in their code and behavior.

Then, the *SystemC-AMS generation methodology* has been applied, as outlined in column *SystemC-AMS* of Table 6.1. Same considerations about the C++ code generation outlined above can be drawn also in the case of SystemC-AMS code generation for the non-determinism internal to each automaton (i.e., the only one presenting some is the valve).

6.4.1.2 Effectiveness of the Proposed Approaches

Ptolemy is one of the main model-based design frameworks for the design of heterogeneous embedded systems [6]. Its strict top-down flow allows to achieve code generation through a set of refinement steps and by applying ad-hoc design choices. Ptolemy has thus been adopted to build a top-down implementation of the water tank system to build a reference in terms of accuracy and of execution time.

The C++ and SystemC-AMS generated code is therefore compared w.r.t. the Ptolemy implementation and w.r.t. the SystemC implementation generated by applying Di Guglielmo et al. [5], to test the advances w.r.t. the current state of the art of UNIVERCM-based flows. Table 6.2 outlines the results of this comparison. Column *Code gen. time (ms)* outlines the overall code generation time, while column *Code (loc)* reports the number of lines of code. Column *Execution time* (s) shows the time necessary to perform the same functionality in all code versions, while column *Error (%)* shows the accuracy w.r.t. the Ptolemy implementation.

In terms of execution time, all the UNIVERCM-based implementations are faster than the Ptolemy-based simulation (ranging from 2.24× to 217.03×). The C++ generated code is in average as fast as the SystemC code generated with Di Guglielmo

Table 6.1 The water tank system: characteristics of the components and generated UNIVERCM automata

Component	Language	UNIVERCM automaton		Conversion time (ms)	C++			SystemC-AMS	
		States	Transitions		Switches/branches	Implem. (loc)		Implem. (loc)	
						Det.	Non-det.	Det.	Non-det.
Tank	CIF	1	1	47	2/1	17	17	67	67
Evaluator	CIF	4	11	61	5/3	68	68	178	178
Valve	CIF	4	11	64	5/4	67	251	203	365
Controller	VHDL	2	6	110	3/3	36	36	113	113
Driver	C	5	10	219	6/2	42	42	107	117

Table 6.2 The water tank system: comparison of the SystemC-AMS and C++ code w.r.t. the Ptolemy top-down implementation and the SystemC implementation

Version	Code gen. time (s)	Code (loc)	Execution time (ms)	Error (%)
Ptolemy	Two person days	–	16.060	–
SystemC	171	750	0.074	0.45
Deterministic C++	160	377	0.098	0.34
Non-deterministic C++	167	561	0.154	0.35
Deterministic SystemC-AMS	201	831	7.040	0.36
Non-deterministic SystemC-AMS	206	1002	7.152	0.36

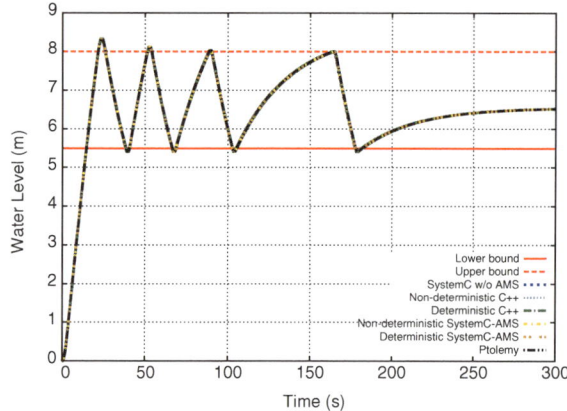

Fig. 6.7 Simulation trace of the code generation alternatives for the water tank system

et al. [5]. Indeed, the SystemC scheduler is substituted by the C++ management function, in charge of activating automata and of updating the system status. On the other hand, in SystemC automata are activated only when fresh data is available thus avoiding the activation of all automata at any execution time. SystemC-AMS is slow (40× slower than the C++ implementation), as its execution is affected both by the SystemC scheduler and by the analog evolution manager. Still, it is 2× faster than the Ptolemy implementation.

Table 6.2 highlights that all the versions of code generated from UNIVERCM have a high accuracy w.r.t. Ptolemy (max. error is 0.45 %). This is highlighted also by Fig. 6.7, where the simulation traces of all code versions are shown. The adherence of all the implementations to the Ptolemy evolution proves that UNIVERCM preserves the correctness of the starting heterogeneous components, and that the transformation and code generation steps proposed in this paper preserve the UNIVERCM semantics also in the generated code. Code generation time is much higher for Ptolemy, as it implied to redesign the whole system from the specifications. These considerations highlight the strength of the proposed automatic code generation approaches, that

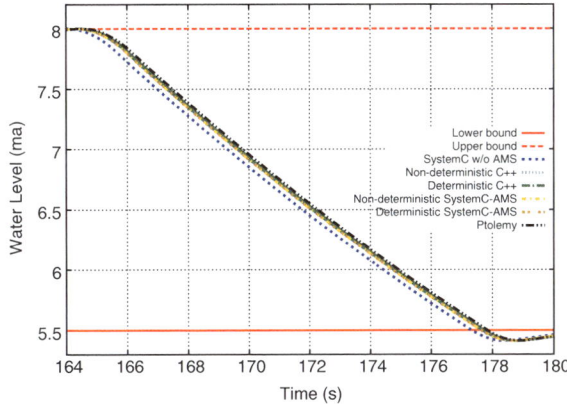

Fig. 6.8 Blow up of a segment of Fig. 6.7 to highlight the precision of the various approaches w.r.t. the Ptolemy implementation

start from a reconciliation of the starting heterogeneity, both in terms of accuracy and of performance.

Figure 6.8 blows up a section of Fig. 6.7 to highlight the differences in terms of accuracy between the different code generation flavors. It is possible to note that the C++ generation methodology achieves a higher precision ($\sim$0.34 %) than SystemC ($\sim$0.45 %), mainly due to the automata activation mechanism, that fully reflects the UNIVERCM semantics. On the other hand, SystemC-AMS is comparable to the C++ implementations ($\sim$0.36 %). However, SystemC-AMS is heavily impacted by a heavy scheduling overhead. Thus, it is important to find a reasonable trade off between accuracy and execution time. In order to achieve a better precision than the C++ versions, SystemC-AMS had to be configured accordingly and the whole simulation took 601.345 ms. Accepting a negligible increase of the error (+0.01 %), it was possible to reduce simulation time to 7.100 ms.

Finally, non-determinism has little impact on the performance and accuracy of the generated code, as preferring one transition to another may slow down reaction to error notifications. However, this is likely due to the low level of non-determinism provided by the water tank system. All variables are written by one single automaton and the only non-determinism occurs with the management of synchronization labels. Thus, the impact of non-determinism will be deepened as future work.

These considerations highlight that the methodologies proposed in the chapter allow to determine a good trade off between accuracy and execution speed, according to the target code generation. Furthermore, the methodologies allow to generate multiple versions and to evaluate a set of alternatives, in order to tune the final code.

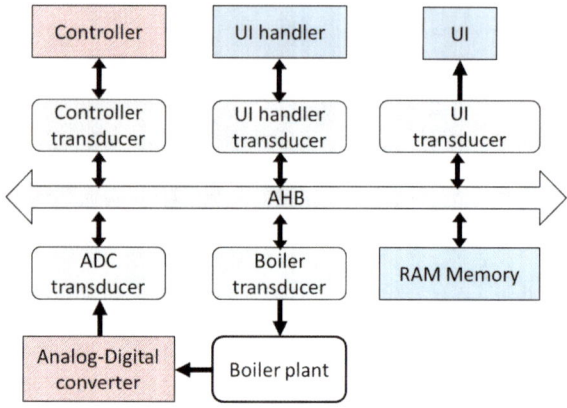

Fig. 6.9 The boiler system

6.4.2 The Boiler System

The starting point of the system is the *bang-bang*-control boiler model provided with
the Mathworks Simulink tool [10]. Such a system is composed of a discrete bang-
bang controller (i.e., Controller) and a boiler plant (dashed in Fig. 6.9). Intuitively, the
controller controls when the boiler is turned on and off. When the boiler's temperature
is lower than the specified set point, the boiler is activated to warm up. Otherwise,
the boiler is switched off to reduce its temperature.

The boiler system has been realized by adding reused components and manually
defined components that complete the system functionality (solid in Fig. 6.9). The
1. *Controller* component has been implemented in SystemC starting from the bang-
bang Stateflow model, while the behavior of 2. the *Boiler plant* has been formalized
into a CIF description. It was then necessary to add 3. an AMBA bus (AHB) SystemC model, 4. a simple generic RAM module and 5. an Analog-to-Digital Converter
(ADC), both modeled by using VHDL.

Finally, the system includes two software components written in C (i.e., UI handler
and UI). The 6. UI handler is a device driver that accesses the RAM memory to retrieve
the value of the digital temperature used by the controller for turning on and off the
boiler, while the 7. UI software prints on a file such values.

Transducers have been introduced to allow binding of components following a
communication protocols different from the standard AMBA one.

6.4.2.1 Application of Code Generation Methodologies

In the starting heterogeneous system, components are implemented in a variety of
languages, ranging from CIF to C and VHDL. All the components have been automatically translated into UNIVERCM automata, combined together and then trans-

Table 6.3 The boiler system: characteristics of the components and generated UNIVERCM automata

Component	Language	UNIVERCM automata	Conversion time (ms)	C++ (loc)	SystemC-AMS (loc)
Controller	SystemC	2	76	384	342
Boiler plant	CIF	1	31	104	185
AMBA AHB bus	SystemC	10	112	1486	1304
RAM	VHDL	4	71	495	403
ADC	VHDL	1	25	158	140
UI handler	C	3	27	164	88
UI software	C	1	29	98	73
Controller transd.	SystemC	3	108	145	180
Boiler transd.	SystemC	1	30	196	179
ADC transd.	SystemC	1	29	142	129
UI handler transd.	SystemC	3	97	164	88
UI SW transd.	SystemC	1	32	157	142

Table 6.4 The boiler system: comparison of the SystemC-AMS and C++ code w.r.t. the Ptolemy top-down implementation and the SystemC implementation

Version	Code gen. time (ms)	Code (loc)	Execution time (ms)	Error (%)
Ptolemy	Two man-days	–	112.896	–
SystemC	1390	3059	7.880	0.34
Deterministic C++	1542	3693	7.948	0.33
Deterministic SystemC-AMS	1422	3166	8.227	0.30

lated to C++ and SystemC-AMS, by following the methodologies presented in this chapter. The main characteristics of the generated code are outlined in Table 6.3.

The UNIVERCM implementation of the system highlighted that the system is deterministic, i.e., it never occurs that the same system status allows the activation of two different transitions with the same priority. This implies that the non-deterministic version of the code would be identical to the deterministic implementation, and it is thus omitted.

6.4.2.2 Effectiveness of the Proposed Approaches

Ptolemy has been adopted to build a top-down implementation also of the boiler system to build a reference in terms of accuracy and of execution time.

The C++ and SystemC-AMS generated code is thus compared w.r.t. the Ptolemy implementation and w.r.t. the SystemC implementation generated by applying Di Guglielmo et al. [5]. Table 6.4 outlines the results of this comparison.

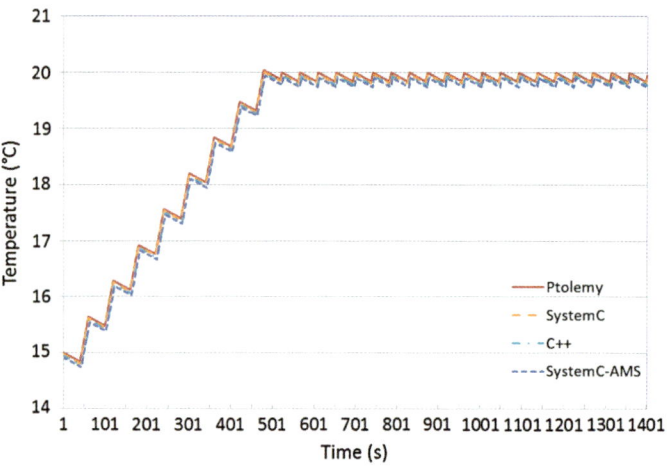

Fig. 6.10 Simulation trace of the code generation alternatives for the boiler system

Also in this case study all UNIVERCM-based implementations are one order of magnitude faster than the Ptolemy implementation (in average the speed up is 14×). The C++ implementation speed is comparable with the one of the SystemC generated code, as functionality is the same and the SystemC scheduler overhead is replaced by the management function overhead. The SystemC-AMS implementation is once again slower than the SystemC and C++ implementations, due to the analog solver overhead. However, this case study features only one simple analog component (i.e., the boiler plant). Thus, the AMS computational weight is by far balanced by the remaining of the system functionality and execution time is increased only by 4 %.

Table 6.4 shows that all the generated versions of the system preserve a high level of accuracy w.r.t. Ptolemy. This is highlighted also by the simulation trace reported in Fig. 6.10, that compares the evolution of temperature during the simulation for all code versions. Figure 6.11 focuses on a subset of execution time (from 450 to 550 s) to blow up the differences between the different implementations. All the code versions achieve a precision comparable with Ptolemy (with an error rate ranging from 0.30 to 0.34 %). This occurs because the analog part is limited w.r.t. the overall functionality and it thus does not interfere on precision and on synchronization.

The adherence of all the implementations to the Ptolemy evolution proves that UNIVERCM preserves the correctness of the starting heterogeneous components, and that the transformation and code generation steps proposed in this paper preserve the UNIVERCM semantics also in the generated code.

Code generation time is much higher for Ptolemy, as it implied to redesign the whole system from the specifications. These considerations highlight the strength of the proposed automatic code generation approaches, that start from a reconciliation of the starting heterogeneity, both in terms of accuracy and of performance.

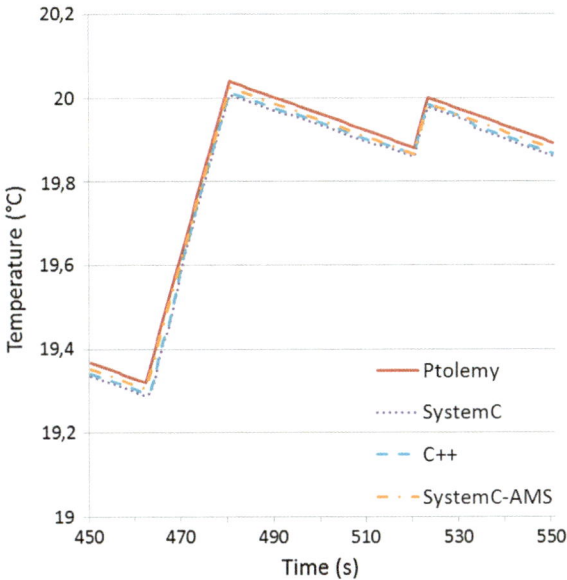

Fig. 6.11 Blow up of a segment of Fig. 6.10 to highlight the precision of the various approaches w.r.t. the Ptolemy implementation

6.5 Conclusions

The chapter enhances reuse and eases design of heterogeneous embedded systems with two code generation flows, targeting C++ and SystemC-AMS. The chapter compares then the behavior of the generated code both in terms of performance and accuracy, and shows the effectiveness of the approach on complex case studies. Future work will investigate the relationship between the supported degree of non-determinism and modeling errors.

Acknowledgments This work has been partially supported by the European project SMAC FP7-ICT-2011-7-288827.

References

1. Balarin F, Watanabe Y, Hsieh H, Lavagno L, Passerone C, Sangiovanni-Vincentelli A (2003) Metropolis: an integrated electronic system design environment. IEEE Comput 36:45–52. doi:10.1109/MC.2003.1193228
2. Bouchhima F, Briere M, Nicolescu G, Abid M, Aboulhamid E (2007) A SystemC/Simulink co-simulation framework for continuous/discrete-events simulation. In: Proceedings of the IEEE international workshop on behavioral modeling and simulation (BMAS) 2007, pp 1–6. doi:10.1109/BMAS.2006.283461

3. Butcher JC (2003) Numerical methods for ordinary differential equation. Wiley, New York
4. Cescirio W, Baghdadi A, Gauthier L, Lyonnard D, Nicolescu G, Paviot Y, Yoo S, Jerraya A, Diaz-Nava M (2002) Component-based design approach for multicore SoCs. In: Proceedings of the IEEE/ACM design automation conference (DAC) 2002, pp 789–794. doi:10.1145/513918. 514115
5. Di Guglielmo L, Fummi F, Pravadelli G, Stefanni F, Vinco S (2012) A formal support for homogeneous simulation of heterogeneous embedded systems. In: Proceedings of the IEEE international symposium on industrial embedded systems (SIES) 2012, pp 211–219. doi:10. 1109/SIES.2012.6356587
6. Eker J, Janneck JW, Lee EA, Liu J, Liu X, Ludvig J, Neuendorffer S, Sachs S, Xiong Y (2003) Taming heterogeneity—the ptolemy approach. Proceedings of the IEEE 91:127–144. doi:10. 1109/JPROC.2002.805829
7. Di Guglielmo L, Fummi F, Pravadelli G, Stefanni F, Vinco S (2013) UNIVERCM: the UNIversal VERsatile Computational Model for heterogeneous system integration. IEEE Trans Comput 62(2):225–241. doi:10.1109/TC.2012.156
8. Keppel D (1993) Tools and techniques for building fast portable threads packages. Technical Report UWCSE 93–05-06. University of Washington, Seattle
9. Lee E, Sangiovanni-Vincentelli A (2011) Component-based design for the future. In: Proceedings of the ACM/IEEE design automation and test in Europe conference and exhibition (DATE) 2011, pp 1–5. doi:10.1109/DATE.2011.5763168
10. MathWorks (2008) Simulink—Dynamic System Simulation for MATLAB, Version 7. MathWorks. http://www.mathworks.com/products/simulink/
11. OSCI AMS Working Group (2010) Standard SystemC AMS extensions Language Reference Manual. Open SystemC Initiative. http://www.systemc.org/

Part IV
Digital Hardware/Software Embedded System Design

Chapter 7
SystemC Modeling with Transaction Events

Bastian Haetzer and Martin Radetzki

Abstract Today, in the design of embedded systems several abstraction levels are used ranging from register transfer level to transaction level. In this contribution we will introduce a new SystemC modeling technique with transaction events for communication modeling. Transaction events extend traditional timed events with a communication payload, thus combining state update and process triggering. This modeling technique can be used at all abstraction levels to create deterministic simulation models in a conceptual clean way.

Keywords SystemC · Transaction-Level Modeling (TLM) · OSCI TLM-2.0 · Register Transfer Level (RTL) · Bus modeling · Transaction Level (TL) modeling styles · Deterministic simulation · Preemption modeling · Parallel simulation · AMBA High-performance Bus (AHB)

7.1 Introduction

In recent years there is a high interest in system level design, especially modeling at the Transaction Level (TL) [5, 6]. SystemC [2] has become the de facto standard for system level modeling. There are several sub levels defined at the transaction level (Fig. 7.1), which differ by the level of accuracy they provide. Cycle-Accurate Transaction Level Models (CATLMs) are evaluated in each clock cycle and thus provide the same amount of information as Register Transfer Level (RTL) models. In Bus Cycle Accurate (BCA) models, only the communication infrastructure (e.g., bus) model is run in each cycle, whereas masters are executed in chunks, possibly spending several cycles [15]. Another abstraction level is the Cycle Count At Transaction Boundaries (CCATB) level introduced by [16]. Here, not only

B. Haetzer (✉) · M. Radetzki
Embedded Systems Engineering Group, University of Stuttgart,
Stuttgart, Germany
e-mail: haetzer@informatik.uni-stuttgart.de

M. Radetzki
e-mail: radetzki@informatik.uni-stuttgart.de

© Springer International Publishing Switzerland 2015   127
M.-M. Louërat and T. Maehne (eds.), *Languages, Design Methods,*
and Tools for Electronic System Design, Lecture Notes in Electrical Engineering 311,
DOI 10.1007/978-3-319-06317-1_7

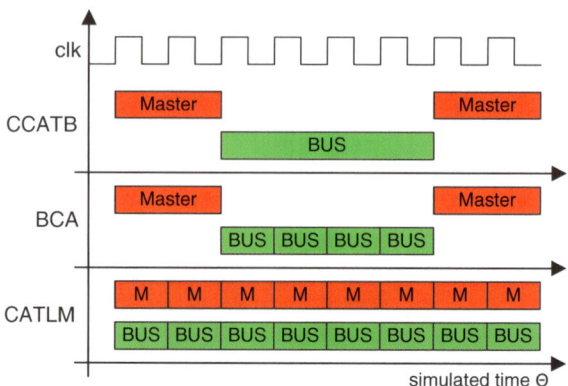

Fig. 7.1 Accurate abstraction levels at the transaction level

masters, but also the communication infrastructure models are executed in chunks whenever possible. Besides these three levels there are other abstraction levels proposed, e.g., untimed and approximately timed TL models [18]. Some approaches also combine several levels in one adaptive model [19]. However, for all such levels there is a common problem in communication modeling, namely to create deterministic models, i.e., make sure that a process is only executed after all inputs have the right (updated) value.

Over the last years, several modeling styles for TLM have been proposed, mainly for bus modeling, which ensure the right order of state update and process execution. In this work, we compare these modeling styles in respect of modeling effort and simulation performance. Furthermore, we propose a new communication modeling style which is based on so-called Transaction Events (TEs). This modeling style elegantly creates deterministic models while hiding the state update details from the models which leads to simpler models and to higher simulation performance. Our new modeling style is compatible to the OSCI TLM-2.0 standard [1] and can be understood as a proposed extension. We will introduce the concept of transaction events at the CCATB level. However, TEs can be used at RTL too, which will also be shown.

The remainder of this chapter is subdivided into the following sections. In Sect. 7.2, related work is given and in Sect. 7.3, the TL modeling styles found in the literature are presented in detail along with a discussion of their pros and cons. Section 7.4 introduces our proposed modeling style. Section 7.5 compares transaction events with payload event queues—a modeling class from OSCI TLM-2.0. Section 7.6 presents extensions of transaction events for preemption modeling. In Sect. 7.7, it will be shown how the concept of TEs can be used for RTL modeling. Section 7.8 shows experimental results for transaction level models. Section 7.9 gives an outlook of TEs in parallel simulation. Finally, Sect. 7.10 concludes the work.

7.2 Related Work

Over the last years, transaction-level modeling has been applied to several communication architectures, e.g., Networks-on-Chip [13, 14]. However, most work concentrates on modeling a bus [7, 16, 18, 22], which is the original domain for TLM. The focus here is often on evaluating the simulation speedup over RTL [17, 23] not on specific modeling guidelines. Nowadays, SystemC [2] can be considered as the most used language for TLM. There are several ways to use language constructs (events, channels, etc.) to create transaction level models. A fundamental modeling decision is about how to implement the state update while maintaining deterministic simulation. An early approach for modeling a bus at the transaction level [11] uses the two phase synchronization principle, which was applied in several other works [4, 7] and was later adapted to another abstraction level [16]. The first TLM proposal by OSCI [20] is using the request-update/update mechanism. Beside these existing approaches, there are other possible techniques, which are presented in detail in the next chapter. The aim of this work is to evaluate the pros and the cons of these modeling styles and to introduce a new solution which simplifies the modeling process.

7.3 Analysis of TL Modeling Styles

A typical SystemC transaction level model consists of masters (also often named as initiators), which are active modules containing a process. These masters are connected through ports to a communication channel model (e.g., a bus model), which provides a communication interface and has a process, too. Slaves (or targets) are normally modeled as passive modules. Such a model is shown in Fig. 7.2.

Normally, we are interested in having deterministic models that guarantee that the simulation output is independent from the simulation run. In SystemC, however, the process execution order is not defined and cannot be specified by the user. This means that, if two processes communicate, there is always the risk that some non-deterministic state update happens. To create deterministic models (or to be more precise, models that create deterministic simulation outputs while using non-deterministic execution), we have to ensure that all processes work with the up-to-date input values. SystemC provides three different language constructs, which can be used to build deterministic models:

- request-update/update mechanism
- delta event notification
- timed event notification.

The request-update/update mechanism is used by so called primitive channels (e.g., `sc_signals`) to establish a signal semantics, i.e., writes to channels are not immediately visible, but are delayed to the next update phase. Delta events are normally used to resolve combinational dependencies, but can also be used to assign

Fig. 7.2 Model of a bus-based
system at the transaction level

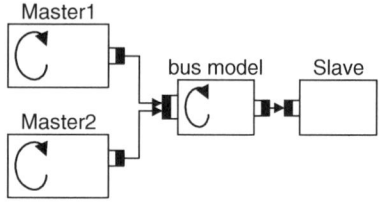

Listing 7.1 Master behavior at the CCATB level

```
1  void master::behavior()
2  {
3    while(run)
4    {
5      computation();
6      wait(time);
7      po_bus->b_transport(payload);
8    }
9  }
```

Listing 7.2 Bus behavior at the CCATB level

```
1  void bus::behavior()
2  {
3    while(true)
4    {
5      wait(m_request_event);
6      // arbitration ...
7    }
8  }
```

process execution to different delta cycles. Timed events can be used to represent
delays in state assignments. All these solutions can be used, but each has its own
advantages and drawbacks.

As an example, we will consider the bus based system from Fig. 7.2. Listings 7.1
and 7.2 show the corresponding master and bus behavior processes. The master
performs some computation for a period of time which is annotated by a wait call.
After that, it initiates a blocking transaction call to the bus representing a read or a
write to the slave, depending on the payload. The bus model implements the interface
method and has its own process responsible for the bus functionality (arbitration
and communication to the slaves). For convenience, in the following we will use
SC_THREADs to describe processes. But SC_METHODs can be used as well because
each SC_THREAD can be transformed to an equivalent SC_METHOD. Next, we will
show different modeling styles for CCATB bus models, which are using the discussed
SystemC constructs to create deterministic models.

Listing 7.3 First bus modeling style (TLM1) at CCATB

```
1  void bus::b_transport(payload_type& payload)
2  {
3    si_payloads[payload.id] = payload;
4    m_request_event.notify(CLK_PERIOD);
5    wait(m_response_events[payload.id]);
6  }
```

7.3.1 TLM1

The first modeling style (Listing 7.3) uses the SystemC request-update/update mechanism to delay the state update to the next SystemC update phase. This ensures that
no state update happens at the time of the rising clock edge and thus all processes
run before any state is updated. SystemC does not allow that a user class derived
from `sc_module` at the same time also inherits from `sc_prim_channel`. That
means to apply the request-update/update mechanism, internal channels have to be
used. In Listing 7.3, we use a signal for every connected master, but other prebuilt
or user defined channels can be used as well. The advantage of this modeling style
is that only events at clock edges are generated. Furthermore, this modeling style is
easy to use, because it is quite similar to modeling at RTL. However, the request-
update/update mechanism creates additional simulation overhead: in each simulation cycle, a list of all channels requiring an update in this cycle has to be created.
This modeling style is proposed by the OSCI TLM-1.0 standard [20], by means of,
e.g., `tlm_req_rsp_channel`.

7.3.2 TLM2

The second modeling style (Listing 7.4) uses delta event notifications to establish the
right sequence of state update and process execution by assigning them to different
delta cycles. Masters do the state update at times of the rising clock edge and the
bus model is triggered at the next delta cycle. There are two main drawbacks of this
modeling style. First, in SystemC, it is not possible to notify an event for a specific
delta cycle at some future time. Consequently, we have to wait for the future time
and only then, we can do the delta notification. This creates additional overhead due
to process activation only for event notification and also makes the modeling more
complicated. Second, this style is not suitable if TL models should communicate
with RTL models, which may take several delta cycles until the final output value is
stable. This modeling style is used by [22] in the Arbitrated Transaction Level Model
(ATLM) of an AMBA AHB bus.

Listing 7.4 Second bus modeling style (TLM2) at CCATB

```
1  void bus::b_transport(payload_type& payload)
2  {
3    wait(CLK_PERIOD);
4    m_payloads[payload.id] = payload;
5    m_request_event.notify(0, SC_NS);
6    wait(m_response_events[payload.id]);
7  }
```

Listing 7.5 Third bus modeling style (TLM3) at CCATB

```
1  void bus::b_transport(payload_type& payload)
2  {
3    wait(DELAY);
4    m_payloads[payload.id] = payload;
5    m_request_event.notify(CLK_PERIOD - DELAY);
6    wait(m_response_events[payload.id]);
7  }
```

7.3.3 TLM3

The third modeling style (Listing 7.5) uses timed event notifications to introduce a delay to the state update. Master processes and the bus process execute at times of the rising clock edge. Somewhere in the open interval between two clock edges, masters update the bus state by waiting a specific time $\Theta_{delay} \in (0, \Theta_{clk})$ before doing assignments. Because of the delay, such models are compatible with RTL models. Furthermore, no channels are required, variables are sufficient for data exchange between processes. The main drawback of this modeling style is that the number of event time points is increased, which leads to an increased number of simulation cycles.

7.3.4 TLM 4

Similar to TLM 3, this modeling style also uses timed event notification. However, compared to TLM 3, the time point of bus execution is changed, i.e., now the masters update the state at the positive clock edge and the negative edge of the clock is used to run the bus process (Listing 7.6). This modeling style is known as the two-phase synchronization principle [11] and is the most commonly used approach for transaction level models [4, 7]. The simple bus example in the SystemC distribution uses this approach to build CATLMs.

This modeling style is also used by [16] to build models at the CCATB level. Here, the bus process is triggered at times of the negative clock edge instead of running each clock cycle. That means the bus model is not cycle accurate, however, the number of cycles the bus model spends is the same as in cycle accurate models leading to

Listing 7.6 Fourth bus modeling style (TLM4) at CCATB

```
1  void bus::b_transport(payload_type& payload)
2  {
3    m_payloads[payload.id] = payload;
4    m_request_event.notify(DELAY);
5    wait(m_response_events[payload.id]);
6  }
```

the name *cycle count accurate*. The bus model is now virtually running at negative clock edges, which has no correspondence to real behavior. Furthermore, by using times of the negative clock edge the bus model is not time accurate anymore, i.e., the start and end times are not the same as with cycle accurate models, but shifted by half of the clock period. This time behavior is the main drawback of TLM4 and is only valid for this modeling style. All other modeling styles presented here are time accurate. In addition, the same drawback as for TLM3 holds, i.e., the increased number of simulation cycles due to increased number of event time points.

In Fig. 7.3, a comparison of the explained modeling styles is given to illustrate how events are used and when state updates are done. The figure describes a sit-

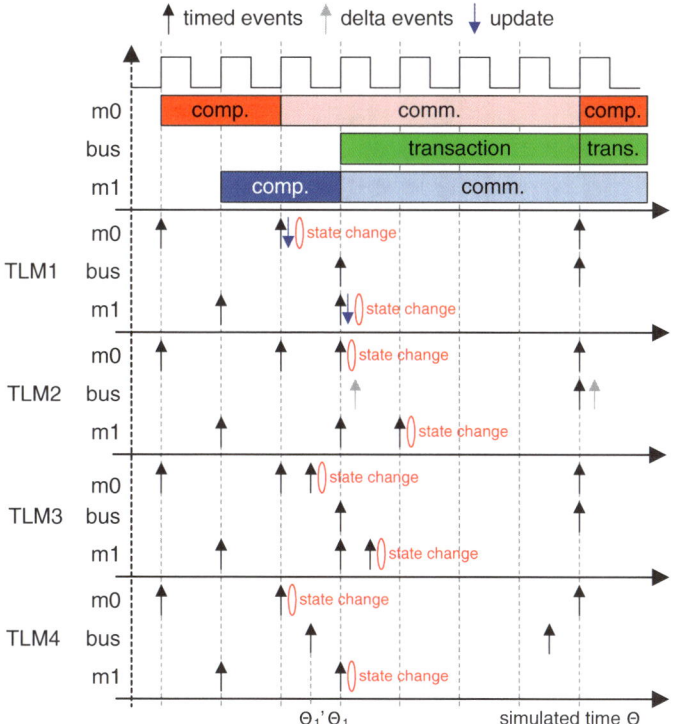

Fig. 7.3 Comparison of CCATB modeling styles

Listing 7.7 Transaction Event class

```
1  template <class T>
2  class transaction_event : public sc_event
3  {
4  public:
5    T m_curr_payload;
6    T m_next_payload;
7
8  protected:
9    virtual void trigger()
10   {
11      sc_event::trigger();
12      m_curr_payload = m_next_payload;
13   }
14 }
```

uation, in which masters issue a communication request after they have finished a computation. The bus process starts the first transaction at the next clock edge at time Θ_1. In TLM1 models, the state update is done during the update phase, after all runnable processes have been executed. In TLM2 models, masters update the state at the rising clock edge and the bus process is executed in the following delta cycle. In TLM3 models, state update happens between two clock edges while the bus is executed at rising clock edges. In TLM4 models, state updates are done at rising clock edges, but the bus runs at negative clock edges (time Θ_1').

7.4 TL Modeling with Transaction Events

As discussed in the previous section, to create deterministic models we have to make sure that all input values of a process are updated before actually running this process. We propose to use Transaction Events (TEs) as a solution to this problem. The idea is to use the event triggering mechanism for performing state updates together with filling the runnable process queue. A transaction_event class, which is derived from sc_event is used for this purpose (Listing 7.7).

This class holds the current active payload together with the desired value after the update (m_next_payload) similar to a sc_signal. A small modification in the SystemC kernel is necessary to make this work: the trigger() function from sc_event has to be declared as a virtual function. This modification guarantees that the trigger() function of the transaction event class is called instead of the function from the base class. There are two things done in the trigger() function: the base function is called so that waiting processes are put in the runnable queue, and finally the state update is done by copying m_next_payload to m_curr_payload. With transaction events, state updates can now be modeled in an elegant way by simply calling the notify method.

Listing 7.8 Bus modeling using transaction events

```
1  void master::behavior()
2  {
3    while(run)
4    {
5      computation();
6      po_bus->b_transport(payload, time);
7    }
8  }
```

Listing 7.8 shows the bus model using transaction events. Masters are not calling `wait()` anymore to annotate the computation time, instead they provide their
computation time as parameter to the interface call. For each master, there is a corresponding transaction event object. The transport call sets the payload as next state of
the transaction event object and calls notify with the request time. The bus process
simply waits on all transaction events (e.g., by a sensitivity list). Because of the
trigger mechanism it is ensured that all transaction events are updated before the
bus process runs. The bus process can now read out the current state of all inputs
(`m_events[i].m_curr_payload`). Typically, in object oriented programming
getter/setter methods are used to access member variables of classes. Such methods
(e.g., `write()` or `read()`) can be added to the transaction event class, similar to
`sc_signals`. In addition, operators (e.g., **operator**=()) can be overloaded to
provide a convenient interface to transaction events.

Figure 7.4 shows how the bus model is processed during simulation using transaction events (cf. to Fig. 7.3). The request from master0 is scheduled for time Θ_1. At
this time the state update happens and the bus process is triggered subsequently. The
request from master1 is scheduled for time Θ_2. As the bus process is not waiting for
any request (because the bus is busy), the event only triggers the state update. It can
be clearly seen from the figure that the number of events necessary to reproduce the
communication sequence is reduced, while maintaining time accuracy.

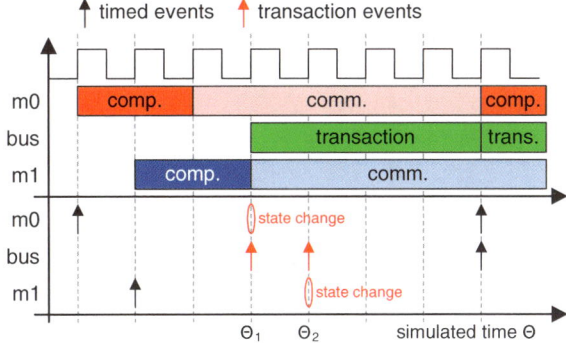

Fig. 7.4 Simulation of transaction events

Fig. 7.5 SystemC scheduler (simplified)

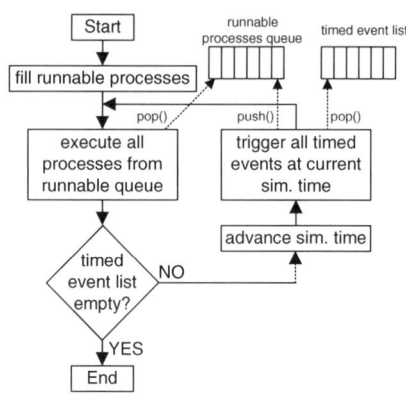

The simulation semantics of transaction events is as follows. As shown in Fig. 7.5, after advancing the current simulation time, the SystemC scheduler triggers all events from the timed event list. Triggering might put waiting processes into the runnable process queue. This procedure is completed before going to the next evaluation phase, where processes are executed. Consequently, the state update with transaction event happens at the notified simulation time Θ_{notify} *before* any process is run.

The example presented so far uses blocking transport interface calls. However, our modeling approach can also be used with a non-blocking interface. In this case a transaction event can be used for the response payload, too. Depending on the time behavior, masters can either wait for the response event or can read the status out of the response payload. If masters can issue several transaction requests without waiting for the response (e.g., bus pipelining), then the transaction event class may be modified according to the modeling requirements. In our proposal, transaction events contain only one payload with variable semantics, but more complex data structures (e.g., queues) can also be incorporated. In this case, the transaction event class has to be extended accordingly by modifying the `trigger()` method with the proper state update function and by providing a suitable interface. Alternatively, new features can also be supported by deriving from the transaction event class and overloading of the `trigger()` method.

In contrast to `sc_ports`, which are not allowed to bind to a `sc_channel` during runtime, transaction events can be created dynamically during the SystemC execution phase. Hence, they can be used to model adaptive systems with communication structures that can change during runtime.

In Hardware Description Languages (HDLs), there are typically modeling constructs for delayed assignments available, which are not directly supported by SystemC. Such delayed assignments can easily be replicated with transaction events. Consider the transport delay mechanism from VHDL:

```
1  y <= transport a and b after 10 ns;
```

With transaction events using the convenience interface, this delayed assignment can be expressed by:

```
1  y = a & b;
2  y.notify(10,  SC_NS);
```

Of course, such behavior can also be modeled in other ways in SystemC, e.g., by a process waiting for a time, but with transaction events, the modeling is more clear.

7.5 Comparison with OSCI TLM-2.0 Payload Event Queues

In OSCI TLM-2.0, there exists a utility class called Payload Event Queue (PEQ), which can be used to schedule transactions for some future simulation time similar to our transaction events. However, the internal mechanism and the usage of the class differs. Transactions can be added to the PEQ, where the payloads are stored together with the scheduled time in a priority list. A process can wait for an event from the PEQ, which will be the earliest event in the queue. Transactions scheduled for the current simulation time can be read out of the queue sequentially by calling `get_next_transaction()`. If there is no more transaction stored with the current simulation time, the earliest transaction in the queue is scheduled again. With PEQ, the same simulation semantics can be expressed, but the state update has to be modeled explicitly, e.g., with a loop extracting all payloads from the PEQ to create the current state (Listing 7.9). This manual recreation of the state increases the modeling effort. With transaction events, the state update details are hidden from the user, which simplifies the modeling (cf. to Listings 7.1 and 7.2). Furthermore, no additional priority lists have to be allocated, all transaction events are simply stored in the global timed event list.

7.6 Extensions for Preemption Modeling

Some bus protocols allow that a higher priority bus request can preempt a currently active burst transfer. Modeling bus preemption needs additional effort, because a simple `wait(time)` in the bus process is not sufficient anymore. If a higher priority request occurs, the bus process has to be triggered again to switch to the new master, what is normally done by a wait on a request event. This event is also scheduled by the bus process for the end time of the transaction to apply for the situation when no preemption occurs. With that technique, the preemption is modeled in the right way, but another difficulty arises: the bus process is triggered by all requests, whether they have lower or higher priority than the active one. If the bus process has to handle such situations, the modeling gets complicated. Furthermore, the unnecessary triggering and execution of the bus process slows down simulation performance. A far better solution will be to run the bus process only for higher priority requests, while lower priority requests do not interrupt bus process execution. With transaction events, this

Listing 7.9 Bus modeling using OSCI TLM-2.0 payload event queues

```
1   void bus::behavior()
2   {
3     while(true)
4     {
5       wait(m_peq.get_event());
6
7       m_payload = m_peq.get_next_transaction();
8       while (m_payload != 0)
9       {
10        m_payloads[m_payload->id] = *m_payload;
11        m_payload = m_peq.get_next_transaction();
12      }
13      // arbitration ...
14    }
15  }
```

Listing 7.10 Extended trigger method for preemption modeling

```
1   virtual void transaction_event::trigger()
2   {
3     if (m_enable)
4       sc_event::trigger();
5     m_curr_payload = m_next_payload;
6   }
```

solution can be easily achieved. For this purpose, we add a boolean field (m_enable) to the member variables of the transaction event class. This variable allows to activate or deactivate the corresponding process triggering. Listing 7.10 shows how the variable is used inside the trigger() function. Now, only if m_enable is true, the corresponding process is triggered, otherwise, only the state update happens. A bus process can use this feature by disabling all lower priority request events at the time of a new granted bus access.

7.7 RTL Modeling with Transaction Events

So far, transaction events have been used for transaction level modeling. However, the concept is independent of the abstraction level. In this section, it will be shown how transaction events can be applied for modeling at RTL. Furthermore, we will introduce a new modeling technique called *querying*, which can be used to resolve combinational dependencies without delta events. Combining these two techniques allows for replacing the request-update and delta event mechanisms. As a consequence, the request-update and delta notification phase can be removed from the

SystemC scheduler while maintaining the same modeling capabilities. This can be useful in parallel simulation, because the number of synchronization between threads can be reduced. This will be illustrated in Sect. 7.9.

7.7.1 SystemC RTL Models

RTL models with SystemC typically contain a sequential process sensitive to the clock edge, which is used for computing the next state based on the current state and the inputs (often represented by a function δ). Additionally, there might be combinational output process(es), which compute the output values based on the current state (function λ). This is especially necessary if output values are of Mealy type, i.e., if they also depend on input values. Outputs that only depend on the current state (Moore type) can be assigned in the sequential process. The local state is stored in `sc_signal`(s). For inputs and outputs, ports with signal interface are used, which are bound to other modules via signals. An example RTL model is shown in Listing 7.11. Here, the combinational process `comb_proc` is used to determine the value of the output signal `po_write`, which depends on the state and on the input `pi_full`. The main mechanism from SystemC used by the RTL models are the request-update mechanism and delta event notification. Both mechanisms are triggered by `sc_signals`.

7.7.2 Signals with Transaction Events

As already introduced for transaction level modeling, transaction events can be used to replace the traditional request-update mechanism from SystemC. Likewise, for RTL modeling, we can provide a new signal class `te_signal` based on TEs, which replaces `sc_signals`. The corresponding class is shown in Listing 7.12. With `te_signals`, the update of the current value happens at times of the clock edge immediately before processes reading from the signal are executed (cf. to Sect. 7.4).

7.7.3 Querying

In traditional SystemC RTL models combinational dependencies are resolved by use of delta events. Updates of the current state or inputs trigger combinational process(es) through a delta event. At the next delta cycle, the new output values are written to the output signals. We call this procedure *writing*. Other components reading from this output signal can read the new value in the next delta cycle through an interface method call of the port (Fig. 7.6a).

Listing 7.11 Example SystemC RTL model

```
1   class rtl_sender : public sc_module
2   {
3   public:
4     sc_port<sc_signal_in_if<bool> >        pi_clk;
5     sc_port<sc_signal_in_if<bool> >        pi_full;
6     sc_port<sc_signal_out_if<bool> >       po_write;
7     sc_port<sc_signal_out_if<unsigned> > po_data;
8
9     virtual void end_of_elaboration()
10    {
11      SC_METHOD(seq_proc);
12        sensitive << pi_clk->posedge_event();
13
14      SC_METHOD(comb_proc);
15        sensitive << lo_reg_state.default_event();
16        sensitive << lo_reg_count.default_event();
17        sensitive << pi_full->default_event();
18    }
19
20  private:
21    sc_signal<state_type> lo_reg_state;
22    sc_signal<unsigned>   lo_reg_count;
23    ...
24  };
```

Listing 7.12 Transaction-Event-based signal

```
1   template <class T, class CLOCK_PERIOD>
2   class te_signal : public
3       transaction_event<T>, public
4           sc_signal_in_if<T>, public
5               sc_signal_out_if<T>
6   {
7   public:
8     void write(const T& value)
9     {
10      m_next_payload = value;
11      notify(CLOCK_PERIOD);
12    }
13
14    const T& read() {return m_curr_payload;}
```

(a)   **(b)**

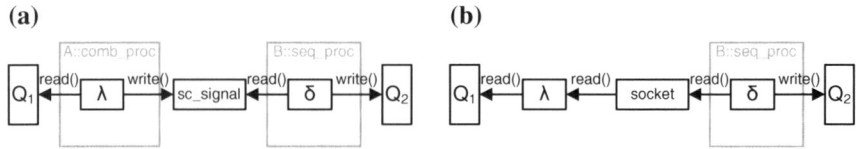

Fig. 7.6 Writing (a) and querying (b) of outputs

Listing 7.13 Socket for querying

```
1   template <class T>
2   class rtl_socket : public sc_signal_in_if<T>
3   {
4   public:
5     virtual const T& read() const;
6     void bind(sc_module* component, method_ptr ⏎
        handler);
7     ...
8   private:
9     sc_module* m_component;
10    method_ptr m_handler;
11  };
12
13  const T& rtl_socket<T>::read() const
14  {
15    return (m_component->*m_handler)();
16  }
```

A completely new point of view is shown in Fig. 7.6b. The idea is to execute the output function λ through the read call of connected components, which we call *querying*. To provide several different outputs with the same signal interface (sc_signal_in_if), an intermediate class similar to sc_export is used, which is called rtl_socket (Listing 7.13). The socket forwards read calls to corresponding member methods that realize the output function λ.

Querying can be used for output functions of Mealy type to remove all combinational processes allowing to resolve combinational dependencies without the need for delta cycles. As a consequence, the delta notification phase is not necessary anymore.

The two techniques, TEs and querying, can be simply applied in RTL models by replacing sc_signals with te_signals and output ports of Mealy type by rtl_sockets. The corresponding combinational process can be removed and the process function is registered at the socket.

7.8 Experimental Results

We have used an AMBA AHB bus [3] based system to evaluate the simulation performance of the presented modeling style. The system consists of several masters and slaves connected to the bus (Fig. 7.7). The slaves are simple models of memory modules. Similar to [7], the masters create generic bus traffic, i.e., bus burst access with a fixed burst length. Each master performs a configurable number of computations and initiates a communication after that.

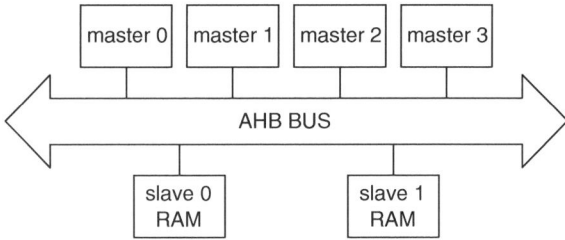

Fig. 7.7 Model of the AMBA AHB bus

Table 7.1 Simulation execution times for AHB bus model

Modeling style	SC_THREAD		SC_METHOD	
	Time [s]	Speedup	Time [s]	Speedup
TLM1	182	1	149	1
TLM2	217	0.8	197	0.8
TLM3	211	0.9	177	0.8
TLM4	169	1.1	140	1.1
OTLM2	152	1.2	133	1.1
TE	131	1.4	99	1.5

We have built a model of the system at the CCATB level using the TL modeling styles presented in Sect. 7.2. We have also implemented a model using our proposed modeling style based on transaction events. The TLx models use the standard SystemC kernel, whereas the transaction event model uses a SystemC kernel with the modification mentioned in Sect. 7.4. Furthermore, a model using the OSCI TLM-2.0 payload event queue was built (OTLM2 model). All experiments were run on a 2.8 GHz Intel Core 2 Quad machine with Linux kernel 2.6. The SystemC 2.2.0 kernel and all models were compiled using GCC 4.1.2 with optimization -O3.

Table 7.1 shows the simulation execution times using a setup of 4 masters and a burst length of four. The reference for comparison is the TLM1 model, because it is fast and time accurate. TLM2 and TLM3 models perform worse than TLM1. The most commonly used approach at transaction level, TLM4, performs slightly better than TLM1, but is not time accurate, because the bus model is running at the negative edge. The model using the OSCI TLM-2.0 payload event queue leads to further performance improvement with a speedup of 1.2 over the TLM1 model. Having a speedup of 1.4, our proposed modeling style with Transaction Events (TEs) outperforms all other modeling styles. We additionally tried using SC_METHODs instead of SC_THREADs for all processes. With SC_METHODs, the simulation execution times are faster, because of the smaller overhead for process context switching. However, the speedup relation remains approximately the same.

We also did experiments with other setups, e.g., more masters, longer computation cycles, different burst length. The speedups in all these experiments are approximately the same as before.

7.9 Outlook Towards Parallel Simulation

Over the last years, there is increasing interest in parallel SystemC simulation
[8, 9, 12, 21, 24]. The approaches so far parallelize the kernel while maintaining compliance to the SystemC scheduler. However, the performance of parallel
approaches that rely on the SystemC simulation phases may be limited. This will be
illustrated with two examples of basic synchronous parallel simulation approaches.

In the approach from [24], only the execution of runnable processes is done in parallel with several threads (Fig. 7.8a). The update and notify phases are processed by
only one master thread. The other threads are waiting during this processing. Through
the centralized processing by the master thread, the synchronization between threads
is minimized. However, the achievable speedup of parallel simulation is limited due
to Amdahl's Law and depends on the ratio of CPU time for sequential processing (update and notify) in relation to CPU time for parallel processing (execute
phase).

In the synchronous approach from [8], each thread processes all simulation phases
of the scheduler (Fig. 7.8b). However, this requires synchronization between threads
after each phase, which is mostly done through a barrier [21].

Barrier synchronization can be a costly operation, especially at low levels
like RTL, where the amount of computation in each cycle typically is low. Furthermore, often there are situations, where some already executed combinational
processes are again triggered for the next delta cycle. In such cases, the number of
runnable processes might be low. In the worst case, only one process is ready to
run, which is dramatically reducing performance. This situation is shown in Fig. 7.9.
Here, the execution of an RTL model is shown where one combinational process
have to be executed twice. It can be easily seen, that such a situation requires a
high amount of barrier synchronization. At higher level models, e.g., at transaction

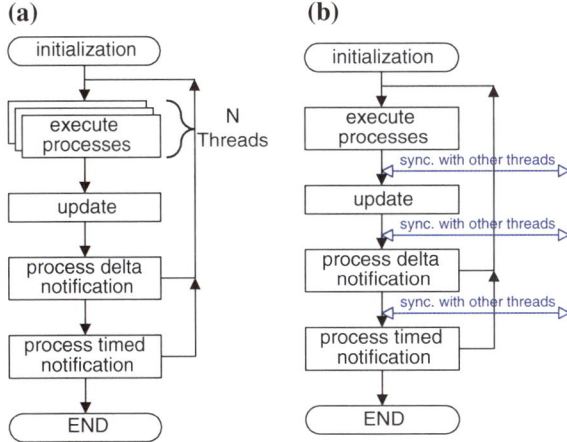

Fig. 7.8 Parallel SystemC simulation kernel approaches: centralized (a) and synchronous (b)

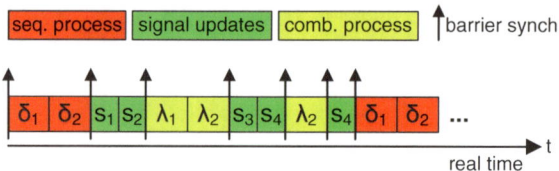

Fig. 7.9 Synchronizations during synchronous parallel simulation

level, delta events and/or the update mechanism are typically not used. Thus, the update phase and delta notification phase is redundant at such levels.

Transaction events are conceptually more elegant than the traditional SystemC simulation cycle, because only timed events exist. In parallel simulation, this may be useful, because communication between threads is only done via timed events and only time synchronization is needed. Additionally, *querying* can be used if some combinational dependencies exist—however, without need for further synchronization. Compared to the basic synchronous parallel SystemC approach (Fig. 7.8b), the number of synchronization can be reduced. Furthermore, different abstraction levels are treated homogeneously, because all modeling is done only with timed events.

Only having timed events is also more in conformance with the concepts of the traditional discrete event simulation field. This makes it easier to adopt other parallel simulation approaches like the promising asynchronous approach [10]. Using transaction events for synchronous and asynchronous parallel simulation will be our next step.

7.10 Conclusion

In this contribution, we have presented the concept of a transaction event, which extends timed events with a communication payload. Transaction events can fully replace the standard SystemC update phase, while providing same modeling capabilities. Furthermore, by a new technique called *querying*, also the delta notification phase is no longer needed. This leads to a clean and homogeneous modeling at all abstraction levels. We have shown how TEs can be used for modeling at the transaction level and the register transfer level. We are currently working on using transaction events in parallel simulation.

References

1. OSCI TLM-2.0 Language Reference Manual (JA32). Open SystemC Initiative (OSCI) TLM Working Group
2. Standard SystemC Language Reference Manual, IEEE Standard 1666–2011. IEEE Computer Society
3. (ARM) ARML (1999) AMBA specification (Revision 2.0), ARM IHI 0011A. http://www.arm.com
4. (ARM) ARML (2003) AMBA AHB cycle level interface (AHB CLI) specification. http://www.arm.com
5. Burton M, Aldis J, Guenzel R, Klingauf W (2007) Transaction level modeling: a reflection on what TLM is and how TLMs may be classified. In: Proceedings of the forum on specification and design languages (FDL) 2007
6. Cai L, Gajski D (2003) Transaction level modeling: an overview. In: Proceedings of the conference on hardware/software codesign and system synthesis (CODES+ISSS) 2007. doi:10.1109/CODESS.2003.1275250
7. Caldari M, Conti M, Coppola M, Curaba S, Pieralisi L, Turchetti C (2003) Transaction-level models for AMBA bus architecture using SystemC 2.0. In: Proceedings of the conference on design, automation and test in Europe (DATE) 2003. doi:10.1109/DATE.2003.1253800
8. Combes P, Caron E, Desprez F, Chopard B, Zory J (2008) Relaxing synchronization in a parallel SystemC kernel. In: Proceedings of the international symposium on parallel and distributed processing with applications (ISPA) 2008, pp 180–187. doi:10.1109/ISPA.2008.124
9. Ezudheen P, Chandran P, Chandra J, Simon B, Ravi D (2009) Parallelizing systemC kernel for fast hardware simulation on SMP machines. In: Proceedings of the workshop on principles of advanced and distributed simulation (PADS) 2009, pp 80–87. doi:10.1109/PADS.2009.25
10. Fujimoto R (2000) Parallel and distributed simulation systems. Wiley, New York
11. Groetker T, Liao S, Martin G, Swan S (2002) System design with SystemC. Kluwer Academic Publishers, Dordrecht
12. Huang K, Bacivarov I, Hugelshofer F, Thiele L (2008) Scalably distributed SystemC simulation for embedded applications. In: International symposium on industrial embedded systems (SIES) 2008, pp 271–274. doi:10.1109/SIES.2008.4577715
13. Indrusiak LS, dos Santos OM (2011) Fast and accurate transaction-level model of a wormhole network-on-chip with priority preemptive virtual channel arbitration. In: Proceedings of the conference on design, automation and test in Europe (DATE) 2011. doi:10.1109/DATE.2011.5763179
14. Kohler A, Radetzki M (2009) A SystemC TLM2 model of communication in wormhole switched networks-on-chip. In: Proceedings of the forum on specification and design languages (FDL) 2009
15. Pasricha S, Dutt N (2010) On-chip communication architectures: system on chip interconnect. Systems on silicon. Elsevier Science, Amsterdam
16. Pasricha S, Dutt N, Ben-Romdhane M (2004) Extending the transaction level modeling approach for fast communication architecture exploration. In: Proceedings of the design automation conference (DAC) 2004
17. Pasricha S, Dutt N, Ben-Romdhane M (2008) Fast exploration of bus-based communication architectures at the CCATB abstraction. ACM Trans Embedded Comput Syst 7(2):217–226. doi:10.1145/1331331.1331346
18. Radetzki M, Salimi Khaligh R (2007) Modelling alternatives for cycle approximate bus TLMs. In: Proceedings of the forum on specification and design languages (FDL) 2007
19. Radetzki M, Salimi Khaligh R (2008) Accuracy-adaptive simulation of transaction level models. In: Proceedings of the conference on design, automation and test in Europe (DATE) 2008. doi:10.1109/DATE.2008.4484912
20. Rose A, Swan S, Pierce J, Fernandez JM et al (2005) Transaction level modeling in SystemC. http://www.systemc.org

21. Roth C, Reder S, Sander O, Hübner M, Becker J (2012) A framework for exploration of parallel SystemC simulation on the single-chip cloud computer. In: Proceedings of the conference on simulation tools and techniques (SIMUTOOLS) 2012, pp 202–207
22. Schirner G, Doemer R (2005) System level modeling of an AMBA bus. Technical report, Center for Embedded Computer Systems, University of California
23. Schirner G, Doemer R (2009) Quantitative analysis of the speed/accuracy trade-off in transaction level modeling. ACM Trans Embedded Comput Syst (TECS) 8(1). doi:10.1145/1457246. 1457250
24. Schumacher C, Leupers R, Petras D, Hoffmann A (2010) parSC: synchronous parallel SystemC simulation on multi-core host architectures. In: Proceedings of the conference on hardware/-software codesign and system synthesis (CODES+ISSS) 2010, pp 241–246

Chapter 8
Automatic Generation of Virtual Prototypes from Platform Templates

Seyed Hosein Attarzadeh Niaki, Marcus Mikulcak and Ingo Sander

Abstract Virtual Prototypes (VPs) provide an early development platform to embedded software designers when the hardware is not ready yet and allows them to explore the design space of a system, both from the software and architecture perspective. However, automatic generation of VPs is not straightforward because several aspects such as the validity of the generated platforms and the timing of the components needs to be considered. To address this problem, based on a framework which characterizes predictable platform templates, we propose a method for automated generation of VPs which is integrated into a combined design flow consisting of analytic and simulation based design-space exploration. Using our approach the valid TLM-2.0-based simulated VP instances with timing annotation can be generated automatically and used for further development of the system in the design flow. We have demonstrated the potential of our method by designing a JPEG encoder system.

Keywords Design automation · Design Space Exploration (DSE) · Predictable platforms · Real-time systems · Simulation · Virtual Prototype (VP) · Mixed-Criticality System (MCS) · Transaction-Level Modeling (TLM) · Constraint programming · Analytical models · Interoperability

8.1 Introduction

Design Space Exploration (DSE) is a key step in platform-based design of embedded multiprocessor systems which aims at finding an optimum platform instance, mapping, and scheduling of application tasks with respect to a set of

S.H. Attarzadeh Niaki (✉) · M. Mikulcak · I. Sander
KTH Royal Institute of Technology, Stockholm, Sweden
e-mail: shan2@kth.se

M. Mikulcak
e-mail: mikulcak@kth.se

I. Sander
e-mail: ingo@kth.se

© Springer International Publishing Switzerland 2015   147
M.-M. Louërat and T. Maehne (eds.), *Languages, Design Methods,*
and Tools for Electronic System Design, Lecture Notes in Electrical Engineering 311,
DOI 10.1007/978-3-319-06317-1_8

performance and cost objectives. The evaluation of design points in the DSE can be performed either analytically, simulation-based, or with a combination of both [9].

Analytic approaches such as scheduling theory [2], dataflow analysis [16, Sect. V] and real-time calculus [17] can be applied while the target architecture is not ready yet and can be used to formulate a correct-by-construction design flow, cutting the extensive verification costs. Unfortunately, these methods mostly assume a deterministic behavior of the system, exact Worst-Case Execution Time (WCET) bounds for tasks, and precise platform characteristics, which are not always valid. On the other hand, simulation-based techniques based on MoCs [8] or TLM [12] do not suffer from these limitations but instead require a longer time to evaluate each design point. Additionally, a careful selection of input stimuli—especially for the corner cases— needs to be performed, which is non-trivial. As a result, combined DSE methods have emerged as a solution to exploit the benefits of these two methods. First, an analytic DSE can be performed to identify a set of Pareto-optimal design points or even optimal solutions for the application tasks and platform components where precise estimates are available. Later, simulation can be performed on a—possibly automatically generated—VP to identify the final solution.

However, the current methods proposed for a combined DSE have considerable limitations. Some approaches are tailored for a small part of the design problem, such as analyzing simulation traces for the investigation of memory architecture parameters [18]. Other techniques generate the simulation models based on additional information from the platform such as a library of virtual prototypes or source level annotations. It is beneficial to exploit the same platform characteristics information used in the analytic DSE phase to generate a rapid prototype of the platform. This prototype can be refined later when a more detailed model of the platform is available.

The present work takes a step towards a combined design flow by *automatically* generating virtual prototypes for simulation-based exploration of the design space based on the output of an analytic exploration phase. In the first phase of the design flow, applications are captured in a framework with well-defined semantics [1] and mapped onto predictable platform templates characterized using a formal framework. Then, for each Pareto-optimal solution a TLM VP is generated which the designer can use for further exploration of the design space. The generated VP is based on the interoperability layer of TLM, which makes it easier to refine it using more accurate platform models.

Using our proposed method the designer can start with implementing the possibly hard real-time applications on predictable platforms in a correct-by-construction flow and later add and test the applications with less time-criticality on a generated VP, hence enabling a mixed-critical design flow.

The contributions of this work are summarized as:

- introducing a flexible method for automatic and rapid generation of virtual prototypes based on characterized predictable platform templates;
- integrating the method into a combined analytic and simulation-based design flow for embedded real-time systems; and
- demonstrating the method in practice by applying it to a JPEG encoder example.

Sections 8.3 and 8.6 serve as the background and related work while the rest of the paper describes our contribution. Section 8.2 motivates the work and sketches the intended design flow while Sect. 8.4 shows how the VPs are generated. Section 8.5 demonstrates the method in action using a case study and Sect. 8.7 concludes the paper.

8.2 Motivation

Using a correct-by-construction design flow for implementing time-critical real-time applications saves the designer from performing extensive verification efforts. An input application model with well-defined timing semantics and a predictable platform which provides timing guarantees enable us to formulate a design-space exploration problem which analytically finds the Pareto-optimal solutions in the design space. However, in design of mixed-critical multiprocessor systems, dynamic exploration of the design space based on simulation methods might still be needed. If there are soft- or non-real-time[1] tasks or applications which are going to be implemented beside the ones with Hard Real-Time (HRT) requirements, we might prefer to get the most performance out of the platform for a representative stimuli, without worrying about occasionally missing a deadline. In such a case, having a fast simulation model of the system which the embedded software designer can use to explore the candidate design points is of great help.

Figure 8.1 shows a two-phase scenario for designing a real-time application with mixed-criticality requirements on top of a multi-core architecture, based on a platform-based design methodology. First, the designer specifies the desired functionality of the application tasks in a platform-independent modeling framework with a well-defined semantics and proper support for the intended abstractions levels, possibly using heterogeneous Models of Computation (MoCs). Examples of such frameworks are Ptolemy [3], HetSC [7], and ForSyDe [1]. A *predictable platform template*, which is mainly a set of flexible and characterized architectural components plus a set of rules stating how to compose them is chosen as the target of the mapping. Together with a set of non-functional system constraints such as the maximum area and minimum throughput of the system, an analytic DSE and mapping tool instantiates a platform and maps the functionality of the application onto it. Depending on the characteristics of the platform components and the constraints, the DSE tool yields different Pareto-optimal solutions.

In the second phase, a VP of each candidate solution is automatically generated to further explore the design-space of the software design. Each VP includes the model of the platform plus the mapping of the input application. The designer, or an automated design flow, can alter the generated VPs, refine the NHRT tasks, execute each VP with the stimuli and pick the best candidate for final implementation. Chapter 9 introduces a similar design flow with mixed-criticality requirements where

---

[1] In other words, Non-Hard Real-Time (NHRT).

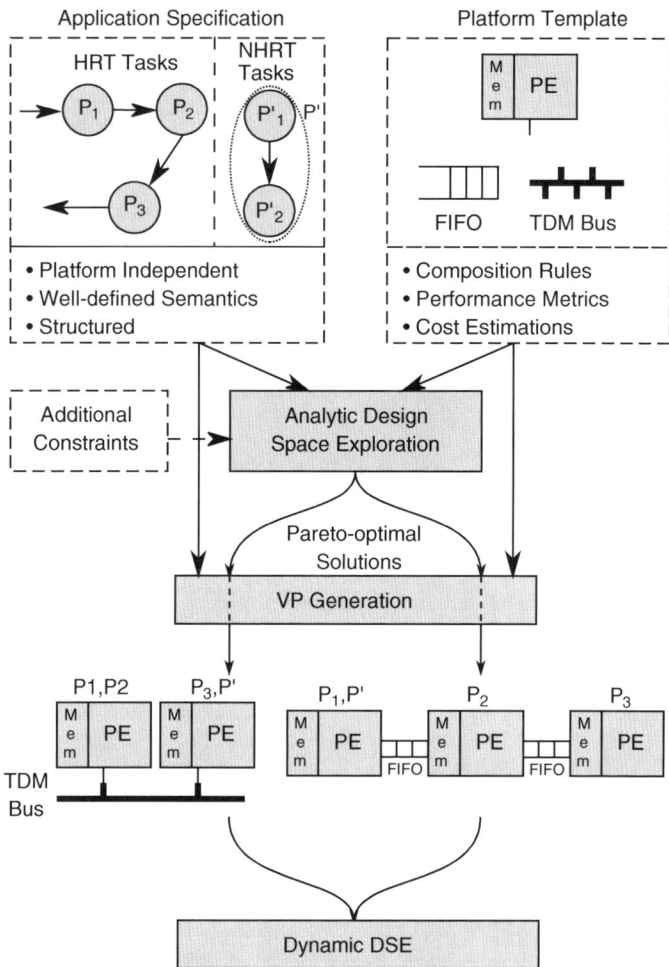

Fig. 8.1 An example of a two-phase system design scenario for real-time MPSoC design with mixed time-criticality. First, starting from a functional specification and a platform template, an analytic design space exploration and mapping tool comes up with solutions that satisfy the given constraints for the hard real-time tasks. Then, a set of virtual platforms are generated for the candidate solution to dynamically explore the design space further for non-hard real-time tasks based on simulation

a simulation-based DSE is preceded with an analytic one which prunes the design space from non-safe design points.

Alternatively, generation of a simulation model out of an analytic model can be motivated by generating a simulation trace of the application mapped to a platform. These traces are used for example in simulation-based cost estimation techniques such as power analysis. Also, the generated VPs can be used to bridge the gap between the formal analytic design methods and the current industrial practice which is based on incremental refinement of a system using available IP blocks, with validation of the system in each refinement step by simulation.

8.3 Background

This section briefly describes the approach we use for characterizing *predictable platform templates*. It is included for clarity and self-containing the paper.

8.3.1 Predictable Component Templates

The framework captures a predictable platform template as a set of *predictable component templates*. Each predictable component template m can provide a specific service, at a given cost.

8.3.1.1 Characterization

The behavior of m can be tuned by setting a set of *variables* V_m upon instantiation. For example, a processing element m_{pe} might be able to run with different frequencies. In this case, the frequency $v_{freq} \in V_{pe}$ is a variable of m_{pe}.

Choosing specific values for variables of a component template m restricts the set of possible values for the other variables. An *instantiation constraint set* C_m captures these restrictions for a component template m. For example, setting a variable of the processing element, which enables its floating point unit, might reduce the maximum allowed value for the frequency variable.

The main purpose of m is to provide a set of *service functions* R_m to the application. In our running example, based on the variable set for the processing element during instantiation, it can provide an execution time function which gives the execution time of each task in the input application.

Each component in an instantiated system comes with its own cost. Depending on the values of a component template variable, the set of component *cost indicators* O_m provide the design space exploration tool with the costs associated with each component template m. The on-chip area usage of the processing element example is one of its cost indicators.

Thus, a predictable component template m is characterized using its variables V_m, instantiation constraints C_m, service functions R_m, and cost indicators O_m.

An *allocation* of a component template m, denoted by A_m, is a copy of its variable set. If m is allocated n times, then $\mathbf{A}_m = \{A_{m,1}, \ldots, A_{m,n}\}$ is called the *allocation set* of m. An *instance* of a component template m, denoted by I_m, is an assignment of values to the variables of one of its allocations while satisfying all of its instantiation constraints. If m is instantiated n times, then $\mathbf{I}_m = \{I_{m,1}, \ldots, I_{m,n}\}$ is the *instantiation set* of m.

8.3.1.2 Examples

Typical components of an embedded platform can be categorized roughly as processors, interconnections, storage, and I/O-elements. We pick a processing element and a TDM bus as an example predictable component templates.

Processing element The guaranteed service that a predictable processor provides for execution of software tasks can be expressed in terms of a Worst-Case Execution Time (WCET). A processing element which is a candidate for running a set of computation tasks K is characterized as a processing component template $m_{pe} = (V_{pe}, \emptyset, R_{pe}, O_{pe})$, where for $\forall k \in K$, we have $V_{pe} = \{v_{freq}\}$, $R_{pe} = \{r_{wcet_k}\}$, and $O_{pe} = \{o_{area}, o_{mem_k}\}$. The frequency of the processor $v_{freq} \in \{f_1, f_2, \ldots, f_n\}$ is a component variable. The processor provides a WCET service $r_{wcet_k} = WCEC_{pe}(k) \times \frac{1}{v_{freq}}$ where $WCEC(k)$ is the worst-case execution cycles for computation task k, derived from the analysis tool. The constant area cost indicator $o_{area} = \texttt{pe\_area}$ and the worst-case memory requirement cost indicator $o_{mem_k} = WCMR(k)$ of each application tasks complete the characterization of our example.

TDM bus A simple TDM bus m_{bus} is characterized as a component template $m_{bus} = (V_{bus}, C_{bus}, R_{bus}, O_{bus})$ where $V_{bus} = \{v_{ports}, v_{tbl}, v_{freq}\}$, $C_{bus} = \{c_{ports,tbl}\}$, $R_{bus} = \{r_{wcct_s}\}$; $\forall s \in 1..v_{ports}$, and $O_{bus} = \{o_{area}\}$ in which the component variables are the number of IP blocks connected to the bus $v_{ports} \in \mathbb{N}$, the TDM table which is an array indicating to which component the access is granted in each time slot $v_{tbl} = \langle v_{tbl,i} \in 1..v_{ports} \mid i = 1..|v_{tbl}| \rangle$, and the frequency of the bus which is the inverse of the length of each slot in terms of global time units $v_{freq} \in \{f_1, f_2, \ldots, f_n\}$. In addition,

$$
\begin{aligned}
c_{ports,tbl} &: \texttt{nvalue}(v_{ports}, v_{tbl}) \\
r_{wcct_s} &= \frac{|v_{tbl}|}{count(v_{tbl},s)} \times \frac{1}{v_{freq}}; \quad \forall s \in 1..v_{ports} \\
o_{area} &= v_{ports} \times (|v_{tbl}| \times \texttt{mem\_area} + \texttt{port\_area})
\end{aligned}
\tag{8.1}
$$

A single instantiated constraint (`nvalue`) ensures that the number of distinct values appearing in the TDM is equal to the number of IP blocks connected to the bus. The WCCT per word is provided by this communication component template is expressed in terms of the number of associated time-slots that a source has access to the bus. The chip area used by the bus is the cost indicator in this example where for each port the amount of logic required for connecting an IP block is assumed to be a constant `port_area` and each memory location for the local TDM tables take `mem_area` units.

8.3.2 Predictable Platform Templates

A platform template $\overline{m}$ introduces a set of template components, their dependencies on each other, services provided by the platform, and overall system costs.

8.3.2.1 Characterization

In an instance of a platform template, the values assigned to variables and the number of instances of different components might be interdependent. We capture this fact using a set of *interdependent instantiation constraints* $\overline{C}$. For example, in a serial interconnection of IP-blocks using simplex links, the number of instances of IPs is constrained to be equal to the number of the links plus one.

A platform template $\overline{m}$ can provide a set of *overall cost indicators* $\overline{O}$ that compute the overall system costs based on the cost indicators of the instantiated component templates. For example, the overall area usage of a typical MPSoC is the sum of the area usage of its individual components plus some additional glue logic.

In short, a predictable platform template $\overline{m}$ presents a set of predictable component templates $M_{\overline{m}}$ and their exposed services, a set of interdependent instantiation constraints $\overline{C}_{\overline{m}}$, and a set of global cost indicators $\overline{O}_{\overline{m}}$.

A platform template is instantiated where at least one of its component templates are allocated and the instantiation set of all of its component templates satisfy the interdependent instantiation constraints.

8.3.2.2 Example

Assuming the availability of component templates for a predictable processor m_{pe}, a simplex point-to-point communication link m_{smp}, and a TDM bus m_{bus} (such as the one presented above), we can characterize a predictable platform template $\overline{m}_p = (M_p, \overline{C}_p, \overline{O}_p)$ with $M_p = \{m_{pe}, m_{bus}, m_{smp}\}$, $\overline{C}_p = \{\overline{c}_{bus,smp}, \overline{c}_{bus,pe}, \overline{c}_{smp,pe}\}$, and $\overline{O}_p = \{\overline{o}_{area}\}$ as

$$
\begin{aligned}
\overline{c}_{bus,smp} &: \ (|\mathbf{A}_{m_{bus}}| = 0) \vee (|\mathbf{A}_{m_{smp}}| = 0) \\
\overline{c}_{bus,pe} &: \ \mathbf{A}_{m_{bus},1} \circ v_{ports} = |\mathbf{A}_{m_{pe}}| \\
\overline{c}_{smp,pe} &: \ |\mathbf{A}_{m_{smp}}| + 1 = |\mathbf{A}_{m_{pe}}| \\
\overline{o}_{area} &= \sum_{m \in M_p} \sum_{A \in \mathbf{A}_m} A \circ o_{area}
\end{aligned}
\tag{8.2}
$$

In the above platform template, the $\overline{c}_{bus,smp}$ restricts the interconnection used in the platform to be either bus-based or composed of simplex links. In the first case, $\overline{c}_{bus,pe}$ ensures that the number of processing elements connected to the bus is equal to the number of bus ports. In serial simplex-based communication, the $\overline{c}_{smp,pe}$ constraint establishes a relation among the number of processing elements and communication links. The overall cost indicator $\overline{o}_{area}$ sums up the individual area cost of the component allocations in the platform.

8.3.3 The DSE Problem

An analytic design-space exploration problem is formulated as a Constraint Satisfaction Problem (CSP), which can be solved using a constraint satisfaction toolkit. A CSP is a set of variables with usually finite domain that are related together by a set of constraints. The goal is to find the possible values for these variables which satisfy these constraints, optionally optimizing for an objective function.

The input application is a task graph, which could be converted from an HSDF graph, and will be mapped to a predictable platform template using the introduced framework. It is modeled as an acyclic task graph with tasks K representing computations and edges E representing the communication. Each task $k_i \in K$ is a triple of decision variables $(\delta_i^k, l_i^k, \rho_i^k)$, where δ_i^k is the time k_i is scheduled to start, l_i^k is the WCET of it and ρ_i^k is the processing element onto which it is mapped. Similarly, each edge $e_{i,j} \in E$ from k_i to k_j is defined as a triple $(\delta_{i,j}^e, l_{i,j}^e, \rho_{i,j}^e)$, which represents the scheduled time, execution time, and the assigned communication element of the communication represented by $e_{i,j}$.

Apart from the constraints coming from the platform, there are three classes of constraints which together form the DSE problem. The first are the constraints coming from the application logic, which for example respect the causal precedence between the nodes and edges of the graph after the mapping and scheduling. The second class are mapping constraints which for example state that not more than a single task can run on a node at any point in time. The third class are the additional constraints provided by the designer, such as the maximum number of available processing cores.

The output of the DSE is the instantiation set of all component templates $\mathbf{I}_m$, mapping of the application tasks ρ_k^k and communication edges ρ_e^e as well as their scheduled start times δ_k^k, δ_e^e.

8.4 Generating Virtual Prototypes

Based on the characterization framework and the analytic design-space exploration problem introduced in Sect. 8.3, this section proposes a flexible method for automatic generation of high-level simulate-able transaction-level models for predictable platforms. The designer is liberated from providing an extra library of transaction-level models for the target platform. Model generation and their timing annotation is only based on the information already provided in the characterization framework.

8.4.1 Overview

In this section, we describe different aspects of the generated VPs shortly followed by the method used to generate the VP.

8.4.1.1 Abstraction

Virtual prototypes can be in different levels of abstraction each for a specific purpose. TLM-2.0 suggests two modeling styles the Loosely Timed (LT) and Approximately Time (AT). The LT style typically uses the blocking transport interface with timing annotations and is sufficient for early software development. AT models use the non-blocking transport interface, include more phases in addition to timing annotations, and can be used for more detailed cycle-accurate architectural exploration. This work focuses on generation of LT models.

8.4.1.2 Integration into the Design Flow

As shown in Fig. 8.1, the VP generation happens after the analytic DSE. The inputs to this step are the functionality of the application, abstract platform characteristics, and the output of the analytic DSE which consists of an instantiation of the platform, mapping of the applications tasks to it and a valid schedule for their execution. For our purpose, only the ordering of the tasks running on a processor is sufficient as the schedule.

Once the LT virtual prototypes are generated they can be used for implementing the dummy tasks (if any) which represent non-hard real-time tasks, and obtaining other simulation-related information such as execution trace of the system. Also, LT VPs can be refined further by replacing the system components with more accurate ones, moving towards an AT model and finally the implementation of the systems.

8.4.1.3 Semantics

The platform executes the input task graph of the application periodically, where each node runs its sub-graph in a self-timed fashion [10]. This means that the tasks on each node are invoked based on the order defined by the schedule and the availability of their input tokens. Computations on each node are performed sequential, but are parallel to inter-node communication, possibly implemented using Direct Memory Access (DMA) cores on each node.

Fig. 8.2 A virtual processor
and its software layers. The
shaded layers are emulated
by the SystemC engine in our
generated VPs

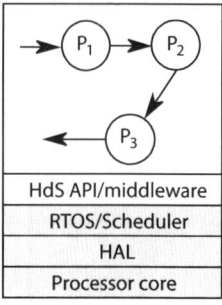

8.4.2 Virtual Processors

A virtual processor is a simulation primitive which implements execution services provided to the system applications. It is used to simulate the component templates which provide the WCET service to the application and are treated as processing nodes in the design flow. It represents a processing node in the hardware architecture consisting of a processing core such as a RISC, VLIW, DSP, or a HW accelerator core together with local memory, and peripherals such as interrupt controller, DMA engine, etc.

Figure 8.2 represents a virtual processor and its software layers. On the top, the application tasks defined in the original application model which are mapped to this node and scheduled in the analytic DSE phase are provided for execution. These applications interface the hardware platform via a Hardware-dependent Software (HdS) API, or middleware layer which provides high-level and architecture-independent communication primitives such as `send_token` and `recv_token`. An RTOS/scheduler layer provides the basic services required for defining the application tasks, providing communication mechanisms between the tasks, and performing context switch among them. The Hardware Abstraction Layer (HAL) or device drivers provides simple communication and I/O facilities. In our generated VPs, all the layers below the middle-ware layer (shaded boxes in Fig. 8.2) are abstracted and implemented using SystemC primitives.

A SystemC process is used to run each application task, `sc_fifos` with blocking semantics implement inter-task communication, and the SystemC kernel performs the context switching. The processes contain a loop in which, following the semantics of the application model, they read their inputs before and write their results after their execution from/to the corresponding buffers. They are also annotated with their WCET by advancing the local processor time. Application tasks mapped to a processor must run sequentially, so, appropriate wait statements for synchronization with local processor time are added after execution of each task and before writing out its results to mimic sequentiality. To model the overlapping computation and communication, a separate SystemC process is added which transmits the computed results and runs in parallel with other tasks. Data receiving logic is implemented by

```
1 Procedure: GenVP(I_proc, I_comm, K, E)
2 forall the I_i ∈ I_proc do
3   └ GenProc(m_i, I_i, {k|ρ_task^k = I_i});
4 forall the I_i ∈ I_comm do
5   └ GenComm(m_i, I_i, {e|ρ_task^e = I_i});
6 BindSockets;
```

```
1 Procedure: GenProc(m, I_m, K_m)
2 GenVirtualProcSkeleton;
3 forall the k_i ∈ K_m do
4     add sc_process;
5     add sc_fifos;        // ∀
      edges
6   └ annotate WCET;       // l_{k_i}^k
7 add DMA sc_process;
  // DMA engine
8 add b_transport;
  // blocking trans. if.
```

```
1 Procedure: GenComm(m, I_i, E_m)
2 init_socket ← targ_socket ← ∅;
3 forall the e_i ∈ E_m do
4   └ add src(e_i) to init_socket, dst(e_i) to
      targ_socket;
5 GenInterconnectSkeleton(init_socket,
    targ_socket);
6 add b_transport;         // blocking
  trans. if.
7 annotate WCCT;           // l_{e_i}^e
```

Fig. 8.3 Pseudo-code for generating the virtual prototype

implementing the b_transport method as part of the blocking transport interface of TLM-2.0. This method decodes the address from the generic payload of the
incoming transfers and puts the data on the appropriate input buffer of the corresponding task input. Each VP has a simple initiator and a target socket provided by
the convenient sockets of TLM-2.0.

8.4.2.1 Generation

The GENPROC procedure in Fig. 8.3 shows the pseudo-code for generating a virtual
processor. Having a characterized processing component template m, its instantiation I_m, and all application tasks K_m mapped to it, the skeleton for a virtual processor is generated first. For each mapped application task an sc_process and for
each of its communication signals an sc_fifo is added. The WCET of each task,
derived from the characterization framework and the instantiation of the processor, is
annotated. The encoding logic for outgoing transfers and the decoding logic for the
incoming transfers are added to the DMA process and the b_transport method,
respectively, using generic payloads.

8.4.3 Interconnection

As stated before, we use the interoperability layer of TLM-2.0 for the communication between the components. A generic interconnection model is used to simulate the component templates which provide the WCCT service to the application and are treated as communication links (e.g., TDM bus, dedicated simplex link) in the design flow. All predictable interconnections are modeled in the generated virtual platforms as modules which implement the blocking interfaces via initiator and target sockets. Based on the communication service delay that they imply, they annotate the blocking transport calls on the forward path with the worst-case execution time that each transport suffers. The tagged simple sockets provided by the convenient sockets and the target buffer address encoded in the generic payload are used to distinguish different transport paths.

8.4.3.1 Generation

The GENCOMM procedure in Fig. 8.3 shows the pseudo-code for generating an interconnection component. For each communication link mapped to this component the processing components to which the source and target nodes are mapped are extracted and added to the set of source and target components. The skeleton code for the interconnection component is generated. For each processing component in the source/target set a simple tagged initiator/target socket with the id of the processing component is added to the skeleton. The blocking transport method of the target sockets are added with the logic to decode the incoming transports and forward it with the WCCT of the link annotated.

8.4.4 Top Level

The pseudo-code for generating a virtual prototype is presented in the GENVP procedure of Fig. 8.3. This step is trivial and comprises the individual generation of the virtual processors and interconnections. Binding of the sockets is based on mapping of the communication links and follows the same logic described in the generation of interconnections.

The generated code compiles against the TLM-2.0 library and simulates the systems with the semantics of the application specification and timing characterization of the framework introduced in Sect. 8.3.

8.5 Case Study

The following section will present the application of the introduced design flow using the example of a real-time image compression algorithm to be implemented on a predictable hardware platform.

8.5.1 Realization of the Design Flow

The platform characterization is captured as an XML+mzn format. The XML part includes all the characterization elements except the implementation of the constraints, including the service and cost functions. The mzn file captures the constraints as predicates of the constraint programming language MiniZinc [11].

   The above format is used to produce a constraint-based formulation of a DSE problem as a MiniZinc model. A generic constraint programming solver kit Gecode [5] is used as the backend to solve the DSE problem.

8.5.2 The Application

The implemented example is a parallel implementation of a JPEG compression algorithm, modeled as an HSDF process network using ForSyDe-SystemC [1]. A possible application, shown in Fig. 8.4a, is in the real-time compression of a collection of live camera images. As can be seen in the example, individual camera feeds occasionally turn black if the sending camera is shut off. This behavior is simulated via a separate NHRT task that is disconnected from the encoder task graph whose functionality and implementation will be refined on the generated VP.

   The encoder consists of a source task (RB) reading blocks of 16×16 pixels of the combined input images which it sends to a variable number of PBs that simultaneously encode these input blocks into JPEG bit streams. To compress the image blocks, the processing elements employ color conversion, chroma down-sampling and subsequent compression of each color component using a discrete cosine transformation and entropy coding, making them the most computationally intensive tasks. These compressed bit streams are then sent to a sink task (WJ) concatenating them into the compressed JPEG output image. This structure, using four block encoding tasks, can be seen in Fig. 8.4b. Due to the implementation of the application as a network of side-effect free processes using ForSyDe, computation and communication are independent from each other and can be analyzed separately.

   The developed process network has been functionally verified by simulation of the ForSyDe-SystemC model and converted into a task graph and the accompanying C code describing the functionality of each block.

(a)   **(b)**   **(c)**

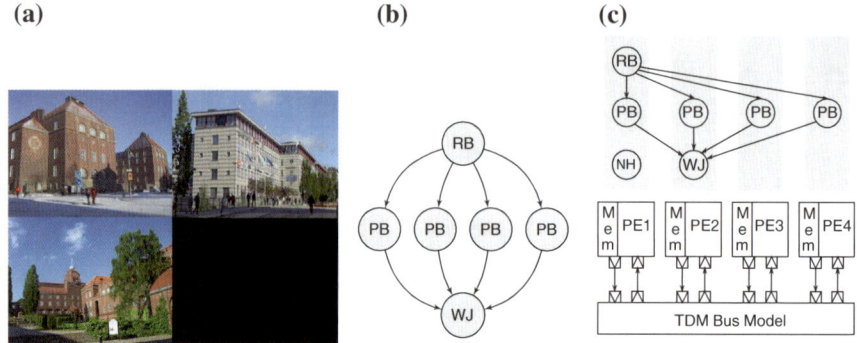

Fig. 8.4 Real-time JPEG encoder example: **a** The collection of four camera input streams to be compressed by the real-time JPEG encoder example. **b** The task graph of the developed parallel JPEG encoder application. **c** The mapping of the application to the target platform

8.5.3 The Platform

The target platform is derived from an FPGA-based predictable hardware architecture and has been captured as a platform template, as explained in Sect. 8.3. Its individual components are a set of processing elements and two possible forms of interconnection structures: a TDM bus, connecting all processing elements, and simplex FIFO buffers, each connecting a pair of processing elements. Listing 8.1 shows the XML file used for the experiments performed in this section.

In this example, the component cost indicator `pe_area` of the processing elements is 1200 Logic Elements (LEs) at a frequency of 50 MHz while the cost indicators `port_area` and `mem_area` of the TDM bus are 100 LEs and 10 LEs, respectively. As it is possible to tightly predict execution time estimates for applications running on the platform, the SWEET tool suite [4] has been used to analyze the worst-case execution time bounds of each task in the application graph. These timing bounds of the application relating to the platform template can be found in Table 8.1.

8.5.4 Results

The parameters of the platform, the task graph as well as the task metrics given in Table 8.1 have been converted into a DSE problem using the developed script and solved using an analytic DSE tool while aiming for a solution yielding minimal FPGA resource usage.

Listing 8.1 The captured platform used in the case study. Constraints appear in a MiniZinc file

```
1  <platform name="simple">
2    <component name="pe">
3      <comp_parameter name="cycle" ↵
           domain="{10,20,40}" />
4      <comp_service service="wcet">
5        <application_arg name="mapped_task" ↵
             domain="tasks"/>
6      </comp_service>
7      <comp_cost cost="area" />
8    </component>
9    <component name="smp">
10     <comp_parameter name="cycle" ↵
           domain="{10,20,40}" />
11     <comp_parameter name="depth" domain="1..128" />
12     <comp_service service="wcct">
13       <application_arg name="mapped_edge" ↵
             domain="tasks"/>
14     </comp_service>
15     <comp_cost cost="area" />
16   </component>
17   <component name="bus">
18     <comp_parameter name="cycle" ↵
           domain="{10,20,40}" />
19     <comp_parameter name="ports" domain="1..32" />
20     <comp_parameter name="tbl" ↵
           domain="1..n_max_insts" size="n_max_insts" ↵
           />
21     <comp_inst_cons constraint="ports_tbl" />
22     <comp_service service="wcct">
23       <application_arg name="mapped_edge" ↵
             domain="tasks"/>
24     </comp_service>
25     <comp_cost cost="area" />
26   </component>
27   <plat_inst_cons constraint="bus_or_smp" />
28   <plat_inst_cons constraint="bus_ip" />
29   <plat_inst_cons constraint="smp_ip" />
30   <plat_service service="wcet">
31     <comp_service service="pe_wcet" />
32   </plat_service>
33   <plat_cost cost="area" />
34 </platform>
```

Table 8.1 Worst-case
execution cycles of the
three tasks in the application

Task	WCEC
read_bitmap (RB)	450,000
process_block (PB)	3,600,000
write_jpeg (WJ)	90,000

8.5.4.1 TDM-Bus-Based Solution

The first proposed solution allots four processing elements to execute the application with a mapping shown in Fig. 8.4c, connected via a TDM bus. The task reading the input frames and one image block compression task are mapped to the first processing element while the remaining three conversion tasks are each mapped to a single processing element. The output writing task is mapped to the second processing element to make use of the parallelism of the application. The cost of this platform instance is 5,360 LEs. Next to this mapping solution, an execution schedule has been generated by the DSE tool, which predicts an execution time of 81,950 μs of a single iteration of the task graph.

Virtual Prototype

Finally, a virtual prototype of the platform instance has been generated, onto which the functionality of the application has been ported. Using this solution, it is possible to simulate the execution of the application using the timing parameters of the specific platform instance suggested by the DSE tool and, if necessary, modify elements to improve upon the solution. The execution of the application on the VP using an input image collection of 425 × 640 pixels yields a network graph iteration length of 72,900 μs while performing 1,564 TLM transfers across the TDM bus.

8.5.4.2 FIFO-Based Solution

The second proposed solution again allots four processing elements to execute the application which are connected by three FIFO buffers.

Mapping and Execution Time

The proposed mapping solution is the same as in the result discussed above, while the cost of the platform instance is 5,280 LEs and the schedule predicts an application iteration time of 79,000 μs. Table 8.2 summarizes the results for comparison.

Table 8.2 Worst-case predicted and measured results of running the application

Scenario	Predicted (μs)	Measured (μs)	Transactions
Bus-based	81,950	72,900	1,564
FIFO-based	79,000	71,500	1,564

Virtual Prototype

Simulation of the application on the VP yields a network graph iteration length of 71,500 μs while performing the same number of 1,564 TLM transfers across the FIFO instances. For both of the generated VPs the simulation time is within the predicted time by the analytic DSE. The observed timing difference is due to the necessary pessimism is the DSE logic and also potential corner cases not revealed by the example's stimuli. This allows for further refinement of timing and functionality of both the application itself as well as the NHRT tasks in the system. For reader's reference a simplified version of the generated code for the bus interconnection is presented in Listing 8.2. The TDM table and processor frequency variables have been assigned and the WCCT service characterization is used to annotate the timing of the blocking transports (Sect. 8.3).

8.6 Related Work

Several approaches for generating MPSoC VPs based on high-level specifications have been proposed [15, 19]. Popovici et al. [14] introduce four abstraction levels to be used in the process of programming MPSoC. Starting from a functional description of the application and a global view on the architecture, a system architecture model is generated after partitioning and mapping. By mapping the communication to hardware resources using a HdS API, a virtual architecture model is obtained as the second step. Adapting the software to the selected communication architecture and adding the OS functionality results to a transaction accurate architecture while the adaptation to specific processor and memories by adding the hardware abstraction layer and instruction-set simulators yields a virtual prototype. The introduced flow requires available simulation models of each hardware type and the software is adapted and verified in each step by simulation.

Kempf et al. [9] propose a design flow, which combines analytic design-space exploration with a simulation-based approach. After an early DSE based on statistical timing information of the tasks, architectures with their temporal and spatial mapping and schedule is generated. Afterwards, abstract simulation models are generated and execution time of software blocks are annotated to them in different levels such as statistical, source-level, or implementation-based to be able run the software on virtual processing units. In contrast to the above approaches, our VP generation does

Listing 8.2 Generated VP for the predictable bus

```
1  const unsigned int NR_OF_INITIATORS = 4;
2  const unsigned int NR_OF_TARGETS = 4;
3
4  class bus1 : public sc_module
5  {
6  ...
7  public:
8   target_socket_type ⏎
        target_socket[NR_OF_INITIATORS];
9   initiator_socket_type ⏎
        initiator_socket[NR_OF_TARGETS];
10
11  SC_HAS_PROCESS(bus1);
12  bus1(sc_module_name name) : sc_module(name)
13  {
14   for (unsigned int i = 0; i < NR_OF_INITIATORS; ⏎
        ++i)
15    target_socket[i].register_b_transport(this, ⏎
        &bus1::initiatorBTransport, i);
16  }
17  ...
18  void initiatorBTransport(int SocketId,
19                           transaction_type &trans,
20                           sc_time &t)
21  {
22   initiator_socket_type *decodeSocket;
23   unsigned int dest_id = ⏎
        decode(trans.get_address());
24   decodeSocket = &initiator_socket[dest_id];
25   trans.set_address(trans.get_address() & ⏎
        getAddressMask(dest_id));
26
27   t += trans.get_data_length() / 4 *
28        sc_time(wcct(SocketId,dest_id>>8), SC_NS);
29   (*decodeSocket)->b_transport(trans, t);
30  }
31  private:
32   vector<unsigned int> v_tbl = {0,1,2,3};
33   unsigned int v_cycle = 20;
34   ...
35   int wcct(unsigned int s, unsigned int d)
36   {
37    return v_tbl.size() * v_cycle /
38         count(v_tbl.begin(), v_tbl.end(), s);
39   }
40  };
```

not require any available TLM model of the components and the generation can be fully automated. Also, valid instances of the platform are instantiated automatically during the analytic DSE phase based on the instantiation constraints introduced in the characterization framework.

The Sesame approach [13] is a simulation-based design-space exploration with the simulation models composed of applications as Kahn process networks, architecture as a discrete-event model, and a dataflow-based mapping layer which drives the architecture based on simulation traces of the application. Unlike Sesame, our approach is based on the interoperability layer of TLM and the generated VPs can be connected to/replaced by the available IPs without extensive adaptation effort.

A highly relevant work is the SystemCoDesigner approach [6]. Starting from a SystemC-based executable formal application model, multi-objective evolutionary optimization is used for analytic DSE. Similar to our approach, generation of VPs for simulation-based exploration is also supported which uses virtual processing components for processing nodes. There is no explicit support for generation of other IPs, including interconnections. In contrast, we permit a more complex modeling of the services and costs for each IP.

The readers are encouraged to consult Chap. 9 for detailed presentation of a design flow for mixed-critical systems which can benefit from the automatic generation of simulation models presented in this work.

8.7 Conclusion

Virtual prototypes are a precious tool for validation of embedded real-time systems during their design flow but they are time-consuming to generate. We have proposed a method for automatic generation of VPs in a combined analytic/simulation-based design flow for predictable platform templates. The core enabler of our method is a framework for characterization of predictable platform templates which captures a set of flexible predictable component templates, their instantiation rules, the services that they provide to the application and their costs. Starting from a platform template captured by the characterization framework and a real-time application captured as a task graph, an analytic design-space exploration is performed which gives a (set of) valid platform instance(s) together with the mapping and schedule of the application on it. For each instance, a simulatable VP is generated which can be used for:

1. execution of the applications on the platform to get simulation traces;
2. development of software code for non-hard real-time tasks; or
3. further refinement to a lower-level simulation models and implementation.

These VPs are valid instances of the platform because they have already satisfied the instantiation constraints in the analytic DSE phase and they are annotated with the timing information captured as service functions in the platform characterization template. The generated VPs are based on the interoperability layer of TLM-2.0 and can be connected to, or refined by partially replacing available IP.

A limitation of this approach is that it can be used for predictable platforms, which provide static timing guarantees to the applications.

References

1. Attarzadeh Niaki SH, Jakobsen M, Sulonen T, Sander I (2012) Formal heterogeneous system modeling with SystemC. In: forum on specification and design languages (FDL) 2012. pp 160–167
2. Davis RI (2011) A survey of hard real-time scheduling for multiprocessor systems. ACM Comput Surv 43(4): doi:10.1145/1978802.1978814
3. Eker J, Janneck J, Lee E, Liu J, Liu X, Ludvig J, Neuendorffer S, Sachs S, Xiong Y (2003) Taming heterogeneity—the Ptolemy approach. Proc IEEE 91(1):127–144
4. Engblom J, Ermedahl A, Sjdin M, Gustafsson J, Hansson H (2003) Worst-case execution-time analysis for embedded real-time systems. Int J Softw Tools Technol Transfer 4(4):437–455
5. Gecode Team (2014) Gecode: generic constraint development environment. http://www.gecode.org/
6. Haubelt C, Falk J, Keinert J, Schlichter T, Streubhr M, Deyhle A, Hadert A, Teich J (2007) A SystemC-based design methodology for digital signal processing systems. EURASIP J Embed Syst 1:15–15
7. Herrera F, Villar E (2008) A framework for heterogeneous specification and design of electronic embedded systems in SystemC. ACM Trans Des Autom Electron Syst 12(3):22:1–22:31. doi:10.1145/1255456.1255459
8. Jantsch A (2004) Modeling embedded systems and SoCs. Morgan Kaufmann, San Francisco
9. Kempf T, Ascheid G, Leupers R (2011) Multiprocessor systems on chip, design space exploration. Springer, Berlin
10. Lee E, Ha S (1989) Scheduling strategies for multiprocessor real-time DSP. In: Proceedings of the IEEE global telecommunications conference and exhibition 'communications technology for the 1990s and beyond' (GLOBECOM) 1989, vol 2, pp 1279–1283
11. Nethercote N, Stuckey P, Becket R, Brand S, Duck G, Tack G (2007) MiniZinc: towards a standard CP modelling language. Principles and practice of constraint programming., Lecture notes in computer scienceSpringer, Berlin, pp 529–543
12. OSCI TLMWG (2008) Transaction level modeling (TLM) library 2.0. http://www.systemc.org/
13. Pimentel A, Erbas C, Polstra S (2006) A systematic approach to exploring embedded system architectures at multiple abstraction levels. IEEE Trans Comput 55(2):99–112
14. Popovici K, Rousseau F, Jerraya AA, Wolf M (2010) Embedded software design and programming of multiprocessor system-on-chip. Embedded systems. Springer, London (Limited)
15. Ptrot F, Gerin P, Hamayun M (2012) On software simulation for MPSoC. In: Nicolescu G, O'Connor I, Piguet C (eds) Design technology for heterogeneous embedded systems. Springer, Netherlands, pp 91–113
16. Stuijk S, Geilen M, Theelen B, Basten T (2011) Scenario-aware dataflow: modeling, analysis and implementation of dynamic applications. In: international conference on embedded computer systems (SAMOS) 2011, pp 404–411
17. Thiele L, Chakraborty S, Naedele M (2000) Real-time calculus for scheduling hard real-time systems. In: Proceedings of the IEEE international symposium on circuits and systems (ISCAS) 2000, vol 4, pp 101–104
18. Uhlig RA, Mudge TN (1997) Trace-driven memory simulation: a survey. ACM Comput Surv 29(2):128–170. doi:10.1145/254180.254184
19. Wieferink A, Meyr H, Leupers R (2008) Retargetable processor system integration into multiprocessor system-on-chip platforms. Springer, Netherlands

Chapter 9
Combining Analytical and Simulation-Based Design Space Exploration for Efficient Time-Critical and Mixed-Criticality Systems

Fernando Herrera and Ingo Sander

Abstract In the context of the design on time-critical systems, analytical models with worst case workloads are used to identify safe solutions that guarantee hard timing constraints. However, the focus on the worst case often leads to unnecessarily pessimistic and inefficient solutions, in particular for mixed-critical systems. To overcome the situation, the paper proposes a novel design flow integrating analytical and simulation-based Design Space Exploration (DSE). This combined approach is capable to find more efficient design solutions, without sacrificing timing guarantees. For it, a first analytical DSE phase obtains a set of solutions compliant with the critical time constraints. Search of the Pareto optimum solutions is done among this set, but it is delegated to a second simulation-based search. The simulation-based search enables more accurate estimations, and the consideration of a specific (or an average-case) scenario. The chapter shows that this can lead to different Pareto sets which reflect improved design decisions with respect to a pure analytical DSE approach, and which are found faster than through a pure simulation-based DSE approach. This is illustrated through an accompanying example and a proof-of-concept implementation of the proposed DSE flow.

Keywords Design Space Exploration (DSE) · Joint Analytical and Simulation Based DSE · Time critical systems · Mixed-Criticality System (MCS) · Embedded system design · Real time system design · Constraint-programming · Predictable systems · Performance estimation · Time Division Multiplex (TDM) bus · Optimization techniques · Formally-based design

9.1 Introduction

Real-time systems are ubiquitous and appear in an important set of practical application domains. A main characteristic of real-time systems is that their correct behaviour depends on the fulfilment of constraints on their time performance,

F. Herrera (✉) · I. Sander
KTH Royal Institute of Technology, Stockholm, Sweden
e-mail: fernanhc@kth.se;fherrera@teisa.unican.es

I. Sander
e-mail: ingo@kth.se

© Springer International Publishing Switzerland 2015
M.-M. Louërat and T. Maehne (eds.), *Languages, Design Methods, and Tools for Electronic System Design*, Lecture Notes in Electrical Engineering 311,
DOI 10.1007/978-3-319-06317-1_9

167

e.g., after considering metrics like response time, throughput, etc. The criticality of such constraints depends on the application domain. In aeronautics, automotive, or medical domains [8], time constraints can be *safety-critical*, and related to a high *Safety Integrity Level* (SIL) [16], while, in consumer electronics, real-time constraints might not guarantee safety, but a minimum Quality of Service (QoS) to ensure the survival of the system in the market. Moreover, mixed-criticality systems have appeared as a consequence of the integration of applications and functionalities with different criticality on their associated performance constraints [8, 14].

Design Space Exploration (DSE), a central design activity of embedded system design in charge of finding an efficient solution given the design constraints and optimisation goals, has to consider these levels of criticality on performance constraints. DSE has become a high and active research field. In [19], two main approaches to DSE are distinguished, *analytical* and *simulation-based* DSE. In an analytical approach the design problem is abstracted into a formal problem, in order to enable the application of analysis techniques [21]. Analytical approaches are convenient for a first and fast filtering of the design space [19]. However, analytical approaches have also limitations. Expressing the design problem of complex systems might not be straightforward, and for real-time design, the consideration of pessimistic workloads can lead to inefficient solutions. Simulation-based performance analysis can tackle more complex platform models and performance metrics, and provide more accuracy for typical use cases, provided a good model of the environment. Moreover, simulation-based performance analysis considers the dynamism of the application. In order to combine the advantages from both approaches, *Joint Analytical and Simulation-based Design Space Exploration* (JAS-DSE) has been proposed in [19]. In most cases, a two-phase approach is proposed, where in a first step simulation is used, to get traces and/or perform a calibration before a second analytical phase [19].

This chapter revises the novel JAS-DSE, or *combined-DSE*, for efficient design of time-critical systems proposed in [13]; updates the related work to consider a recent and close proposal [17]; extends the experimental section to show the speed-up; and adds a section where the limitations of the shown proof-of-concept and thus the possible extensions of the proposed framework are introduced.

The proposed flow consists of a first analytical DSE phase, relying on constraint programming and worst-case estimations. The goal of this phase is to filter the design space and get a set of *safe solutions*. These solutions are safe in the sense that the fulfilment of the critical time constraints is ensured at a certain level of criticality. The actual safety level depends on the predictability of the platform executing the critical part, and on the technique for evaluating the worst-case workloads. In a second phase, the set of *safe solutions* is evaluated through an executable performance model of the system, which is stimulated by input traces reflecting the most likely or a specific working scenario for the system. The consideration of that scenario, and the fine granularity in the performance assessment thanks to simulation and basic-block level annotations enables an accurate finding of the Pareto solution set. This Pareto set is in general different to the one that would be found only by means of the analytical phase based on worst-case workloads.

The rest of the chapter is structured as follows. Section 9.2 discusses the related work. Section 9.3 formulates the problem and Sect. 9.4 states the generic flow. Then, Sect. 9.5 presents an illustrative DSE problem and a proof-of-concept instance of the combined-DSE flow based on open frameworks and easy to reproduce. Section 9.6 discusses the main limitations of the proof-of-concept example in order to introduce future extensions of this work. Finally, Sect. 9.7 provides the main conclusions.

9.2 Related Work

Simulation-based methods have been extensively applied for performance analysis and so for DSE. A summary of them can be found in [19]. DSE frameworks such as *Sesame* [27, 36] or *Multicube* [31] have proposed a clear separation among application, platform architecture, and mapping in the system description, and moreover, the separation of the system simulatable performance model from the exploration tool, e.g., *Multicube Explorer* [39], which steers the search. Well defined interfaces and configurable performance models which avoid re-modelling and re-compilation every DSE iteration, and fast performance assessment technologies, such as native simulation [28, 37], have enabled simulation speed-ups of orders of magnitude, which has been exploited in frameworks like Multicube-SCoPE [25]. The UML/MARTE COMPLEX methodology [15] enabled a model-driven front-end for DSE.

Several analytical approaches have been proposed so far. They have abstracted applications and platform resources to different formalisms, such as *network calculus* [35], *conditional process graphs* [7], or *workload models* using formal scheduling analysis techniques and symbolic simulation, such as Sym/Ta [12]. Real-Time (HRT) theory [6] has traditionally addressed the analysis of *time critical* systems by relying on worst-case workloads, e.g., Worst-Case Execution Times (WCET) [38]. Similarly, the consideration of hard-real time constraints in analytical DSE approaches has been based on worst-case workloads. Due to the NP-complex nature of the mapping and scheduling problem, earlier approaches tackled the problem in separated phases [32], which prevented completeness. In [33] each implementation found is suited to a specific *scenario*. Constraint-based Programming (CP) has been proposed in [3] to achieve completeness, and to separate the modelling of the DSE problem from the solver engine, which can be independently optimised. In [22], an analytical DSE is performed at design time to find a set of Pareto optimum mappings considering a throughput-energy trade-off, and which are used by a dynamic scheduler at run-time.

Relying on sound modelling formalisms has been common to both types of DSE approaches. Most of the aforementioned analytical approaches [3, 22, 32] have relied on the *Synchronous Data Flow* (SDF) or the *Homogeneous Synchronous Dataflow* (HSDF) [24]. The SDF MoC provides functional determinism by construction and analysability for deciding deadlock protection and other properties, such as boundeness. Later approaches [33] have relied on more generic models, e.g. the *Scenario-Aware SDF* (SA-SDF) MoC [34]. However, simulation-based approaches are capable to cope with more general formalisms, e.g., Kahn Process Networks (KPNs), as in [27].

In order to achieve the advantages from analytical and simulation based appro-aches, e.g., evaluation speed and accuracy, JAS-DSE approaches have been pro-posed [19]. A common JAS-DSE approach is to use simulation traces to compute the workloads which are later used in an analytical phase. In [23], simulation is used to obtain traces of the computation and communication events of the application which are abstracted as an acyclic communication analysis graph (CAG), and which is later used for performance analysis. Big speed-ups suitable for DSE are obtained. However, the approach does not link to an automatic exploration and/or optimisation mechanism. The methodology in [9] achieves sub-optimal configurations of sepa-rated cache memory by using a sensitivity-based optimisation. The approach relies on a system-level simulation environment and on an analytical model of the energy consumption for obtaining the traces. The approach in [21] couples network calculus *event streams* with events in a SystemC simulation environment through a set of *con-verters*. The approach speeds-up the evaluation of a configuration, since not all the system has to be simulated, making it more suitable for DSE. This approach is focused on the combination of analytical and simulation-based techniques for enhancing per-formance estimation, e.g., [21, 23], and not in defining a DSE flow, stating how to integrate search and optimization techniques with performance assessment.

A recent work [17] has got much closer to [13], since it also proposes a two-phases approach. A relevant distinction with the aforemetioned JAS-DSE approches is that the proposed approach places the analytical phase in first place. In fact, in [17], the pruning of the design space is totally delegated to the analytical phase, which relies on a set of heuristic-based algorithms for producing the potential HW/SW partition, task clustering and mapping. All the solutions derived from the analytical phase are later assessed by means of a simulation framework. The methodology in [17] supports the modelling of a virtual plaform with a shared-memory based mechanism.

In contrast, the methodology proposed in [13], and revised in this chapter, tar-gets the design of more efficient real-time systems or systems where certain time constraints are critical. For it, the main objective of the analytical phase is to dis-card any solution which cannot fulfill *hard* or *critical* time constraints. At the same time, the analytical phase will perform a fast pruning of the design space. However, it might not prevent a large set of solutions compliant with the critical constraints, making unfeasible an exhaustive simulation of every safe solution. On one side, this is an issue, on the other side, getting a large set of solutions compliant with the critical time constraints raises the chances for finding efficient solutions in the later simulation-based phase. This explains the convergence of two specific characteristics in the proposed approach, which are at the same time two main differences versus the approach in [17], namely: an analytical phase based on constraint-programming, which preserves the possibility to obtain a complete set of compliant or safe solutions; and an exploration tool in the simulation-base phase, to cope with the simulation-based exploration of large safe solution sets. To feed the former analytical phase with worst-case workloads, a third and previous phase, based on simulation traces, as in

[9, 23], or as it is later shown in Sect. 9.5 in the proof-of-concept example, can be added. However, the proposed approach does not fix a specific method.

9.3 Problem Formulation

Let a system S be composed of a set of applications A running on a set of platform resources R; a set of performance metrics P associated to the system; and a set of constraints on performance and resources $Cr(P) \cup Cr(R)$. Let consider the spliting of P into a specific set of time-related metrics $T | T \in P$, plus a remaining set of performance metrics $P_1 \cup P_2$, such $P = T \cup P_1 \cup P_2$. Notice that $P_1 \cup P_2$ can contain also time-related metrics, and that the set of constraints on T, $Cr(T)$, fulfills $Cr(T) \subseteq Cr(P)$. Finally, let consider also a set of performance goals $G(P)$. S is *time critical* because $\exists Cr(T)$ which fulfillment has associated a high criticality, denoted $\chi(Cr(T)) = \chi_{HI}$. Moreover, S is a *mixed-criticality* system if there are additional performance constraints with a lower criticality, denoted $\chi(Cr(P_1)) = \chi(Cr(P_2)) = \chi_{LO}$. Then the DSE problem stated consists in finding an efficient set of solutions (ESS), ideally a Pareto set, according to $G(P)$, ensuring the fulfillment of $Cr(T)$ for a χ_{HI} criticality level and of $Cr(P_1) \cup Cr(P_2)$ for a χ_{LO} criticality level.

9.4 Proposed Flow

The combined-DSE approach proposed is sketched in Fig. 9.1. In a first step, a *Worst Case Analytical Design Space Exploration* (WCA-DSE) filters the original design space, to get the set of *Safe Solutions* (SSs), i.e., which fulfills $Cr(T)$ at the χ_{HI} criticality level. That means an assessment of the performance based on worst-case workloads associated to a high criticality level, $WCW(\chi_{HI})$. Here, $WCW(\chi)$ denotes a Worst-Case Workload for a given criticality χ_{HI} level. Typically, the higher the criticality, the more pessimistic the WCW [1]. The criticality strongly determines the performance estimation technique used to obtain the WCW (PET_1 in Fig. 9.1), which has to take into account possible variations or refinements on the platform architecture, and the dynamism, that is, the data and state dependency of the workload figures to be associated to a piece of code.

An analytical description of the DSE problem (CP in Fig. 9.1) is captured in a declarative way, by relying on a constraint-programming language. Such description contains different constraints which define valid implementations of the system by considering, for instance dependencies on the applications; the limitations on available resources, $C(R)$; and on the feasible architectural mappings. The CP-model will also capture an analytical performance model for time-related metrics, e.g., *throughput* and *response times*; and their associated constraints $Cr(T)$. Support of constraints on different types of time performance metrics, such as throughput and

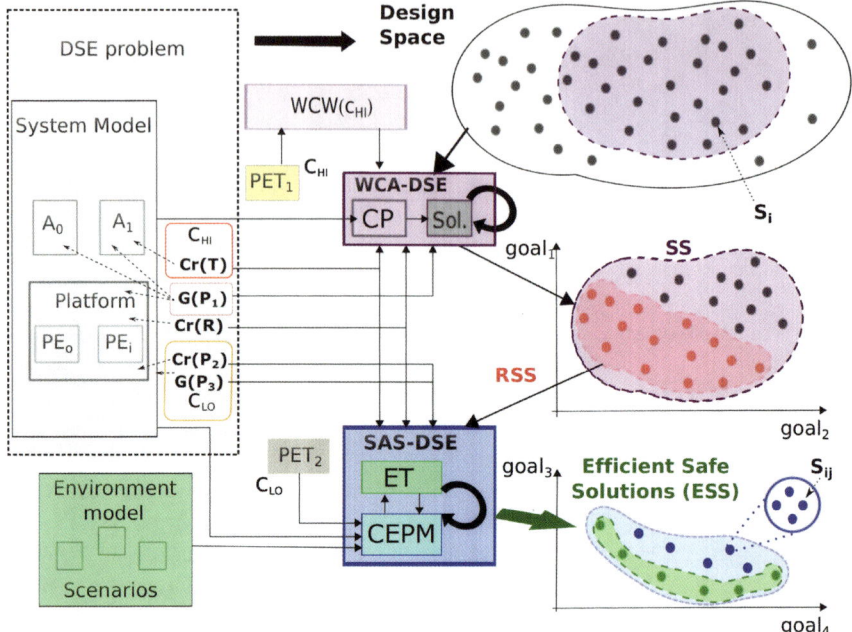

Fig. 9.1 Proposed JAS-DSE flow

response times, enable to tackle in an unified way design problems which have been separately addressed so far, e.g., throughput optimisation and hard-real-time scheduling.

A solver (Sol. in Fig. 9.1) provides all the solutions compliant with the analytical model, that is, the Safe Solution (SS) set. If the SS set on the original $Cr(T)$ is too big for the second phase of the proposed approach, a *reduced set* of safe solutions (RSS) of size size$(RSS) <$ size(SS) can be produced. For it, the original time-critical constraints $Cr(T)$ can be tightened, settling a more demanding $Cr'(T)$ set. Furthermore, more constraints on an additional set of performance metrics P_1, can be used, whenever such constraints can be also stated in an analytical form and reflected in the CP program. The additional constraint set $Cr(P_1)$ can be tuned to yield the targeted size(RSS). Constraint-based programming frameworks enable optimisations, which can be used to guide the production of the RSS.

In the proposed approach, a second *Scenario-Aware Simulation-based Design Space Exploration* (SAS-DSE) phase is applied in order to obtain the Pareto optimum solutions among the SS set, or among the RSS if it is the case. In its simplest form, the SAS-DSE consists in performing a simulation-based performance estimation of each solution of the SS/RSS set. For it, a configurable and executable performance model (CEPM) of the system, coherent with analytical model, is built up, and linked to an executable environment model. The environment model is in charge of feeding the most likely stimuli (scenario) which the system will receive in its working environment. The CEPM is *configurable* in the sense that the generation of a simulatable model of a specific solution only requires to read the specific values of the explored variables which define such solution, without requiring manual re-coding and re-compilation. This speeds-up each iteration of the DSE loop in the SAS-DSE phase.

The methodology is not tied to a specific simulation-based performance assesment technique (PET_2 in Fig. 9.1). For fast simulation vs accuracy trade-off, the use of native source simulation integrating performance annotations at basic block level [28] is proposed. Resource constraints $Cr(R)$ and critical constraints $Cr(T)$ used in the WCA-DSE phase are also an input to the SAS-DSE phase, which has to validate them. The result of the SAS-DSE is a new estimation for the performance metrics associated to the solutions of the SS (or RSS) set. In general, these new estimations will be less pessimistic and will involve a shift in the position of each solution in the Pareto diagram. Therefore, the application of the SAS-DSE phase is capable to show more accurately and for a given scenario the most efficient and yet safe solutions. Such set of solutions is called *Efficient Safe Solutions* (ESS) set. Moreover, additional goals on the T set, $G(T)$, can be stated, which would make more likely the consideration of solutions which were not optimum after the WCA-DSE phase. The simulation approach facilitates the consideration of performance metrics $P_2|P_2 \in P$ of different nature (time, energy, power, etc.), and therefore a wider set of constraints $Cr(P) = Cr(T \cup P_1 \cup P_2)$, and more general goal functions $G(P)$ in the SAS-DSE phase. Finally, the proposed flow contemplates the integration of an Exploration Tool (ET) in the SAS-DSE phase. An ET in the SAS-DSE phase makes it feasible to cope with potentially large safe solution sets, and thus to apply the WCA-DSE analytical phase considering only the highly critical constraints. This way, the chances for finding efficient solutions in the SAS-DSE phase are maximized.

9.5 Proof-of-Concept DSE Problem and Flow Implementation

9.5.1 DSE Problem Model

Fig. 9.2 sketches a proof-of-concept system model. The application model abides a Synchronous Dataflow (HSDF) MoC [24]. An application A is partitioned into a set of $N \in \mathbb{N}$ side-effect free functions f_i, which are associated to corresponding tasks or actors a_i (nodes in Fig. 9.2). A $N \times N$ dependency matrix D reflects the dependencies among actors (shown as solid directed edges in Fig. 9.2). In D, $d_{ij} = 1|d_{ij} \in D$ means that a_i has to execute before a_j, while $d_{ij} = 0$ means that there is no dependency. Such dependencies can reflect either a data-dependency or an explicit synchronization, but the model does not capture or involves a specific implementation.

For instance, Fig. 9.2 application can be represented as:

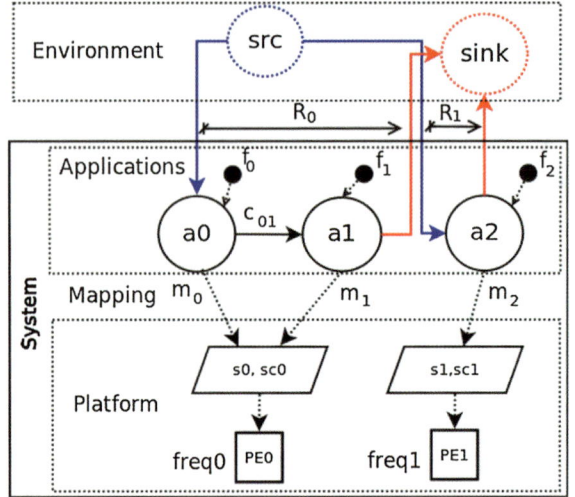

Fig. 9.2 System model

$$A = \{a_0, a_1, a_2\} = \{0, 1, 2\} \tag{9.1}$$

$$D = \begin{vmatrix} d_{00} & d_{01} & d_{02} \\ d_{10} & d_{11} & d_{12} \\ d_{20} & d_{21} & d_{22} \end{vmatrix} = \begin{vmatrix} 0 & 1 & 0 \\ 0 & 0 & 0 \\ 0 & 0 & 0 \end{vmatrix} \tag{9.2}$$

Eq. (9.1) assigns unique integer identifiers to the actors. Equation (9.2) reflects a dependency matrix showing a dependency between a_0 and a_1.

The platform model comprises a HW layer and a SW layer. The HW layer consists of a finite set of processing elements (PE_j), where the working frequency of each PE ($freq_j$) can be customized. A frequency vector $freq$ is used to model it, and a finite range of frequencies is assumed for each $freq_j$ for simplicity. The SW platform is composed of a finite set S of static schedulers (s_j). In this example, we will assume that each PE will run only one static scheduler associated to a single PE. Therefore, there is a one-to-one association among schedulers and PEs, which match in number ($N_S = N_{PE}$). For instance, the HW/SW platform of Fig. 9.2 can be represented as follows:

$$S = \{s_0, s_1\} = \{0, 1\} \tag{9.3}$$

$$PE = \{PE_0, PE_1\} = \{0, 1\} \tag{9.4}$$

$$freq = \{freq_0, freq_1\} = \{100, 50\} \tag{9.5}$$

Eqs. (9.3) and (9.4) assign unique integer identifiers to scheduler and PE instances. Equation (9.5) configures PE working frequencies. The architectural mapping defines the association of application tasks to schedulers, or equivalently in this case, to processing elements. The mapping can be expressed as a matrix M, where $m_{ij} = 1 | m_{ij} \in M$ represents allocation of an actor $a_i \in A$ to a scheduler $s_j \in S$, while $m_{ij} = 0$ means that there is no allocation. Once the mapping is defined, the set of actors assigned to a scheduler is specified, and a design solution has to be completed by defining the static schedule sc_j for each scheduler s_j. For a HSDF Graph (HSDFG), the size of sc_i matches the number of actors allocated to the scheduler. For the Fig. 9.2 example, mapping and static schedule solutions can be represented as follows:

$$M = \begin{vmatrix} m_{00} & m_{01} \\ m_{10} & m_{11} \\ m_{20} & m_{21} \end{vmatrix} = \begin{vmatrix} 1 & 0 \\ 1 & 0 \\ 0 & 1 \end{vmatrix} \equiv proc = [0, 0, 1] \tag{9.6}$$

$$sc = \{s_0, sc_1\} = \{\{0, 1\}, \{2\}\} \equiv next = [1, -1, -1] \tag{9.7}$$

Eqs. (9.6) and (9.7) also show the equivalent compact forms through the `proc` and `next` vectors shown in [2]. The application model is annotated with *Worst-Case Executed Instructions* (WCEI). WCEI is a characterization of the computational effort required by a functionality, independent of the Cycles Per Instruction (CPI) (we will assume $CPI = 1$) and of the processor frequency. Then, WCETs are calculated as $WCET_{ij} = WCEI_i * (1/freq_j)$, where i refers to the i-th functionality/actor and j to the j-th processing element.

To complete the statement of the DSE problem, as well as the system model, we need to state the performance constraints, that is, boolean conditions stated in terms of the performance metrics of interest. In the Fig. 9.2 example, we will handle two types of performance metrics. Two *response times*, R_0 and R_1, for the corresponding connected task sets, $\tau_0 = \{a_0, a_1\}$ and $\tau_1 = \{a_2\}$, are considered. Then, assuming that the system triggers after receiving input data and will not process more data until both task sets have completed, the *throughput* for this system can be defined as:

$$Th = 1/max\{R_0, R_1\} \tag{9.8}$$

Th_{min} denotes a constraint on throughput, such $Cr(Th) = \{Th > Th_{MIN}\}$. A deadline D_k denotes a constraint on each k-th response time, such $Cr(R_k) = \{R_k < D_k\}$. Real-time theory focus on deadline constraints, i.e., $Cr(T) = \cup_k D_k$, while other approaches [3, 11, 22] focus only on the throughput constraint, i.e., $Cr(T) = Cr(Th)$. However, the proposed approach and this proof-of-concept example covers both kinds of constraints on time-metrics in a holistic way, for all possible constraints, $Cr(T) = Cr(R_0) \cup_k D_k$, or for a subset of them, e.g., $Cr(T) = Cr(R_0) \cup Cr(Th)$.

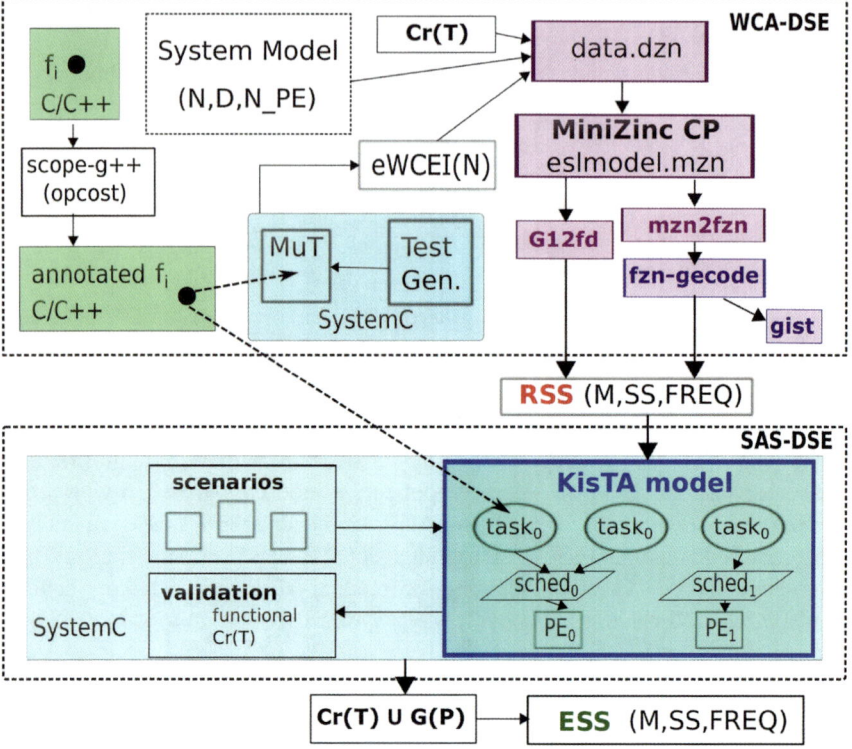

Fig. 9.3 Implementation of the combined DSE flow

9.5.2 Implementation of the Combined-DSE Flow

An implementation of the combined-DSE flow capable to tackle the DSE problem shown in the previous section has been done. It is sketched in Fig. 9.3.

For tackling the WCA-DSE phase, the DSE problem of Sect. 9.5.1 has been generically described as a constraint satisfaction problem using the *MiniZinc* CP language [26]. Listing 9.1 shows a simplified version of the generic MiniZinc program (without frequency parametrisation) for the DSE problem introduced in Sect. 9.5.1.

First, a set of *MiniZinc parameters* are declared, i.e., the number of tasks N, the number of processing elements N_PE, the D (dependency matrix) and the vector of worst-case execution times wcet. They are stated (in the data.dzn file) before solving the MiniZinc program, and behave as constants during the analytical DSE. The Minizinc program contains a set of *decision variables*, e.g., M (mapping matrix), s (starting job times), e (end job times), and end. While all of them are subject to the tuning of the solver, only a group of them represent a solution variable, namely, M, plus s (or equivalently e). Notice that from the vector of starting times s, and from M, the set of scheduling vectors *sc* (or the next vector) can be extracted, and that

Listing 9.1 MiniZinc code of the Sect. 9.3 DSE problem

```
1  int: N; int: N_PE;
2  array [1..N] of int: wcet;
3  array [1..N,1..N] of par 0..1: D;
4  array [1..N,1..N_PE] of var 0..1: M;
5  int: seq_t = sum(i in 1..N) (wcet[i]);
6  int: digs = ceil(log(10.0,int2float(seq_t)));
7  array [1..N] of var 0..seq_t: s;
8  array [1..N] of var 0..seq_t: e;
9  var 0..seq_t: end;
10 % end times (no preemption)
11 constraint forall(i in 1..N)
12            (s[i] + wcet[i] = e[i] );
13 % precedence constraint
14 constraint forall(i,j in 1..\,N where i!=j) (if ↵
       D[i,j]>0 then e[i] <= s[j] else true endif);
15 % jobs to same processor sequentialized
16 constraint forall(i,j in 1..\,N where i!=j)
17    (forall(k in 1..N_PE) (
18    (M[i,k]!=0 /\ M[j,k]!=0 /\ (e[i] <= s[j] \/ ↵
       e[j] <= s[i]))
19    \/ M[i,k]==0 \/ M[j,k]==0));
20 % mapping constraint
21 constraint forall(i in 1..N) (
22       sum(j in 1..N_PE) (map[i,j]) == 1);
23 % global performance constraint
24 constraint forall(i in 1..N)
25       (e[i] <= end);
```

the one-to-one association between scheduler instances and PEs have been captured in a compact way, through a single PE vector.

A set of constraints are associated to the system model. The constraint in Listing 9.1, line 11, reflects the relation among starting and end times, assuming non-preemptive scheduling. Line 14 constraint states the precedence among jobs obliged by the dependency matrix. For instance, in the Fig. 9.2 example, $D[1, 2] > 0$ means that a_o job has to be executed before a_1 of the same HSDFG firing, or equivalently, that the start time of a_1 has to to be bigger or equal than the end time of a_o, i.e., $s[1] >= e[0]$. Notice that this constraint relies on the fact of handling an HSDFG, so there will be one job per periodical firing of the HSDFG. The constraint in line 16 states that any pair of jobs assigned to the same processor cannot execute at the same time. Line 21 constraints task migration and ensures that any valid solution maps every task to a PE. Finally, line 24 constraint bounds the set of interesting solutions to those where the execution time of the HSDFG is bounded by the fully sequential case.

Modelling performance constraints can be done also as additional constraints, as shown in Listing 9.2, added to the Listing 9.1 code. The constraint in line 1 of Listing 9.2 states the deadline constraints for individual tasks. For it, a deadline vector Dd is used, where $Dd[i] = 0$ means that there is no deadline for the i-th task. Line 2 sets

Listing 9.2 MiniZinc code of the Sect. 9.3 DSE problem

```
1  constraint forall(i in 1..N) ((bool2int(Dd[i])*e[i] ↵
       <= Dd[i]);
2  constraint forall(i in 1..N) ((bool2int(th_inv)*e[i] ↵
       <= th_inv);
```

Listing 9.3 MiniZinc code for stating the analysis target

```
1  solve satisfy
2  solve minimise th_inv
```

the throughput constraint by stating the maximum response time. If the constraint is set to 0, the condition holds, and thus no throughput constraint is stated. In case that no performance constraint is set, the aforementioned Listing 9.1, line 24 constraint limits the set of interesting solutions.

The simple CP shown in Listing 9.1 can be extended with additional constraints to focus on subsets of solutions and to involve a more efficient search. Section 9.5.3 discuss and reports the effects of adding additional constraints to the CP introduced in this section.

Line 1 of Listing 9.3 shows how the solver can be asked to provide all compliant solutions (what is required in the proposed flow). Line 2 illustrates how to ask for an optimisation (of throughput in the shown case).

Getting back to Fig. 9.3, two integer solver back-ends have been used to show the decoupling of the DSE problem description from the solver engine. One is the g12fd solver available in the *G12 Minizinc distribution*. The other is an efficient solver included in Gecode [10], an open C++-based CP-toolkit. Gecode also includes tools like gist, which has been used to visualise the exploration as a decision tree. Both solvers have been used to provide all safe solutions. Each solution is defined by a mapping, a static scheduling for each PE, and the configured frequencies, that is, the tuple (M, sc, freq), or equivalently (proc, next, freq). Each of these solutions is used to configure the executable performance model in the WCA-DSE phase.

The proposed flow does not fix a specific technique for the estimation of worst-case workloads feeding the analytical phase (PET1 in Fig. 9.1). In the proof-of-concept flow implementation, a simulation-based technique has been used to calculate a vector with the number of *Maximum Observed Executed Instructions* (MOEI). In such PET1 implementation, each functionality associated to a task has been enclosed by a *Module under Test* (MuT) in Fig. 9.3, which has been stimulated by a module which generates random test vectors. Each vector accounts for a valid set of inputs enabling a single fire of the functionality. The response time of the MuT functionality is measured for each vector, and the maximum one reported as a MOEI. Therefore, the MOEI is to the WCEI, as the *Maximum Observed Executed Time* (MOET) to the WCET [38]. That means that, strictly speaking, a MOEI is not a bound to the execution time and therefore, it is not what we propose to feed to the analytical phase. However, as it is reasoned in Sect. 9.6, using MOEI will be sufficient for the demonstration purposes of this chapter.

Moreover, a main advantage is that it enables that the performance estimation techniques PET1 and PET2 of the presented proof-of-concept implementation rely on a common source code annotation technique which facilitates comparisons and the illustration of the advantages of the combined-DSE approach. The `scope-g++` utility of the SCoPE performance estimation framework [30] has been used. By compiling source code with the SCoPE `scope-g++` utility (enabling the `opcost` option), each basic block, is automatically and transparently annotated. The annotation takes into account the different types of source-level operations performed, e.g., conditional branches, additions, logical comparisons, etc. [28]. The `scope-g++` utility introduces several annotations. Specifically, an annotated variable which accounts for the number of instructions has been used. The number of instructions executed by a piece of code depends on the processor architecture. In the `scope-g++` utility, this is tuned by changing the configuration of an operator cost file. We have used the default operator costs, tuned to an ARM architecture. If the measured piece of code has several paths traversing several basic blocks, the measured number of executed instructions will depend on the input data and state.

For building the executable performance model used for the SAS-DSE phase, a SystemC-based infrastructure for simulation and time analysis called KisTA [20] has been used. KisTA supports the generation of executable and configurable performance models (CEPM) of the system, and its linking to a SystemC description of the scenarios, which provide the stimuli and which retrieves the performance data and performs functional validation. For simulation and assessment of the performance of a solution under a specific scenario, it is only necessary to pass to the KisTA CEPM, the parameters (frequency vector, mappings and schedulings) found by the WCA-DSE phase defining a safe solution.

A complete explanation of the KisTA framework is out of the scope of this paper. For the setting of the experiment, it has been exploited that KisTA supports the modelling of schedulers, and different scheduling policies, specifically of static schedulings. It also supports the annotation of worst-case workloads, to validate the results of the analytical phase. What it is more interesting for the experimental set shown in this chapter, KisTA has a hook to integrate code annotated with `scope-g++`. This way, the dynamism of the model, and thus the different time behaviour of the model can be exposed when different stimuli can be applied. The use of the same annotation technique for estimating worst case workloads and the simulated workloads provides a common reference to enable a fair comparison between the proposed JAS-DSE method, and aforementioned methods relying only on either analytical or simulatable performance models. Moreover, KisTA provides reports of metrics, such as PE usage, which can be incorporated to the DSE in the simulation-base phase. The simulation-based framework also enables to perform a functional validation. By using KisTA synchronization facilities, the KisTA model has been ensured to be coherent with the dependencies stated in the analytical model.

Table 9.1 Number of solutions found by the WCA-DSE for different versions of the CP model

th^{-1}	TS	TS+M	OS+FS	OS	TS+D1	OS+D1
6	152	76	38	5	76	4
5	58	29	17	2	36	2
4	16	8	6	1	16	1
3	2	1	1	1	2	1

9.5.3 Experiments

9.5.3.1 Experiments on the WCA-DSE

A first set of experiments was done for analysing the effect of constraints on the number of solutions found. For it we ran the solver for four different versions of the CP program to find all satisfying solutions given a single constraint on maximum latency (th^{-1}). Table 9.1 reports the number of solutions found for different values of the constraint. All the experiments considered $N = 3$, $N_{PE} = 2$, fixed frequencies and fixed $wcet = \{1, 2, 3\}$. The TS column reports the results for the first version of the CP program, basically reflected by the Listing 9.1, which is a *timed scheduling* version, where two schedules with jobs in the same order, but different time tags, define two different solutions. The TS+M column reports the number of solutions found by a the TS version when adding a symmetry breaking condition to avoid considering solutions which are equivalent if processor names are swapped. The OS column corresponds to an *order scheduling* version, which distinguishes schedules only by their ordering. The third column (OS+FS) refers to an order scheduling version flexible in the time that a processor can start its assigned computation. Although every solution set was obtained in less than 70 ms with any of the solvers, the results can be scaled to more complex cases and they show the importance to transfer into constraints all the knowledge of the design problem for an efficient search. In the example, the symmetry breaking condition, proper of homogeneous architectures, reduced the TS solution set to the half. Considering if order scheduling, instead of timed scheduling, is sufficient for the design purpose is also important since it avoids the much larger set of timed-scheduled solutions.

Additionally, we checked the capability of the implemented WCA-DSE phase for merging constraints. We were able to run the WCA-DSE for a design problem, which included a throughput constraint and a deadline constraint ($D_1 = 4 \Rightarrow R_1 < 4$), both, for the TS version, (TS+D1), and for the OS version (OS+D1). For it, we only had to add both constraints to the input `data.dzn` file. The results show that posing all the performance constraints available from the design problem, throughput and one deadline here, makes more efficient the WCA-DSE, being noticeable for the TS case.

Table 9.2 Maximum Observed Executed Instructions (MOEI)

Function →	f_0 (ALU)	f_1 (averager)	f_2 (sort)
MOEI (instr.)	257	118	776

9.5.3.2 Exercising the Combined DSE

For exercising the whole combined flow, a specific code for f_0, f_1, f_2 has been considered. For f_0, a three input ALU functionality has been implemented. The ALU applies either an addition or a product operation on two input integer vectors of size 5. The operation applied is configured by a third mode input. The operations have saturation, to limit values to a maximum value 255. For f_1 and f_2, an averager of a size 5 input vector and a bubble sort of the same type of vector have been implemented, respectively. We consider a design case, where there is a critical throughput constraint $Th = 1/(16000\,\text{ns})$ and where it is known that the system will work in an scenario where inputs to the sorting functionality (f_2) are partially preordered, and where the ALU mostly performs multiplications.

The MOEIs of each functionality was obtained and reported in Table 9.2. It was obtained through the performance estimation technique PET1, explained in Sect. 9.5.2, using 10 million random input vectors for each functionality.

These MOEIs were used to feed the order scheduling version of the MiniZinc description. Two values of frequency were considered, 50 MHz and 100 MHz, which here are labelled L and H, respectively. For this simple example, 70 implementation solutions can be considered if the combinations of possible clusterings, scheduling orders (without considering dependencies), and frequency assignations are taken into account. When the flow was fed with a single throughput constraint $Th = 1/(16000\,\text{ns})$, the WCA-DSE phase produced 12 solutions in less than 3 ms (using the gecode solver). The *simulation time* of each solution took 8 ms in average. The simulation of each solution was minimized by exploiting that KisTA has features to limit the required *simulated time* based on the task set properties and considering a sequential execution of the task set. Taking into account the measured simulation times, the JAS-DSE exploration takes 99 ms (3 ms of WCA-DSE plus $12 \cdot 8$ ms of SAS-DSE), while the full simulation of each variant for this simple case would have

Table 9.3 Solutions found by the WCA-DSE for $th\_inv = 16000\,\text{ns}$

Solution	$freq_i$	$proc_i$	$next_i$	cost	L_{WCA}/ns	L_{SAS}/ns
1	H,H	0,0,1	1,−1,−1	6	7760	3620
2	L,H	0,0,1	1,−1,−1	5	7760	7260
3	H,H	0,1,1	−1,−1,1	6	8940	4350
4	L,H	0,1,1	−1,−1,1	5	8940	6080
5–7	H,L	0,0,0	*	3	11510	6800
8	H,L	0,0,1	1,−1,−1	5	15520	6340
9	L,L	0,0,1	1,−1,−1	4	15520	7260

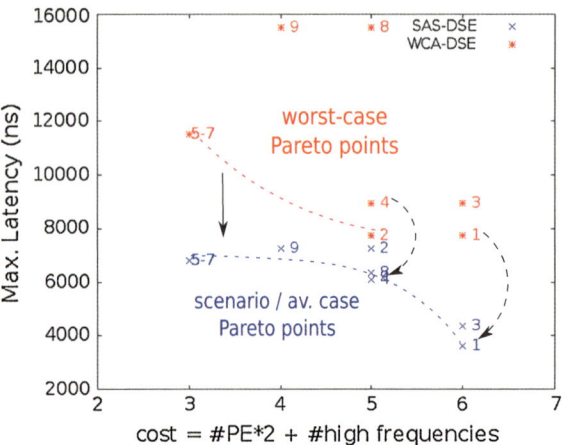

Fig. 9.4 Translation of WCA-DSE solutions after the SAS-DSE

taken 560 ms (70·8 ms). Therefore, the speed-up of the JAS-DSE flow on the example versus an approach based on exhaustive simulation of each variant was 82.3 %.

The set of *safe solutions* is summarised in Table 9.3. A solution is identified by the first column, while the next three describe the solution through the frequencies assigned to the processor, the mapping ($proc_i$) and the schedules ($next_i$) ([2] notation). From the 12 solutions, 6 were fully sequential, but only three of them, solutions 5–7, are reported, since the other three refer to the same solution, but applying the high frequency to the unused processor. A cost was assigned to each solution calculated as `cost=2*#PE+#H`, where `#PE` and `#H` are the number of processors and high frequencies respectively. The L_{WCA} column reports the latency of the solution, where $L_{WCA} = th_{WCA}^{-1}$. The solutions found by the WCA-DSE phase have been represented in red with the $\star$ symbol and placed by their the cost and latency in the Pareto diagram of Fig. 9.4. The WCA-DSE phase reported solutions 1 and 2 are equivalently optimum in terms of latency. Considering the cost, solution 2 was better. In order to let the solver obtain solution 2 as the optimum one, the cost criteria has to be added to the constraint program.

A simulation with KisTA using MOEIs of each solution reported in Table 9.3 confirmed each value of the L_{WCA} column. After this, the SAS-DSE phase was applied. For it, each solution of the Table 9.3 was simulated with KisTA, configured to execute the aforementioned functionalities, and estimating the workload with the `scope-g++` annotations. The WCA-DSE solutions moved to the ones reported in the column L_{SAS} of Table 9.3, represented in blue with the x symbol in Fig. 9.4.

Figure 9.4 clearly shows that the configuration of Pareto optima solutions changes. Interestingly, the simulation-based phase exposes new solutions, i.e., solutions 4 and 1, as part of the Pareto set, while other solutions, i.e., solution 2, disappear from the Pareto set. The simulation-based phase also enabled to expose for the given scenario trade-offs which did not exist after the WCA-DSE analysis, i.e., between solutions with cost 5 and 6. Moreover, more interesting information is unveiled after

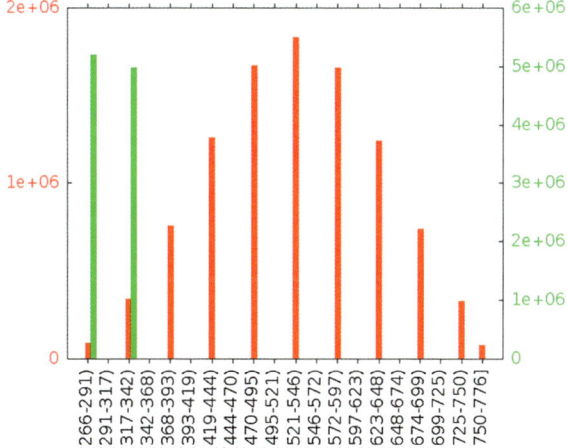

Fig. 9.5 Histogram (for 10^7 input vectors) of instructions executed by function f_2 (*sorter*) for random stimuli (*red*) and for a specific scenario (*green*)

the simulation-based analysis, such as changes in the relevance of the trade-offs. For instance, for the considered example, the application of the SAS-DSE phase reveals that the speed-up vs cost trade-off between solutions 5–7 and solutions 4 or 2 is not as interesting as it was indicated by a pure analytical-based assessment.

A deeper analysis shows the connection of the shifts, as the one between solutions 2 and 4, with the consideration of a specific scenario. In this case, the scenario keeps f_0 and f_1 workloads close to their maximum, but the workload of function f_2, i.e., the sorter, was reduced more than 50 %, as shown in Fig. 9.5. Under such conditions, solution 2 keeps a minimum latency of 7260 ns because of the sequencing of f_0 and f_1 in the first processor, and regardless the shortened execution of f_2 in the second processor. However, the same scenario enables a more drastic reduction of the latency in solution 4 (from 8940 to 6080 ns), which balances better the computation by executing f_1 in the same processor as f_2.

Therefore, taking the same design constraint, throughput in this case, the SAS-DSE phase, attending to a scenario-aware optimisation, may select safe solutions different from the ones that would be selected if the optimisation is left to the WCA-DSE phase.

9.6 Extension of the Approach

The use of MOEIs does not invalidate the results of Sect. 9.5.3 to the effects of showing the over-estimation of the analytical phase. In fact, considering WCEIs will show a more remarkable over-estimation. Moreover, MOEI calculation relies on the same annotation technique used for the dynamic annotation, which enables to get more accurate performance assessment in the SAS-DSE phase. Thus, more efficient

Pareto solutions can be derived and it can be better detected where the important trade-offs for a given scenario are. However, a practical application to a real-time and mixed-criticality design requires the integration to the proposed approach of tools and techniques capable to provide WCETs and/or WCEIs [38]. Further experimental examples to explore the potential of the approach, and the use cases where it better fits are required. Such experiments shall include examples with analysis and simulation times bigger than the proof-of-concept shown, and shall take into account the times for obtaining the worst-case workloads feeding the analytical phase, and the overheads in the connection of the analytical and simulation-based phases.

The JAS-DSE approach proposed in Sect. 9.4 is generic regarding the system modelled. However, there are obvious points where the presented proof-of-concept example and implemented framework can be generalized. For instance, the application model can be extended. An immediate extension is the support of cyclic graphs or more generic computational models, e.g., Synchronous Data Flow (SDF), or the Scenario-Aware SDF (SA-SDF) MoC [33], to take into account *inter-* and *intra*-application dynamism already in the analytical phase. This generalization will not remove sense to a JAS-DSE approach, which can still get better accuracy since the simulation-based assessment can take into account *intra-node dynamism*, that is, the different response times that can appear at the firing of each single node of the application due to input data and the actor state. Similarly, the platform model can be enhanced to target the modelling of a predictable platform. For instance, the communication penalties of a predictable communication resource, e.g., a TDM bus, could be taken into account. Both application and platform model generalizations require the extension of the corresponding analytical and simulation frameworks.

The proof-of-concept example did not deal with advanced throughput calculations [5, 11, 18, 22], for simplicity and to focus and remark better the capability of the proposed approach to cover at once in the analytical phase different types of constraints, which have traditionally lead to separated analysis theories.

Another possible enhancement has to do with providing the user an interface to capture the input to the flow, specifically the system model template, in order to hide to the user the complexities and particularities of constraint-programing. A framework, as the one reported in [29], is a good candidate, since it not only hides such complexities, to ask the user just the required input information, but it already overcomes the aformentioned modelling limitations.

The JAS-DSE approach described in Sect. 9.4 covers the case of using one exploration tool or solver for the analytical model and another one for its executable counterpart. It enables the application of different exploration mechanisms adapted to each phase, analytical and simulation-based. It is also useful for coping with cases where the set of *safe solutions* found by the WCA-DSE phase is yet large for an exhaustive simulation. Notice that, in order to maximixe the chances for finding efficient solutions, it is not interesting to reduce the size of the set of *safe solutions* by analytical means, either with additional non-critical constraints or making the critical ones artificially tighter. To the contrary, the goal is to target a complete set of safe solutions, which maximizes the amount of candidate solutions. The larger the amount of candidate solutions, the higher the chances to find more efficient solutions for a given scenario through accurate simulation-based assesment. The proof-of-concept example and flow implementation shown in this chapter has focused on showing that the approach can actually lead to improved solutions, using for simplicity and clarity a single solver (the one used to run the CP model) in the analytical DSE phase. No solver or heuristic was used in the SAS-DSE phase, since all the safe solutions were simulated. The integration of an exploration tool in the SAS-DSE flow is not straightforward and deserves additional treatment beyond the bounds of this chapter, since the exploration tool in the SAS-DSE phase has to consider the constraint that any valid solution has to fit to a custom and likely non-homogeneous set of solutions. Anyhow, a sucessful integration of a search heuristic tied to that constraint would enable that the presented methodology can be applied to complex time-critical and mixed-criticality systems.

Another possible enhancement is related to the approach presented in Chap. 8. Such an approach introduces *platform template characterizations*, a flexible way to describe a family of predictable platforms. Platorm template characterizations take into account the instantiation rules of the component templates of the platform, the services they provide, and their costs. This approach would enable to increase the amount of potential safe solutions under consideration. Moreover, Chap. 8 deals with automated generation of simulatable models from the platform template characterizations in a combined analytical and simulation-based DSE approach. Therefore, the integration of this work with the presented approach, should facilitate the development of a JAS-DSE framework where not only the automation, but also the coherence of the analytical and simulatable models is ensured.

A complete support of the methodology proposed in Fig. 9.1 will make it suitable for the design of efficient mixed-criticality systems.

These systems require the consideration of platforms, which are *heterogeneous* in terms of predictability, that is, platforms, which merge predictable resources with non-predictable ones. In this sense, the integration of further simulation-based performance assessment technologies and tools would be very interesting. For instance, the integration of native estimation technology, e.g., through the SCoPE framework [30], would enable fast assessment of the non-predictable parts. Moreover, the integration of such performance estimation technology would enable the consideration of additional performance metrics beyond time related ones, e.g., power consumption and temperature.

9.7 Conclusion

This chapter has presented a novel DSE flow, which targets efficient design of time-critical and mixed-criticality systems. Through a first analytical phase based on worst-case workloads, the flow enables the identification of a set of safe solutions. The implementation of this phase relies on constraint-based programming, which enables a holistic and portable description and analysis of the DSE problem, and the derivation of a complete set of safe solutions, taking into account different types of time constraints. Passing such complete solution set to a later simulation-based phase maximizes the chances to find more efficient solutions, thanks to the capability of the simulation-based performance assessment to consider the intra-node dynamism and the specific working scenario. The proof-of-concept simulation-based framework introduced in Sect. 9.5 has shown that the approach can yield solutions more efficient, by considering the system working environment, than the solutions found by a pure analytical phase, without sacrificing certainty on the time constraint fulfilment; and that it can be done significantly faster than through a pure simulation-based approach. Many of the enhancements mentioned in Sect. 9.6 will be tackled in the context of the FP7 CONTREX project [4].

Acknowledgments This work has been partially funded by the Excellence Post-doc Position I-2011-0646 granted by the School of Information and Communication Technology of the KTH Royal Institute of Technology, Sweden.

References

1. Baruah S, Li H, Stougie L (2010) Towards the design of certifiable mixed-criticality systems. In: Proceedings of the real-time and embedded technology and applications symposium (RTAS) 2010
2. Bonfietti A, Lombardi M, Milano M, Benini L (2009) Throughput constraint for synchronous data flow graphs. In: Proceedings of the international conference on integration of AI and OR techniques in constraint programming for combinatorial optimization problems (CPAIOR) 2009, vol 5547. doi:10.1007/978-3-642-01929-6_4
3. Bonfietti A, Benini L, Lombardi M, Milano M (2010) An efficient and complete approach for throughput-maximal SDF allocation and scheduling on multi-core platforms. In: Proceedings of the design, automation and test in Europe conference (DATE) 2010, 3001 Leuven, Belgium, pp 897–902. http://dl.acm.org/citation.cfm?id=1870926.1871143
4. CONTREX EU project website (2014) http://contrex.offis.de
5. Dasdan A, Irani S, Gupta RK (1999) Efficient algorithms for optimum cycle mean and optimum cost to time ratio problems. In: Proceedings of the IEEE/ACM design automation conference (DAC) 1999, pp 37–42
6. Davis RI, Burns A (2011) A survey of hard real-time scheduling for multiprocessor systems. ACM Computing Surveys
7. Eles P, Kuchcinski K, Peng Z, Doboli A, Pop P (1998) Scheduling of conditional process graphs for the synthesis of embedded systems. In: Proceedings of the design, automation and test in Europe conference (DATE) 1998, Washington, DC, USA, pp 132–139. http://dl.acm.org/citation.cfm?id=368058.368119
8. Ernst R, Burns A, Thiele L, Rhun JL (2012) Mixed critical system design and analysis. In: Proceedings of the international conference on embedded software (EMSOFT) 2012
9. Fornaciari W, Sciuto D, Silvano C, Zaccaria V (2001) A design framework to efficiently explore energy-delay tradeoffs. In: Proceedings of the IEEE/ACM/IFIP international conference on hardware/software codesign (CODES) 2001, pp 260–265. doi:10.1109/HSC.2001.924686
10. Gecode website (2013) www.gecode.org
11. Ghamarian A, et al (2006) Throughput analysis of synchronous data flow graphs. In: Proceedings of the international conference on application of concurrency to system design (ACSD) 2006

12. Henia R, Hamann A, Jersak M, Racu R, Richter K, Ernst R (2005) System level performance analysis–the SymTA/S approach. IEE Proc Comput Digital Tech 152(2):148–166
13. Herrera F, Sander I (2013) Combining analytical and simulation-based design space exploration for time-critical systems. In: Proceedings of the forum on specification and design languages (FDL) 2013, pp 1–8
14. Herrera F, Niaki S, Sander I (2013) Towards a modelling and design framework for mixed-criticality SoCs and systems-of-systems. In: Proceedings on the euromicro conference on digital system design (DSD) 2013, pp 989–996. doi:10.1109/DSD.2013.112
15. Herrera F, et al (2012) A MDD methodology for specification of embedded systems and automatic generation of fast configurable and executable performance models. In: Proceedings of the IEEE/ACM/IFIP international conference on hardware/software codesign and system synthesis (CODES+ISSS) 2012
16. IEC61508 (2010) IEC/EN 61508. Functional safety of electrical/electronic/programmable electronic safety related systems
17. Jia Z, Nez A, Bautista T, Pimentel A (2014) A two-phase design space exploration strategy for system-level real-time application mapping onto MPSoC. Microprocess Microsyst 38(1):9–21. doi:10.1016/j.micpro.2013.10.005
18. Karp RM (1978) A characterization of the minimum cycle mean in a digraph. Discrete Math 23(3):309–311. doi:10.1016/0012-365X(78)90011-0
19. Kempf T, Ascheid G, Leupers R (2011) Multiprocessor systems on chip, design space exploration. Springer, New York. doi:10.1007/978-1-4419-8153-0 Please check the inserted location of publisher is correct in Ref. [19].
20. KisTA website (2014) http://people.kth.se/~fernanhc/KisTA_website/KisTA_index.html
21. Knzli S, Poletti F, Benini L, Thiele L (2006) Combining simulation and formal methods for system-level performance analysis. In: Proceedings of the design, automation and test in Europe conference (DATE), pp 236–241
22. Kumar A, Srikanthan T (2012) Accelerating throughput-aware run-time mapping for heterogeneous MPSoCs. ACM transactions on design automation of electronic systems (TODAES). doi:10.1145/0000000.0000000
23. Lahiri K, Dey S, Raghunathan A (2000) Performance analysis of systems with multi-channel communication architectures. In: Proceedings of the international conference on VLSI design (VLSI) 2000
24. Lee E, Messerschmitt D (1987) Synchronous data flow. Proc IEEE 75(9):1235–1245. doi:10.1109/PROC.1987.13876
25. Multicube-SCoPE website (2013) http://www.teisa.unican.es/gim/en/scope/multicube. Accessed 13 May 2013
26. Nethercote N, Stuckey P, Becket R, Brand S, Duck G, Tack G (2007) Minizinc: towards a standard CP modelling language. In: Proceedings of the international conference on principles and practice of constraint programming (CP). Springer, Heidelberg, pp 529–543
27. Pimentel A, Erbas C, Polstra S (2006) A systematic approach to exploring embedded system architectures at multiple abstraction levels. IEEE Trans Comput 55(2):99–112. doi:10.1109/TC.2006.16
28. Posadas H, lvaro Daz, Villar E (2012) Embedded systems–theory and design methodology. In: Tanaka K, Posadas H, Díaz A, Villar E (ed) SW annotation techniques and RTOS modelling for native simulation of heterogeneous embedded systems. doi:10.5772/38000
29. Rosvall K, Sander I (2014) A constraint-based design space exploration framework for real-time applications on MPSoCs. In: Proceedings of the design, automation and test in Europe conference (DATE) 2014
30. Scope Website (2013) http://www.teisa.unican.es/scope
31. Silvano C, et al (2010) MULTICUBE: multi-objective design space exploration of multi-core architectures. In: Proceedings on the IEEE computer society annual symposium on VLSI (ISVLSI) 2012, pp 488–493. doi:10.1109/ISVLSI.2010.67
32. Stuijk S (2007) Predictable mapping of streaming applications on multiprocessors. PhD thesis, Faculty of electrical engineering, Eindhoven University of Technology, Eindhoven, The Netherlands. http://www.es.ele.tue.nl/~sander/publications/phd-thesis.php

33. Stuijk S, Geilen M, Basten T (2010) A predictable multiprocessor design flow for streaming applications with dynamic behaviour. In: Proceedings on the euromicro conference on digital system design (DSD) 2010, IEEE computer society, Washington, DC, USA, pp 548–555. doi:10.1109/DSD.2010.31

34. Theelen B, et al. (2006) A scenario-aware data flow model for combined long-run average and worst-case performance analysis. In: Proceedings of MEMOCODE '06, pp 185–194. doi:10.1109/MEMCOD.2006.1695924

35. Thiele L, Chakraborty S, Naedele M (2000) Real-time calculus for scheduling hard real-time systems. In: Proceedings on the IEEE international symposium on circuits and systems (ISCAS) 2000, vol 4, pp 101–104. doi:10.1109/ISCAS.2000.858698

36. Thompson M, et al (2007) A framework for rapid system-level exploration, synthesis, and programming of multimedia mp-socs. In: Proceedings of the IEEE/ACM/IFIP international conference on hardware/software codesign and system synthesis (CODES+ISSS) 2007, ACM, New York, NY, USA, pp 9–14. doi:10.1145/1289816.1289823, http://doi.acm.org/10.1145/1289816.1289823

37. Wang Z, Henkel J (2012) Hycos: hybrid compiled simulation of embedded software with target dependent code. In: Proceedings of the IEEE/ACM/IFIP international conference on hardware/software codesign and system synthesis (CODES+ISSS) 2012, ACM, New York, NY, USA, pp 133–142. doi:10.1145/2380445.2380471, http://doi.acm.org/10.1145/2380445.2380471

38. Wilhelm R et al (2008) The worst-case execution-time problem—overview of methods and survey of tools. ACM Trans Embed Comput Syst (TECS) 7(3):1–53

39. Zaccaria V, et al. (2010) Multicube explorer: an open source framework for design space exploration of chip multi-processors. In: Proceedings of the international conference on architecture of computing systems (ARCS) 2010

Chapter 10
Bridging Algorithm and ESL Design: MATLAB/Simulink Model Transformation and Validation

Liyuan Zhang, Michael Glaß, Nils Ballmann and Jürgen Teich

Abstract MATLAB/Simulink is today's de-facto standard for model-based design in domains such as control engineering and signal processing. Particular strengths of Simulink are rapid design and algorithm exploration. Moreover, commercial tools are available to generate embedded C or HDL code directly from a Simulink model. On the other hand, Simulink models are purely functional models and, hence, designers cannot seamlessly consider the architecture that a Simulink model is later implemented on. In particular, it is not possible to explore the different architectural alternatives and investigate the arising interactions and side-effects directly within Simulink. To benefit from MATLAB/Simulink's algorithm exploration capabilities and overcome the outlined drawbacks, this work introduces a model transformation framework that converts a Simulink model to an executable specification, written in an actor-oriented modeling language. This specification then serves as the input of a well-established Electronic System Level (ESL) design flow, enabling Design Space Exploration (DSE) and automatic code generation for both hardware and software. We also present a validation technique that considers the functional correctness by comparing the original Simulink model with the generated specification in a co-simulation environment. The co-simulation can also be used to evaluate different quality numbers of implementation candidates during DSE. As a case study, we present and investigate a torque vectoring application from an electric automotive vehicle.

Keywords Electronic System Level (ESL) · MATLAB/Simulink · Model transformation · Model validation · Design Space Exploration (DSE) · SystemC · Model-Based Design (MBD) · Code generation · SysteMoC · Torque vectoring

L. Zhang (✉) · M. Glaß · N. Ballmann · J. Teich
Hardware/Software Co-Design, Department of Computer Science,
Friedrich-Alexander-Universität Erlangen-Nürnberg (FAU), Erlangen, Germany
e-mail: liyuan.zhang@cs.fau.de

M. Glaß
e-mail: glass@cs.fau.de

J. Teich
e-mail: teich@cs.fau.de

© Springer International Publishing Switzerland 2015   189
M.-M. Louërat and T. Maehne (eds.), *Languages, Design Methods,*
and Tools for Electronic System Design, Lecture Notes in Electrical Engineering 311,
DOI 10.1007/978-3-319-06317-1_10

10.1 Introduction

Driven by the rapid development of microelectronics technology, the functionality and, thus, the design complexity of modern distributed embedded systems are continuously increasing. To cope with these challenges, *Electronic System Level* (ESL) [21] design methodologies introduce higher abstraction and model the complete embedded system as an *executable specification* at system level. At this level, the decisions such as the partition of functional units to software or hardware are yet to be made. Therefore, *Design Space Exploration* (DSE) [20] allows for an early evaluation of design decisions and searches for optimized implementation alternatives. After DSE, various tools, cf. to Gerstlauer et al. [11], are available to (semi-)automatically synthesize the implementation into software and hardware. Overall, ESL design helps the designer to deliver optimized systems and to shorten the design cycle.

In domains such as control engineering and signal processing, the development of an embedded system typically starts with the application engineer using domain-specific modeling tools such as *MATLAB/Simulink* [25] to build a functional model, e.g., a controller in a feedback control system. The application engineer often also uses Simulink to model the physical environment and to create the test bench for design validation. Simulink allows rapid design and is therefore often used to carry out algorithm optimization in early design stages. By using *Simulink Coder* [24], a Simulink model can be automatically translated into embedded C code for software implementation or HDL code for hardware implementation. However, there is no information about the architecture that a Simulink model is later implemented on. Therefore, considering different implementation alternatives and investigating architectural interferences and side-effects directly within Simulink is not possible.

In this work, we aim at closing the gap between classic ESL design flows and Simulink models by applying 1. *model transformation* and 2. a *system-level validation* technique (see Fig. 10.1). We employ an actor-oriented modeling language (*SysteMoC* [10]), which is based on *SystemC* [12], the de-facto standard for system-level modeling, to serve as the intermediate representation of the Simulink models and the input of an ESL design flow. Representing a Simulink model in an actor-oriented fashion is very suitable due to the nature of Simulink modeling, cf. to Lee and Neuendorffer [19]. Here, we introduce a model transformation framework (Sect. 10.4) that automatically generates an executable specification in SysteMoC from a given Simulink model. This executable specification is transformed with a component library to an exploration model. The exploration model is used within a Design Space Exploration (DSE) to consider different implementation candidates, which are evaluated by the DSE framework with respect to multiple design objectives and constraints. DSE delivers a set of high quality implementation candidates, from which the designer may subsequently choose the best trade-off as the system-level implementation for subsequent design phases. Moreover, we propose to validate the correctness of the automatic generated SysteMoC model using a co-simulation approach. We consider a control application from an electronic automotive vehicle

Fig. 10.1 Proposed design flow from MATLAB/Simulink to prototype via automatic model transformation and validation via co-simulation

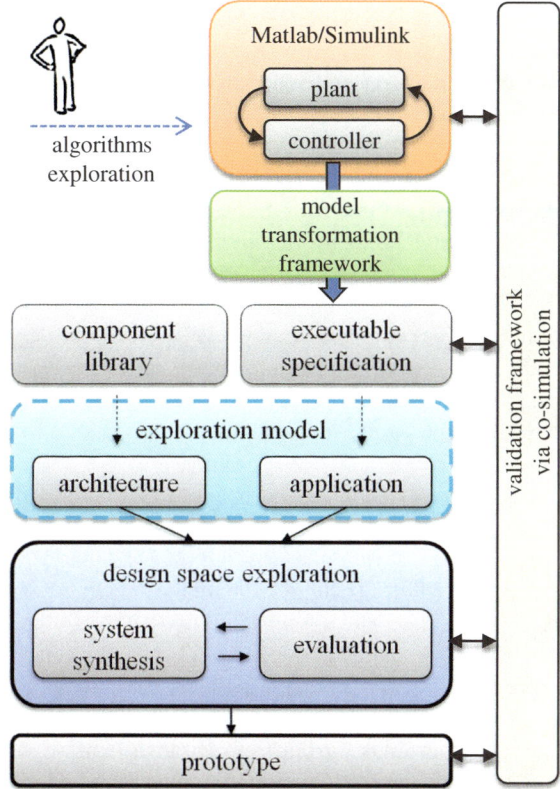

to give evidence of the effectiveness of the proposed approach in Sect. 10.5 before we conclude the chapter in Sect. 10.6.

10.2 Related Work

Several contributions relevant to this work have been made in recent years by various research groups. Caspi et al. [5] focus their work on designing embedded software by translating Simulink models to *SCADE/Lustre* [8]. This intermediate representation is then implemented on the *Time Triggered Architecture* (TTA) introduced by Kopetz [17], which is a platform for running safety critical applications. Czerner and Zellmann [6] try to combine SystemC and MATLAB/Simulink for cycle-accurate hardware modeling and system verification by integrating SystemC modules into Simulink via *S-Functions* [22]. A co-simulation framework between SystemC and MATLAB/Simulink is built by Boland et al. [4] with the purpose of hardware verification of DSP-based designs. There, Simulink is used to model the environment and to generate real-world stimuli to drive the *design under verification* that is implemented in SystemC. MathWorks also has a commercial tool (*HDL Verifier* [23]) to verify hardware designs using HDL simulators and FPGA *hardware-in-the-loop* test-benches. Above works only focus either on software design or on hardware design.

Kai et al. [14] present a Simulink-based MPSoC design flow by composing the Simulink model into a *Combined Algorithm and Architecture Model* (CAAM). The CAAM model can be then implemented at different abstraction levels using a multi-threaded code generator. Atat and Zergainoh [1] propose to refine a Simulink model at three different abstraction levels: transactional model, macroarchitecture model, and microarchitecture model. The system verification is carried out at all abstraction levels. These works try to enable system-level design with Simulink models. However, the partitioning of functional blocks into software or hardware must be performed manually by the designer.

Jersak et al. [13] introduce an approach to transform time-driven Simulink models into *System Property Intervals* (SPIs) to enable system-level timing analysis. They propose to transform time-driven models to data-driven models by combining register and virtual FIFO queues to guard the data exchange in multi-rate systems. Baleani et al. [2] use the *Synchronous Reactive* (SR) model as intermediate layer to enable the model transformation between MATLAB/Simulink, SR, and the model-based development tool set *ASCET* [9], which enables automatic code generation for automotive applications. In contrast to these works, we also consider domain- and application-specific knowledge during model transformation, where either a data-driven or a time-driven Model of Computation (MoC) can be chosen.

Another option to transform MATLAB/Simulink models is offered by employing *SystemC Analog/Mixed-Signal (AMS) extensions* [3], enabling the modeling of continuous systems within a SystemC-based design. In particular, it enables the transformation of *hybrid systems* (i.e., systems containing continuous and discrete state) from Simulink to SystemC. Since Simulink is no longer required to simulate the continuous part of the system, the simulation speed compared to the co-simulation technique increases. However, it may often not be desired or even possible to transform the huge and possibly closed-source continuous environmental models from Simulink to SystemC. Moreover, the result of such transformation does not come with the same benefits with respect to DSE.

In this work, we propose 1. a model transformation framework to automatically generate an executable specification from a Simulink model and 2. a system-level validation technique. Since our executable specification serves as the input for a well-established ESL tool flow introduced by Keinert et al. [15] that enables DSE, highly-optimized system implementations that satisfy given constraints can be achieved automatically. The advantage of our validation technique lies in re-using the test bench (including the environment model) created in Simulink.

10.3 Design Fundamentals

This section introduces the basics of Simulink and the actor-oriented model used to represent a Simulink model as an executable specification.

10.3.1 Simulink

Simulink is a commercial software from MathWorks [25] for modeling, simulation, and analysis of dynamic systems. Simulink is widely used to solve problems in automotive applications, communications, electronics, and signal processing. The basic elements in Simulink are *blocks* and *lines*. Basic blocks are mandatory units to perform computation or display functions, such as *Add, Memory, Scope*, etc. The designer builds hierarchical systems by encapsulating basic blocks into *subsystems*. Lines (also called *edges* or *channels*) are used to connect blocks and have *register* semantics (*non-destructive read, destructive write*). Besides these basic building elements, Simulink also has rich libraries that offer a broad variety of predefined blocks and can be used together with user-defined functions. Simulink provides several *solvers* to compute models that contains continuous and/or discrete states. It is very efficient to use Simulink during early design stages for algorithm exploration. By using *Simulink Coder*, a Simulink model can be automatically translated into highly-optimized C code for software implementation or HDL code for hardware implementation.

10.3.2 Executable Specification

In [10], a library for modeling and simulation of actor-oriented behavioral models termed *SysteMoC* is introduced. SysteMoC is based on SystemC, the de-facto standard for system-level modeling, adding actor-oriented MoCs to form executable specifications. In actor-oriented models, *actors*, which encapsulate the system functionality, are potentially executed concurrently and communicate over dedicated abstract *channels*. Thereby, actors produce and consume data (so-called *tokens*), which are transmitted by those channels.

- An actor (see the example in Fig. 10.2) is a tuple $a = (I, O, F, R)$, containing a set of actor ports partitioned into a set of actor input ports I (e.g., i_1) and a set of actor output ports O (e.g., o_1, o_2), a set of functions F (e.g., *fpositive, fnegative*), and a *Finite State Machine* (FSM) R. Actors can be grouped together to form *graphs*. A graph may also contain other graphs to build a hierarchical system.

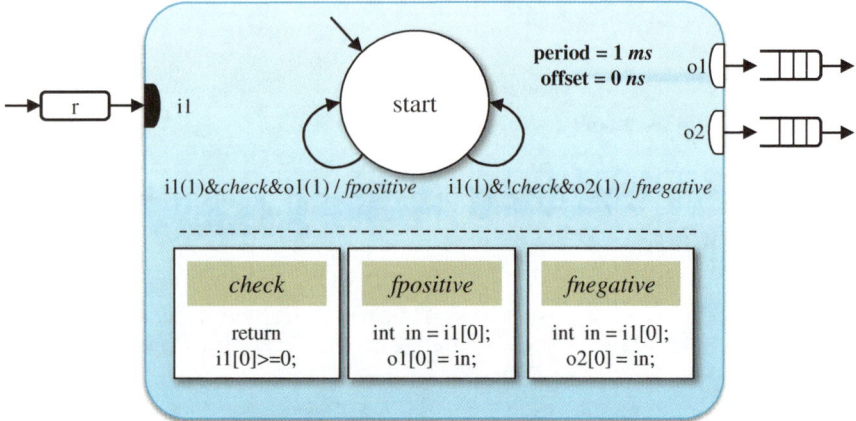

Fig. 10.2 A graphical representation of a SysteMoC actor, which sorts a sequence of input data depending on its algebraic sign

- The functions F encapsulated in an actor are partitioned into so-called *actions* f_{action} and *guards* k and are executed during a *transition* of the FSM R that also represents the communication behavior of the actor (i.e., the number of tokens consumed and produced for each transition).
- A transition is a tuple $t = (q_{src}, k, f_{action}, q_{dst})$ containing the source state q_{src} before the execution of the transition, and the destination state q_{dst} after the execution of the transition. An action f_{action} (e.g., *fpositive*) performs a computation task for the actor and may consume or produce tokens on the channel. A guard k (e.g., *check*) checks the availability of a transition by returning a Boolean value and the assignment of one or several guards to the FSM implements the required control flow. The firing of an actor corresponds to the execution of exactly one transition of the actor. If multiple actors have transitions that can be fired, they are chosen non-deterministically by the SysteMoC runtime system.
- A channel is a tuple $c = (I, O, n, d)$, containing a set of channel ports partitioned into a set of channel input ports I and a set of channel output ports O, its buffer size $n \in N_\infty = \{1, \ldots, \infty\}$, and a possibly empty sequence $d \in D^*$ of initial tokens, where D^* denotes the set of all possible finite sequences of tokens. In SysteMoC, actors are only permitted to communicate with each other via channels, to which the actors are connected by ports. Hence, in a SysteMoC actor, the communication behavior is completely separated from its functionality.

Figure 10.2 gives a graphical representation of an actor, which sorts a sequence of input data arriving on input port i_1 to either output port o_1 or o_2 depending on its algebraic sign. The actor reads its input data from a register r and writes its output data into two FIFOs. The actor has only a *start* state. Transitions of the finite state machine R are depicted as directed edges in the actor. Each transition is annotated with an activation pattern, a Boolean expression which decides if the transition can be taken, and an action which is executed once the corresponding transition is taken. Using parameters *period* and *offset* indicates that this actor is time-triggered.

SysteMoC exploits the *event-driven scheduler* from SystemC to manage the firing sequence of the actors according to their FSMs. The basic SysteMoC implementation uses channels with FIFO semantics to provide a unidirectional point-to-point connection between an actor output port and an actor input port. Using FIFOs as communication channels makes SysteMoC suitable to model data-driven systems (e.g., signal processing applications). In other areas such as automotive systems, tasks are often executed periodically, i. e. they are time-triggered. In order to improve this kind of systems, SysteMoC supports *periodic actors* and *register* channels (see Fig. 10.2). A periodic actor has two additional parameters: *period* and *offset*, which describe the time that the SysteMoC scheduler evaluates the FSM of the actor. A register channel has non-destructive read, destructive write semantics. This extension enables a time-triggered MoC, hence, the modeling ability is greatly enhanced. For example, in a typical embedded control system, the sensors can be modeled using periodic actors and the rest of computation blocks can be data-driven.

We use *actor-based design* to construct the executable models (Fig. 10.3). The key advantage of actor-based design is that the interaction between actors follows some kind of communication pattern, called Model of Computation (MoC) [18]. A certain model of computation is given by a predefined type of communication behavior and a scheduling strategy for the actors. Separating actor computation and actor communication gives the designer the ability to refine the communication at different abstraction levels. Thus, the model is ideal for ESL design, since at the system level, the decision about the hardware software partitioning has yet to be made. Design

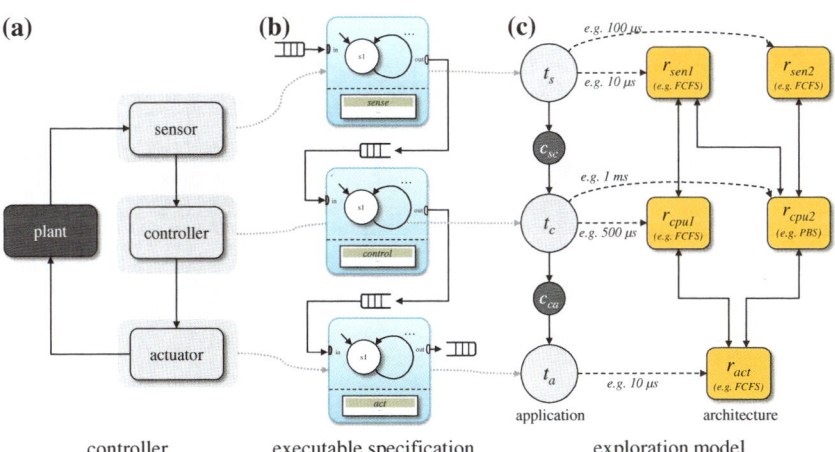

Fig. 10.3 To apply design space exploration for the controller (a), a graph-based exploration model (c) is automatically generated from the executable specification (b), a given architecture, as well as mapping constraints between them

Space Exploration (DSE) follows the commonly accepted Y-chart approach [16]. For each implementation, multi-objective optimization is used to evaluate the implementation quality. In the end, we obtain a set of high quality candidates as optimized implementation solutions.

10.4 Model Transformation

After the initial modeling in Simulink is finished, a block diagram of the system (e.g., containing controller and plant) is at hand. The Simulink model is verified via simulation. If the simulation results meet the design goals, the model transformation can be started.

10.4.1 Model Transformation Preparation

The first step of model transformation is *transformation preparation*, which changes the interface of a model that is going to be transformed (see Fig. 10.4). The part that is going to be transformed to SysteMoC (e.g., the controller) is connected with the environment model (e.g., the plant). Before model transformation, the designer must disconnect the chosen part from the environment (e.g., disconnect the controller from the plant) and then re-connect every input at the top level of the chosen part with an *Inport* block and every output at the top level with an *Outport* block. These I/O blocks form the interface of the chosen part. After model transformation, these I/O blocks are mapped to special actors that enable the data exchange between Simulink and SysteMoC.

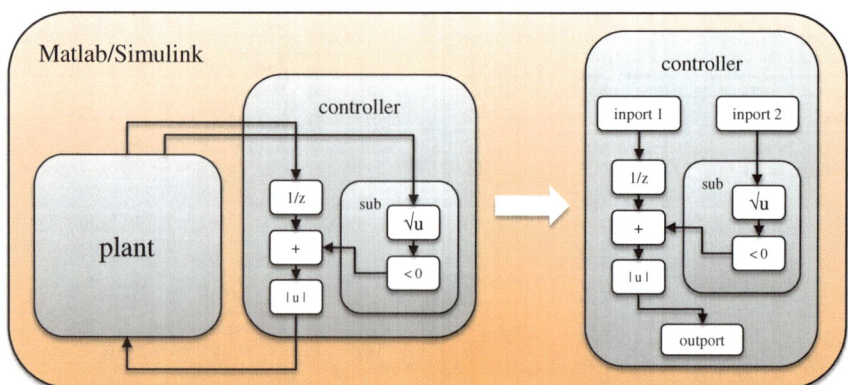

Fig. 10.4 The automatic model transformation requires a preparation step, in which those parts that shall be transformed (e.g., the controller) are disconnected from the other parts (e.g., the plant) and re-connected via I/O blocks

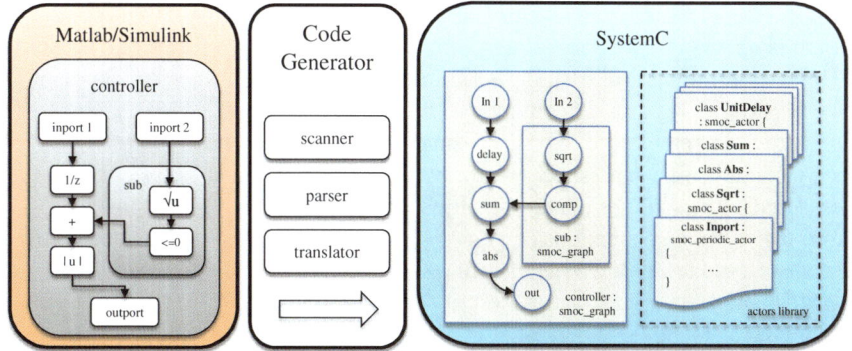

Fig. 10.5 The model transformation framework consists of an actor library and a Code Generator that maps Simulink basic blocks and lines to SysteMoC actors and FIFO channels

10.4.2 Model Transformation Framework

The core of the model transformation framework is a *Code Generator* that takes Simulink as input language and produces SysteMoC as output language. The structure of the Code Generator is shown in Fig. 10.5. The Code Generator has the most common operations in compiler design, such as lexical analysis, parsing, and code generation. There are three building parts:

- The *Scanner* reads the Simulink block diagram and filters the basic information elements.
- The *Parser* analyzes and identifies the semantics of the elements (e.g., basic blocks, lines). All the necessary information needed for model transformation is determined here (see Sect. 10.4.3), which includes the hierarchy and topology of the Simulink block diagram.
- The *Translator* determines the targeted MoC for the current Simulink block diagram (see Sect. 10.4.4). Additionally, the translator is responsible for the SysteMoC code generation.

10.4.3 Evaluating Simulink Block Diagrams

The Code Generator reads the source code of a Simulink block diagram in the form of an *mdl*-file and abstracts all the necessary information by using the *scanner* and the *parser*. The elements in an *mdl*-file currently supported by the Code Generator are *atomic blocks*, *subsystems*, *reference*, *user-defined blocks*, *lines*, and *branches*. We do not consider Simulink models with *algebraic loops*. An atomic block (i.e., Simulink basic block) represents a basic computation or display function. A subsystem contains multiple Simulink atomic blocks as well as subsystems. A reference is used to link an atomic block or a subsystem. It contains, therefore, only a path to the implementation, which is normally stored in a library file (e.g., *simulink.mdl*). The Code Generator supports user-defined blocks, e.g., *S-Function* written in *C/C++*, by generating a wrapper in the SysteMoC model (as a placeholder). However, the designer has to integrate the implementation of the S-function into the wrapper.

A Simulink line represents a communication channel and has register semantics. Each line can have branches, which represent the multiple destinations. The multi-driver for a connection (i.e., one line, multiple source blocks) is forbidden in Simulink, but multicast of signals is allowed (i.e., one line, multiple target blocks).

All the necessary information described above are parsed by the Code Generator. The Code Generator also determines the *I/O data types* of each Simulink block. All signals by default have *double* types except 1. the types of the signals are implied from the block type or 2. the types are specified by the designer.

10.4.4 Model Transformation

The main task of the *translator* is to transform the Simulink model to SysteMoC with the targeted MoC. If the Simulink model is a *single-rate system*, all blocks share the same sampling rate. For this kind of models, it is straightforward to represent them as data-driven SysteMoC models. The Code Generator maps each Simulink block to a SysteMoC actor, whose function code is stored in the *actors library* (see Fig. 10.5). Each Simulink subsystem is mapped into a SysteMoC graph. Each Simulink line that has a point-to-point connection is mapped to a SysteMoC FIFO, which is a unidirectional point-to-point connection. For each Simulink line that enables multicast, a *multicast actor* is added between the source actor and the destination actors. This additional actor only serves as a relay for transmitting the data to all destination actors.

If the Simulink model is a *multi-rate* system, the blocks are sampled at different rates. For example in Fig. 10.6a, two Simulink blocks $b1$ and $b2$ are connected through a line l. Block $b1$ has the sampling rate λ and block $b2$ has the sampling rate λ/n. Block $b1$ and block $b2$ are transformed to actor $a1$ and actor $a2$, respectively. There are several options to represent this Simulink model in the SysteMoC environment:

10.4.4.1 Rate Transition Actors

Adding a *rate transition actor* c allows to explicitly coordinate the data exchange between $a1$ and $a2$ (see Fig. 10.6b). Two FIFOs are used to connect $a1$ and $a2$ with c. Per activation, $a1$ produces one token on the left FIFO, c reads one token from the left FIFO and produces n tokens on the right FIFO, $a2$ reads one token from the right FIFO. This solution is easy to implement but consumes additional memory space and causes extra delay.

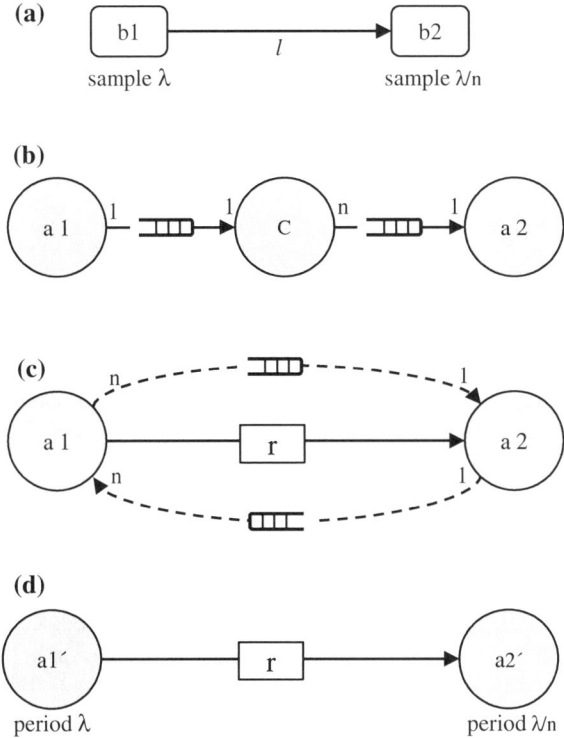

Fig. 10.6   A Simulink multi-rate model, as shown in (a), can be represented in SysteMoC by: adding a rate converter actor (b); transforming time-driven models to data-driven models (c); or creating time-driven SysteMoC models (d)

10.4.4.2   Data-Driven Transformation

Applying the technique introduced in [13] indicates adding a register channel r and two *virtual* FIFOs to govern the activation of $a1$ and $a2$ (see Fig. 10.6c). Activation of $a1$ and $a2$ is enabled by the availability of tokens on the two virtual FIFOs. This solution replaces the absolute periodic timing in Simulink models with relative execution rates. Thus, the time-driven Simulink MoC is transformed into a data-driven model. However, adding two virtual FIFOs per connection may increase the complexity of the design.

10.4.5   Time-Driven Transformation

Here, we propose to transform Simulink multi-rate models to time-driven SysteMoC models in order to preserve the simulation semantics of Simulink models. Blocks $b1$ and $b2$ are mapped to periodic actors $a1'$ and $a2'$. Line l is mapped to a SysteMoC register channel r (see Fig. 10.6d). The sample rate of the Simulink block corresponds to the period of the periodic actor. Here, the activation of $a1$ and $a2$ is no longer governed by the availability of data on the communication channel. Because SysteMoC is originally designed to mainly model data-driven applications, the data dependencies (or topology dependencies in Simulink) are automatically preserved by the FSMs and FIFOs. But, this is no longer given when time-driven semantics are applied: If the two periodic actors have the same period, the SysteMoC runtime scheduler will not consider the partial order dependencies, hence, either actor that can be first simulated. Thus, the simulation behavior of SysteMoC may differ from Simulink. As a remedy, we propose to introduce artificial *offsets* for the periodic actors to reflect partial-order dependencies in time-driven SysteMoC. The Code Generator assigns a proper offset automatically for each periodic actor by running an analysis of the topology dependencies. In summary, the proposed model transformation technique can be divided into three parts:

1. using periodic actors to enable time-driven simulation,
2. using register channels to preserve the Simulink communication semantics,
3. using artificial timing offsets to include partial-order dependencies into SysteMoC models.

No matter which MoC is applied, each Inport block added during transformation preparation (Sect. 10.4.1) is mapped to a periodic actor. These periodic actors can be seen as hardware sensors (fetching the states of the environment model in Simulink). The sample rate t_a in a sensor actor is specified by the designer. Each Outport block added during transformation preparation is mapped to an actor, which is a representation of an actuator (sending the computation results back to Simulink). These sensor and actuator actors are typically grouped together to form a *co-simulation interface* for the auto-generated SysteMoC model.

10.5 Case Study: Torque Vectoring

In this section, an automotive application is used to evaluate the accuracy and efficiency of the proposed model transformation framework. Torque Vectoring (TV) is a new driver assistance system that distributes torque sent to each wheel to suit driving conditions and road surface in order to get more traction in curves. In this work, the Automotive Simulation Models (ASMs) of dSPACE [7] are used for modeling an electric rear-wheel drive vehicle and the environment in Simulink. A torque vectoring differential is realized by modifying the ASM to contain two basic engines (Fig. 10.7), so the left-side engine controls the torque sent to the left rear wheel, and the right-side engine for the right rear wheel, respectively. These two engines are controlled by a *torque vectoring controller*, which is implemented by an application engineer. A simple testing maneuver and the driving conditions are configured in ASM. The driving scenario is as follows: 1. the vehicle first remains motionless at position [0, 0] for 2 s; 2. the vehicle starts to accelerate and keeps a straight cruise without any steering;

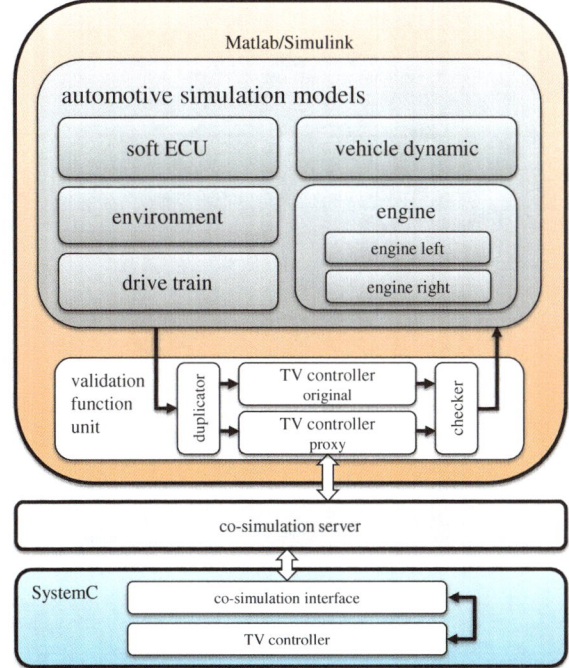

Fig. 10.7 Overview of the co-simulation between torque vectoring controller in SysteMoC and environment in MATLAB/Simulink for validation

3. the acceleration lasts for 20 s and the speed of the vehicle reaches 90 km/h at the new position [400, 0]; 4. the driver performs a *step steering*. This action lasts for 1s and causes the steering wheel to turn left for 100°; 5. the steering wheel keeps its position for the rest of the scenario.

Simulating this maneuver in Simulink (with a continuous solver) first with TV enabled and then with TV disabled, the changes of the vehicle's position (Fig. 10.8) show that using TV controller shortens the radius while turning.

After the initial modeling and validation in Simulink, the TV controller is transformed to SysteMoC by the model transformation framework. Since TV is a single-rate system, the data-driven SysteMoC is used. The auto-generated torque vectoring controller contains 10 graphs, 98 actors, 110 FIFOs, and 770 lines of code (excluding library code).

After converting the Simulink model to an executable specification, a validation function unit is created in Simulink. Next, the co-simulation server and co-simulation interface are configured. The co-simulation checks whether the auto-generated TV controller will perform as intended in its operational environment. The testing maneuver configured in ASM is reused to evaluate the generated TV controller. The result of the co-simulation is given in Fig. 10.9: While the vehicle turns, in order to get more traction to shorten the curve, each engine is applied

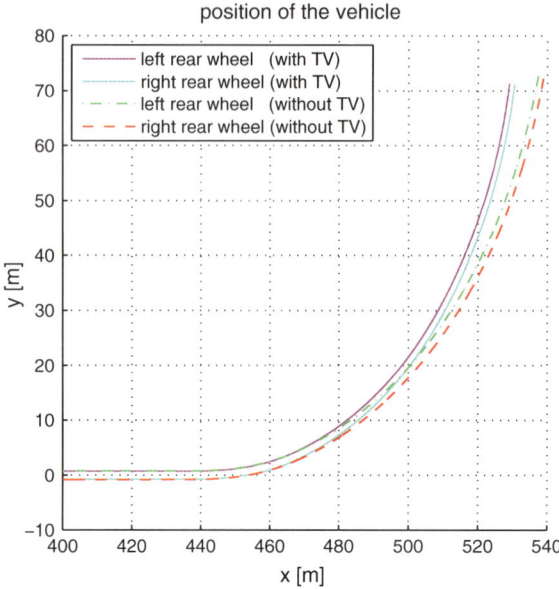

Fig. 10.8 Effect of torque vectoring on the vehicle's position: simulation results with TV controller enabled (*solid lines*) and disabled (*dashed lines*)

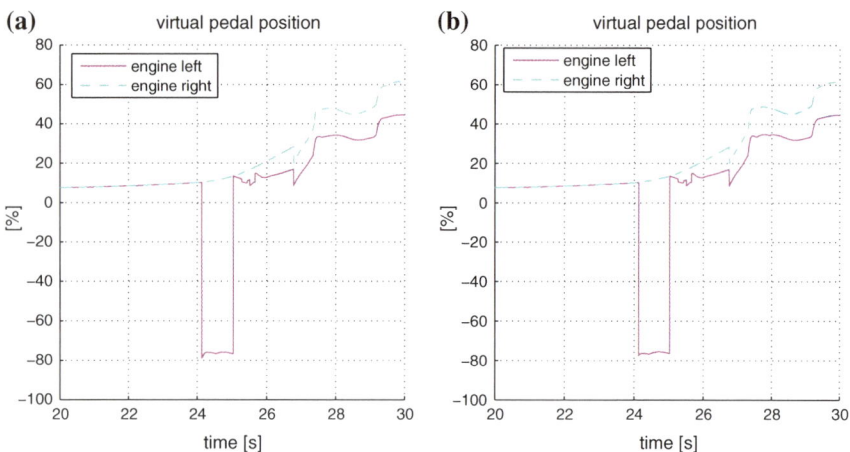

Fig. 10.9 Validation via co-simulation of Simulink and SysteMoC using the virtual pedal position seen by the engine: the proposed co-simulation shown in (**b**) delivers almost identical results compared to the plain Simulink simulation depicted in (**a**)

with an individual pedal signal calculated by the TV controller based on the current physical pedal position. These pedal signals are interpreted by the engines as *virtual pedal positions*, with a positive position for acceleration and a negative position for deceleration. Figure 10.9a shows how these two pedal signals change while using the Simulink TV controller. When the vehicle is turning, the engine that generates torque for the inner wheel (in this case the left-side engine) actually receives a signal from the TV controller indicating a negative pedal position, this causes the inner wheel to brake while the outer wheel is still experiencing acceleration. As depicted in Fig. 10.9b, the SysteMoC TV controller delivers almost identical simulation results. The co-simulation shows a minor simulation deviation, which is observed by the validation function unit through comparing the pedal signals (see Fig. 10.9) calculated by the original TV controller and the generated TV controller. The small deviation shown in Fig. 10.10, which is always below 0.09 %, indicates a high accuracy of the proposed model transformation.

The development time for the presented case study is shown in Fig. 10.11: The development time consists of (I) implementing the initial model in MATLAB/ Simulink (mandatory); (II) validating the initial model within the Simulink environment (mandatory); (III) automatic model transformation (proposed); and (IV) validation of the auto-generated model via co-simulation (proposed). The first two mandatory phases of modeling and validation in Simulink consume almost 70 % of the complete development time. The proposed automatic model transformation that converts the Simulink TV controller into SysteMoC requires 20 % of the overall development time—a high value at first glance. The reason is that the actors library did not yet include all the atomic blocks used in the Simulink TV model. Thus, the designer had to implement those Simulink atomic blocks and add them to the actors library. Note that this effort has to be invested only once for an atomic block. It is expected that the time consumption of this phase reduces dramatically

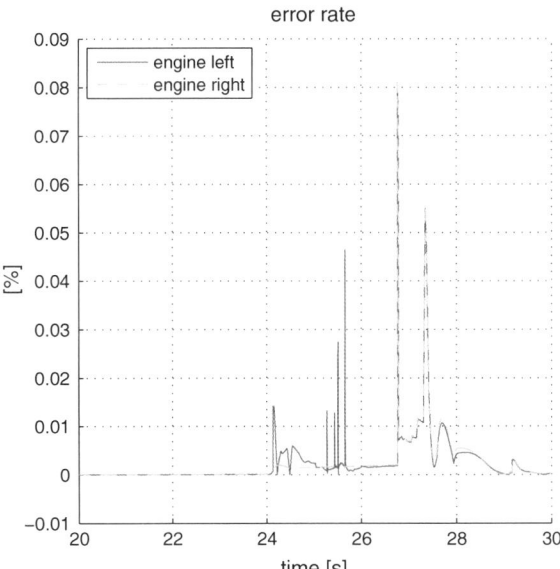

Fig. 10.10 The relative simulation error for the virtual pedal position remains below 0.09 %

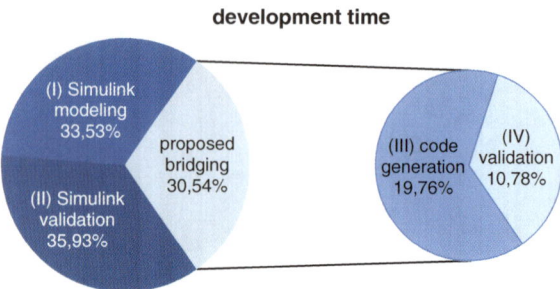

Fig. 10.11 The proposed methodology reduces the time from the Simulink model to the ESL model
to only ≈30 % of the overall development time

for future applications (ideally to 0 %), given the constant extension of the actors
library. But, this phase may include the integration of user-defined S-functions in the
ESL model such that an application-dependent amount of development time remains
to be taken into account. The last phase of system-level validation consumes about
10 % of the overall development time, resulting from the creation of the validation
function unit and the configuration of the co-simulation framework. For the presented case study, it can be concluded that bridging algorithm design to automatic
ESL design flows requires less than 50 % of extra development time compared to the
mandatory MATLAB/Simulink part. Note that a huge amount of the extra development time arose from implementing basic blocks and configuring the co-simulation.
Both aspects do not scale with the complexity of the modeled application, such
that for future large and complex Simulink models, the extra effort for bridging to
ESL design flows will become almost negligible.

10.6 Conclusion

In this chapter, we presented a framework that enables an automatic model transformation from MATLAB/Simulink to an actor-oriented design language (SysteMoC),
which enables Design Space Exploration (DSE). This framework is integrated into
an ESL design flow to further reduce development efforts. The automatic generation of an executable specification from Simulink has freed system designers from
converting Simulink functional models into implementation models manually. On
the other hand, by applying the proposed system validation technique, which is
based on the co-simulation of the original Simulink model and SysteMoC model,
the designer can easily validate the correctness of the executable specification.
Furthermore, combining this work with DSE allows the designer to automatically get
a first-hand evaluation on the performance of different implementation alternatives.

Acknowledgments The work has been partially supported by EFRE funding from the Bavarian Ministry of Economic Affairs (Bayerisches Staatsministerium für Wirtschaft, Infrastruktur, Verkehr und Technologie) as a part of the "ESI Application Center" project.

References

1. Atat Y, Zergainoh NE (2008) Automatic code generation for MPSOC platform starting from SIMULINK/MATLAB: new approach to bridge the gap between algorithm and architecture design. In: Proceedings on the international conference on information and communication technologies: from theory to applications (ICTTA) 2008, IEEE, pp 1–6
2. Baleani M, Ferrari A, Mangeruca L, Sangiovanni-Vincentelli AL, Freund U, Schlenker E, Wolff HJ (2005) Correct-by-construction transformations across design environments for model-based embedded software development. In: Proceedings of the design, automation and test in Europe conference (DATE) 2005, IEEE, pp 1044–1049
3. Barnasconi M, Einwich K, Grimm C, Maehne T, Vachoux A (2011) Advancing the SystemC analog/mixed-signal AMS extensions. Open SystemC Initiative (OSCI)
4. Boland JF, Thibeault C, Zilic Z (2005) Using MATLAB and Simulink in a SystemC verification environment. In: Proceedings of the design and verification conference (DVCon) 2005
5. Caspi P, Curic A, Maignan A, Sofronis C, Tripakis S, Niebert P (2003) From Simulink to SCADE/Lustre to TTA: a layered approach for distributed embedded applications. ACM Sigplan Not ACM 38:153–162
6. Czerner F, Zellmann J (2002) Modeling cycle-accurate hardware with Matlab/Simulink using SystemC. In: The European SystemC users group meeting (ESCUG) 2002
7. dSPACE (2013) Automotive simulation models (ASM). dSPACE. http://www.dspace.com/
8. ESTEREL (2013) SCADE suite™: control and logic application development. ESTEREL technologies. http://www.esterel-technologies.com/
9. ETAS (2013) ASCET. ETAS. http://www.etas.com/
10. Falk J, Haubelt C, Teich J (2006) Efficient representation and simulation of model-based designs in SystemC. In: Proceedings of the forum on specification and design languages (FDL) 2006, pp 129–134
11. Gerstlauer A, Haubelt C, Pimentel AD, Stefanov TP, Gajski DD, Teich J (2009) Electronic system-level synthesis methodologies. IEEE Trans Comput Aided Des Integr Circ Syst 28(10):1517–1530
12. Grötker T, Liao S, Martin G, Swan S (2002) System design with SystemC. Kluwer Academic Publisher, Boston
13. Jersak M, Cai Y, Ziegenbein D, Ernst R (2000) A transformational approach to constraint relaxation of a time-driven simulation model. In: Proceedings of the international symposium on system synthesis (ISSS) 2000, IEEE Computer Society, Washington DC, pp 137–142. doi:10.1145/501790.501820
14. Kai H, Sang-il H, Popovici K, Brisolara L, Guerin X, Li L, Yan X, Chae SI, Carro L, Jerraya A (2007) Simulink-based MPSoC design flow: case study of motion-JPEG and H.264. In: Proceedings of the ACM/IEEE design automation conference (DAC) 2007, pp 39–42
15. Keinert J, Streubühr M, Schlichter T, Falk J, Gladigau J, Haubelt C, Teich J, Meredith M (2009) SYSTEMCODESIGNER—an automatic ESL synthesis approach by design space exploration and behavioral synthesis for streaming applications. ACM Trans Des Autom Electron Syst 14(1):1–23
16. Kienhuis B, Deprettere E, Vissers K, van der Wolf P (1997) An approach for quantitative analysis of application-specific dataflow architectures. In: Proceedings of the IEEE international conference on application-specific systems, architectures and processors (ASAP) 1997, pp 338–349. doi:10.1109/ASAP.1997.606839

17. Kopetz H (2011) Real-time systems: design principles for distributed embedded applications. Springer, Hermann
18. Lee EA (2000) What's ahead for embedded software? Computer 33(9):18–26
19. Lee EA, Neuendorffer S (2004) Actor-oriented models for codesign. Formal methods and models for system design. Kluwer Academic Publishers, Norwell, pp 33–56
20. Lukasiewycz M, Streubühr M, Glaß M, Haubelt C, Teich J (2009) Combined system synthesis and communication architecture exploration for MPSoCs. In: Proceedings of the design, automation and test in Europe Conference (DATE) 2009. IEEE Computer Society, Nice, pp 472–477
21. Martin G, Bailey B, Piziali A (2010) ESL design and verification, a prescription for electronic system level methodology. Morgan Kaufmann, San Francisco
22. MathWorks T (2013) S-Function. http://www.mathworks.com/
23. The MathWorks (2013) HDL Verifier™: Verify HDL and Verilog using HDL simulators and FPGA-in-the-loop test. The MathWorks. http://www.mathworks.com/
24. The Mathworks (2013) Simulink Coder: Generate C and C++ code from Simulink and Stateflow models. The MathWorks
25. The Mathworks (2013) Simulink: Simulation and model-based design. The MathWorks

Chapter 11
Software Allocation in Automotive Networked Embedded Systems: A Graph-Based Approach

Yasser Shoukry, M. Watheq El-Kharashi, Sherif Hammad, Ajay Kumar
and Ghada Bahig

Abstract Complex automotive networked embedded systems require novel algorithms for exploring different design decisions at early stages of the design flow. The problem of allocating the software components on electronic control units lies at the core of these design decisions. This chapter formalizes this allocation problem using graph theory. The proposed formalism allows the designer to use a wide variety of graph-theoretic optimization algorithms, which are capable of minimizing more than one criterion simultaneously. The proposed algorithm is then proven, by means of numerical examples, to find the same solution as mathematical optimization, but it is 15 times faster in computation time.

Keywords Automotive networked embedded systems · Automotive Electronic Control Units (ECUs) · Automotive SoftWare Component (SWC) · AUTomotive Open System ARchitecture (AUTOSAR) · Graph-theoretic optimization · Multi-level graph partitioning · Partition-and-map algorithm · Recursive mincut heuristic · Task allocation problem · Vehicle engineering

Y. Shoukry (✉) · M.W. El-Kharashi · S. Hammad
Computer and Systems Engineering Department, Ain Shams University, Cairo, Egypt
e-mail: yasser_shoukry@eng.asu.edu.eg

M.W. El-Kharashi
e-mail: watheq.elkharashi@eng.asu.edu.eg

S. Hammad
e-mail: sherif.hammad@eng.asu.edu.eg

A. Kumar
Mentor Graphics Corporation, Wilsonville, OR, USA
e-mail: ajay_kumar@mentor.com

G. Bahig
Mentor Graphics Egypt, Cairo, Egypt
e-mail: ghada_bahig@mentor.com

© Springer International Publishing Switzerland 2015   207
M.-M. Louërat and T. Maehne (eds.), *Languages, Design Methods,
and Tools for Electronic System Design*, Lecture Notes in Electrical Engineering 311,
DOI 10.1007/978-3-319-06317-1_11

11.1 Introduction

The growth of electronic systems led to an increase in number of functionalities and SoftWare Components (SWCs) in vehicle engineering. Current vehicles have more than 80 Electronic Control Units (ECUs) and thousands of SWCs. This large number of ECUs and SWCs (or functions) complicates the task of efficient allocation of SWCs onto the different ECUs, especially when considering the impact of this allocation on the overall system performance.

The problem above is studied in Hardung [6] and Honnavara [7]. Hardung [6] proposed to use search-based optimization techniques. These techniques suffer from the exploding solution-space. He showed an example, which needs 400,000 years on an ordinary computer to evaluate all possible solutions. Honnavara [7] developed a binary quadratic programming model representation of the problem. The model applies a branch-and-cut algorithm to obtain the optimal solution. Although his mathematical formulation of the problem ensures a Pareto-optimal solution, the proposed mathematical model suffers from a scalability problem in terms of expanding the model to newer minimization criteria (e.g., bus schedulability). Similar problems are studied in other contexts. For example, the work reported in Leonardi et al. [11] proposes a methodology to realize a distributed embedded system, in the context of smart buildings, given dataflow specifications. It utilizes mathematical optimization techniques aiming to minimize a quadratic objective. The general problem of design space exploration for networked embedded systems is also reported in Fummi et al. [5]. The objective was to use a mathematical language to model a distributed application in terms of tasks, hosting nodes, and interactions with the environment.

In this chapter, we model the SWC allocation problem using graph theory. Graph representation allows more flexibility in adding new minimization criteria without affecting the main algorithm. Graph-based optimization techniques are widely proven to be efficient in many problems, like allocation of components in VLSI [13, 14] as well as task scheduling and allocation in operating systems [10, 12, 15]. Accordingly, the proposed representation allows making use of familiar algorithms in graph theory in order to achieve better results based on such formalization. This chapter draws inspiration from the "task Allocation by Recursive Mincut (ARM)" heuristic scheme [3]. We adapt the ARM heuristic to the problem under consideration.

The proposed algorithm is implemented as a tool, which utilizes the AUTomotive Open System ARchitectures (AUTOSARs) description [2]. The information available in the AUTOSAR model is parsed and fed to the algorithm. The output of the algorithm is then written back automatically into the AUTOSAR model files. The performance of the proposed algorithm is then compared to a mathematical optimization done using the binary quadratic programming model. The proposed recursive scheme is proven to get the same results, as obtained by the mathematical optimization, but in shorter computation times for the same examples.

The rest of this chapter is organized as follows. A brief review of current multilevel graph partitioning algorithms is presented in Sect. 11.2. The main contribution of this chapter, which is the graph representation of the SWC allocation problem along with the "Partition-and-Map" algorithm used to solve the allocation problem, are then developed in Sects. 11.3 and 11.4. Different evaluation criteria are discussed in Sect. 11.5. Three examples of running the proposed algorithm are provided in Sect. 11.6. Finally, Sect. 11.7 concludes this work.

11.2 Multilevel Graph Partitioning

Consider a weighted graph $G = (V, E)$ as a set of vertices V and a set of weighted edges E, where each vertex v and edge e has an associated weight, denoted by $w(v)$ and $w(e)$, respectively. A decomposition of V into k disjoint subsets $V_1, V_2, ..., V_k$, is called a k-way partitioning of V. The cut of a k-way partitioning of V is equal to the sum of weights of the edges that contain vertices from different sub-domains or partitions. Many algorithms were developed in order to partition a graph into two domains while obtaining the minimum cut size. Recently, a new class of graph partitioning algorithms has been developed based on the multilevel paradigm. In these algorithms, a sequence of successively smaller graphs is constructed. A partitioning of the smaller graph is computed and this partitioning is then successively projected to the next level of finer graphs. At each level, an iterative refinement algorithm (e.g., KL [8] or FM [4]) is used to further improve its quality. Experiments have shown that multilevel graph partitioning algorithms can produce substantially better solutions than those produced by non-multilevel schemes [1].

11.3 Graph Model of the Software Allocation Problem

Allocating SWCs onto ECUs represents a provider-consumer problem, where ECUs provide resources in terms of memory bytes and processor cycles, whereas SWCs represent consumers for these resources. The objective of the software allocation problem is not to optimize the utilization of these resources or balance the load on different ECUs. However, these resources provide physical limitations on how the SWCs can be allocated, which the algorithm has to satisfy.

SWCs communicate with each other via signals. If two communicating SWCs are assigned to different ECUs, then all signals between these two SWCs will be transferred over the bus connecting these two ECUs. These signals are either periodic (identified by signal frequency and signal size) or sporadic signals (identified by minimum-start interval and signal size).

The intuition of the proposed SWC-ECU allocation framework can be explained as follows. Given a set of SWCs, ECUs, and signals, we can model our problem as a graph. Graph vertices represent SWCs while graph edges represent the communi-cation between SWCs. Edge weights differ according to the minimization criteria. Edge weights are chosen such that the higher the edge weight, the higher tendency of these two SWCs (connected by this edge) to be allocated on the same ECU.

We start by allocating all SWCs to one ECU. If no ECU can fit all the SWCs (i.e., the resources provided by this ECU are not enough to satisfy the resources required by all SWCs), the algorithm tries to bisect the SWCs into two different clusters while minimizing the objective function. Accordingly, a graph partitioning algorithm is applied on the prescribed graph. The algorithm tries again to find suitable ECUs to fit the smaller SWC graphs. This process is then repeated a number of times until all SWCs are allocated on ECUs. Remember that graph edges correspond to the criteria the algorithm aims to minimize. Hence, by utilizing a graph partitioning algorithm that tends to cut through edges with minimum weights, the resulting partitioning in turn corresponds to the minimization of the required criteria.

We consider two main minimization criteria in our examples, named communication bandwidth and sporadic signal bandwidth. However, the proposed framework can be extended directly to other criteria as well.

11.3.1 Minimizing Communication Bandwidth between ECUs

Highly coupled software components, i.e., those communicating frequently to achieve their goals, are better to be allocated on the same ECU. Accordingly, the first optimization criteria aims to increase the intra-communication bandwidth between the SWCs operating within the same ECU. However, as discussed before, graph weights are assigned as a criterion that needs to be minimized. Hence, it is more convenient to minimize the bus communication bandwidth between different ECUs instead of maximizing the intra-communication bandwidth over the same ECU.

By minimizing the communication bandwidth criterion, the graph edge weight is given by:

$$w(e) = \max \left\{ \frac{1}{\text{signal frequency}}, \text{minimum start interval} \right\} \times \text{signal size}, \quad (11.1)$$

where "signal frequency" is the property associated with periodic signals, whereas "minimum start interval" is the property associated with sporadic signals and reflects the lower bound on the frequency of such signals. "Signal size" measures the number of bytes required by each signal.

11.3.2 Minimizing Sporadic Signal Bandwidth between ECUs

A major task in designing real-time systems is to design a bus scheduling policy. On the first hand, signals with periodic occurrence on the bus are easier to be scheduled. On the other hand, sporadic signals which do not occur periodically, but have a lower bound on their occurrence, are harder in terms of designing a scheduling policy. Hence, sporadic signals decrease the schedulability of the bus and decrease the bus stability. By minimizing the sporadic signal bandwidth, the proposed framework tries to decrease the bandwidth required by sporadic signals. It is obvious to show that minimizing the overall communication bandwidth does not guarantee minimization of sporadic signals' bandwidth. Hence, we are interested in reflecting this behavior in our framework. Accordingly, if the signal is sporadic the graph edge weight is given by Eq. (11.2) and zero otherwise.

$$w(e) = \text{minimum start interval} \times \text{signal size} \qquad (11.2)$$

11.4 Partition-and-Map Algorithm

In this section, we discuss the details of the "Partition-and-Map Algorithm" along with some issues related to the performance and design constraints. The objective of this algorithm is to allocate each SWC onto an ECU while minimizing the overall communication bandwidth and the sporadic bandwidth between ECUs as well.

11.4.1 The Algorithm

For each minimization criterion, we have a preference vector item to weight the effect of each criterion. The proposed algorithm starts by constructing the graph representation for the problem, as discussed in Sect. 11.3. For each minimization criteria, a separate graph is constructed. All graphs are then merged to form one graph with multiple edge weights. After creating and merging graphs, the algorithm creates a queue called "Unmapped Clusters". Then, the algorithm works, as shown in Algorithm 11.1.

The proposed algorithm uses a multi-objective, multi-level partitioning algorithm, like the k-way Fiduccia-Mattheyses (FM) heuristic described in Kumar et al. [9]. The objective of the k-way FM is to bisect the graph into two clusters such that the sum of edge weights between the two clusters is minimal. Since edge weights represent the tendency of two SWCs to be allocated to the same ECU, then by finding the minimum edge cut, the proposed "Partition-and-Map" algorithm ensures finding clusters that satisfy the minimization goal.

As discussed before, the main objective of the proposed algorithm is to allocate the SWCs in order to minimize the communication. The proposed algorithm does not try to balance the loads between different ECUs. Hence, it is more convenient to use a 2-way partitioning instead of k-way partitioning (with $k > 2$), which usually appears in the context of load balancing. The intuition behind the 2-way partitioning is the following. First, the algorithm tries to allocate all SWCs to the same ECU. If it succeeds, then the resulting system will have zero communication bandwidth and zero sporadic signals. If the physical constraints imposed by the available resources prevent such allocation, the algorithm partitions the set of SWCs using 2-way partitioning. This partitioning will result in an increase in the amount of needed ECUs by one and thus communication between the two ECUs is needed. However, a k-way partitioning will result in increasing the number of used ECUs by k and thus more and more communication is needed by those k ECUs.

Algorithm 11.1: The Partition-and-Map algorithm

Input: A list of different ECUs and a weighted graph representing different SWCs and communication between them

Output: A list of all ECUs and SWCs mapped to each ECU

1 Sort all ECUs in an ascending order based on the available resources;
2 Put all SWCs in one cluster and push this cluster onto the "Unmapped Clusters" queue;
3 **while** *the "Unmapped Clusters" queue is not empty* **do**
4   Pop an unmapped cluster from the queue;
5   Use either Algorithm 11.2 or Algorithm 11.3 to map this cluster to one of the ECUs;
6   **if** *the cluster can fit onto one of the ECUs* **then**
7   decrease the resources provided by this ECU;
8   reorder ECUs again;
9   **if** *the cluster cannot fit onto any ECU* **then**
10   Use the multi-objective, multi-level partitioning algorithm described in [9] in order to bisect this cluster into two smaller clusters;
11   Push the resulting two clusters into the "Unmapped Clusters" queue;

Algorithm 11.2: Mapping SWCs to ECUs using the first-fit approach

---

Input: A list of different ECUs, a list of remaining resources for each ECU, and the required resources by the SWC cluster, which needs to be mapped

Output: An ECU to which the SWC is to be assigned

1 Let $i = 1$;
2 **while** $i \neq$ *length of the ECU list* **do**
3   **if** *available resources of* ECU_i > *the required resources by the SWC cluster* **then**
4   **return** ECU_i;
5 **return** *Failure*;

An aside effect of the intuition behind the algorithm is the minimization of the number of used ECUs and hence the system cost. This follows from the fact that the algorithm tries to map all SWCs onto just one ECU and then gradually increase the number of ECUs.

Algorithm 11.3: Mapping SWCs to ECUs using the best-fit approach

---

Input: A list of different ECUs, a list of remaining resources for each ECU, and the required
       resources by the SWC cluster, which needs to be mapped

Output: An ECU to which the SWC is to be assigned

1 Let $i = 1$;
2 **while** $i \neq$ *length of the ECU list* **do**
3   **if** *available resources of ECU$_i$ > the required resources by the SWC cluster* **then**
4 | Calculate r_i = available resources of ECU$_i$ − required resources by the SWC cluster;
5   **else**
6 | $r_i = 0$;

7 **if** $r_i = 0$ *for all i* **then**
8 | **return** *Failure*;
9 **else**
10 | **return** *ECU$_i$ with the maximum r$_i$*;

11.4.2 First-Fit Versus Best-Fit Mapping

Algorithm 11.1 can use either the first-fit mapping algorithm or the best-fit mapping algorithm while trying to fit an SWC cluster onto an ECU. In the first-fit algorithm (Algorithm 11.2), the ECUs are sorted in an ascending order based on the remaining resources and the SWC cluster is mapped to the first ECU that fits the SWC cluster. Despite the fact that first-fit approach (Algorithm 11.1) performs fast in practice (as will be shown in the results Sect. 11.6, Examples 2 and 3), the major drawback of this approach could be stated as follows: As more SWC clusters are mapped, the remaining resources in each ECU become less. After some time, each ECU may have small amount of resources available. The sum of these few resources is adequate to map new SWCs , but since these resources are distributed over multiple ECUs, they are all unused. This is similar to the fragmentation problem that arises in storage devices.

The best-fit approach (Algorithm 11.3) tries to overcome the drawbacks of the first-fit approach at the expense of the execution performance. In best-fit approach, another loop is required, which goes through all ECUs aiming to find the ECU, which results into leaving fewer resources compared to all other ECUs. This ensures fewer resource gaps in the final SWC allocation.

11.4.3 Design Constraints

In many design examples, some constraints need to be satisfied. In the context of mapping SWCs to ECUs, we identify three design constraints, which the "Partition-and-Map" algorithm handles directly. These constraints are named:

1. Pre-mapped SWCs,
2. Mutually inclusion of SWCs, and
3. Mutually exclusion of SWCs.

The first constraint, which is "pre-mapped SWCs" asks that certain SWCs must be assigned to specific ECUs. For example, an SWC, which handles some sensors measurements, must be mapped to the ECU, which contains these sensors. In this case, the "Partition-and-Map" (Algorithm 11.1) have a prior initialization step, where these SWCs are mapped to the required ECU even before the algorithm starts.

The "mutually inclusion of SWCs" asks that certain SWCs must be mapped to the same ECU. In such case, we treat these SWCs as only one "bundled" SWC. The rest of the algorithm remains the same.

On contrast to the second constraint, the "mutually exclusion of SWCs" asks that certain SWCs are not mapped to the same ECU. In this case, we add two modifications to the proposed algorithm. First, while partitioning the SWCs graph into clusters, one needs to add relatively high weight on the edge connecting the two mutually exclusive SWCs. This forces the multi-way partitioning algorithm to cluster these two SWCs into two different clusters. Second, the first-fit and best-fit mapping algorithms (Algorithms 11.2 and 11.3) need to accommodate that constraint. The two mapping algorithms check whether the SWC cluster (which needs to be mapped) has a conflict with the ECU that is selected for mapping. If there exists a conflict, the mapping algorithm discards this ECU and searches for the next first/best-fit ECU.

11.4.4 ECU Heterogeneity

So far, the SWCs are assumed to require a fixed amount of resources regardless the type of the ECU. A more practical case is that each SWC has different resource requirements based on the ECU. For example, the code footprint for the same SWC differs from a 16-bit processor to a 32-bit processor. In order to accommodate such cases, we attach an array of resource requirements to each SWC. The mapping algorithms (Algorithm 11.2 and 11.3) utilize these information while comparing the required resources to the available ones on each ECU.

11.5 Evaluation Criteria

To evaluate results of the proposed algorithm, six metrics are developed to measure the mapping results as given by Eqs. (11.3)–(11.8). The first two metrics evaluate the communication bandwidth criterion. The third metric evaluates the bus-schedulability criterion since, from the scheduling point-of-view, the less the sporadic signals on the bus, the better the schedulability. The other metrics are considered as aside metrics, which can affect the decision of the designer. These metrics reflect the utilization of the available resources as a consequence of different SWC allocations.

11.5.1 Inter-ECU Communication Bandwidth

This metric is calculated for each ECU. It measures the communication sent and/or received over the bus by each particular ECU as a percentage of the total communication in the system:

$$\% \left[\frac{\text{Total intercommunication bandwidth}}{\text{Total communication bandwidth}} \right]. \tag{11.3}$$

11.5.2 Intra-ECU Communication Bandwidth

This metric is calculated for each ECU. It measures the intra-communication inside a particular ECU as a percentage of the total communication in the system:

$$\% \left[\frac{\text{Total intra-communication bandwidth}}{\text{Total communication bandwidth}} \right]. \tag{11.4}$$

11.5.3 Sporadic Communication Bandwidth

This metric is calculated for each ECU. It measures the sporadic communication sent and/or received over the bus by each particular ECU as a percentage of the total communication in the system:

$$\% \left[\frac{\text{Total sporadic communication bandwidth}}{\text{Total communication bandwidth}} \right]. \tag{11.5}$$

11.5.4 Total Normalized Cost

This metric is calculated for the whole system. It measures the cost of the ECUs being utilized as a percentage of the set of the provided ECUs:

$$\% \left[\frac{\sum \text{Cost for all ECUs that contain SWCs}}{\sum \text{Cost for all ECUs}} \right]. \tag{11.6}$$

11.5.5 Processors' Cycles Utilization

This metric is calculated for each ECU. It measures the utilization of the processor computational resources measured as the percentage of processor cycles utilized in each ECU:

$$\% \left[\frac{\sum \text{Processor cycles used}}{\text{Processor cycles provided}} \right]. \tag{11.7}$$

11.5.6 Memory Bytes Utilization

This metric is calculated for each ECU. It measures the utilization of the processor memory resources measured as the percentage of processor memory bytes utilized in each ECU:

$$\% \left[\frac{\sum \text{Memory bytes used}}{\text{Memory bytes provided}} \right]. \tag{11.8}$$

11.6 Implementation and Results

The proposed algorithm is implemented using Java on the Eclipse platform. The input to the developed software is a set of AUTOSAR (version 3.0) XML description files. The software parses these XML files in order to get the information of the ECUs from the `ECUResourceTemplate` nodes in the XML files, SWCs from the `SwcImplementation` nodes, and Signals from the `AssemblyConnector` and the `InternalBehavior` nodes. The distinction between the periodic and sporadic signals is based on the type of the `InternalBehavior` event associated with the corresponding `Runables`. The output of the mapping algorithm is then written to the `SystemImpl` node in the same XML files.

Here, we will show three different examples. The first example is to show that the proposed algorithm tends to minimize more the two criteria proposed earlier based on the given preference vector. The second example is to show that the proposed algorithm can give the same results of the optimal solution obtained by mathematical optimization model, proposed by Honnavara [7], while reducing the computation time by an order of a magnitude. The third example is to show that the proposed algorithm can scale up with the increase in the dimension of the system.

Table 11.1 SWCs configuration used in Example 1

SWC	Cycles required (%)	Memory required (Bytes)
SWC1	10	256
SWC2	10	384
SWC3	10	640
SWC4	20	256
SWC5	30	640
SWC6	40	1,152
SWC7	40	2,176
SWC8	50	1,651

Table 11.2 ECUs configuration used in Example 1

ECU	Cycles provided (%)	Memory provided (Bytes)	Cost ($)
ECU1	70	1,024	50
ECU2	70	2,048	80
ECU3	70	3,072	100
ECU4	70	4,096	120
ECU5	70	5,120	170

11.6.1 Example 1: Performance Results

This example is used to test the performance of the proposed algorithm. Number of SWCs : 8, number of ECUs: 8, number of Signals: 18. Configuration of these ECUs, SWCs, and signals are presented in Tables 11.1, 11.2 and 11.3, respectively.

Applying the first phase of the algorithm, to build the graph, we get the graph shown in Fig. 11.1. Vertices represent different SWCs and edge weights represent {signal frequency, minimum start interval} for each signal connecting two SWCs.

Results of running the proposed algorithm with the following preference vectors [100, 0], [0, 100], and [50, 50], which target minimization for bandwidth only, sporadic communication only, and minimizing both criteria, respectively, are presented in Table 11.4. The first column shows the preference vector, second column shows the ECU, and the third column shows the output of the algorithm, which is the allocation of SWC. The remaining columns show resource and communication statistics resulting from this allocation.

The results of running the proposed evaluation criteria Eqs. (11.3)–(11.8) for the three different runs of the algorithm are shown in Table 11.5. These metrics show that the proposed algorithm tends to minimize the required preference. For instance, rows 1 and 3 show that the minimum values for bandwidth and sporadic

Table 11.3 Signals configuration used in Example 1

Source SWC	Destination SWC	Signal frequency (Hz)	Minimum start interval (s)
SWC1	SWC7	5.0	
SWC1	SWC3		0.0833
SWC2	SWC7	7.0	
SWC2	SWC1	9.5	
SWC2	SWC4	8.0	
SWC3	SWC1	10.0	
SWC3	SWC4		0.1111
SWC3	SWC6	5.0	
SWC4	SWC2	9.0	
SWC4	SWC5		0.0833
SWC5	SWC4	11.0	
SWC5	SWC3	10.0	
SWC5	SWC6	7.0	
SWC6	SWC3	6.0	
SWC6	SWC1		0.333
SWC7	SWC4	2.0	
SWC7	SWC1	4.0	
SWC8	SWC3	9.0	

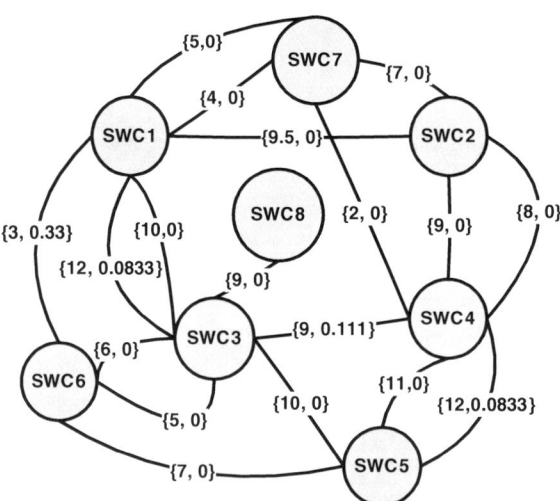

Fig. 11.1 Merged graph used in Example 1. *Vertices* represent different SWCs and *edge weights* represent {signal frequency, minimum start interval} for each signal connecting two SWCs

Table 11.4 Mapping results of Example 1 for three different preference vectors

Preference vector	ECU	Allocated SWCs	Remaining processor cycles (%)	Remaining memory (%)	Inter-ECU bandwidth (%)	Intra-ECU bandwidth (%)	Sporadic bandwidth (%)
[100, 0]	1	1, 2	71.43	37.50	41.88	6.86	10.83
	2	5, 6	0.00	12.50	33.94	5.05	10.83
	3	3, 4	0.00	0.00	77.98	7.94	17.33
	4	8	28.57	59.69	6.50	0.00	0.00
	5		100.00	100.00	0.00	0.00	0.00
[0, 100]	1	4, 5	71.43	37.50	32.49	16.61	6.50
	2	2, 8	0.00	12.50	30.68	0.00	0.00
	3	3, 7	0.00	0.00	57.04	0.00	15.17
	4	1, 6	28.57	59.69	42.24	2.17	8.67
	5		100.00	100.00	0.00	0.00	0.00
[50, 50]	1	4, 3	57.14	12.50	67.87	6.50	17.33
	2	8	28.57	19.38	6.50	0.00	0.00
	3	7, 2	28.57	16.67	27.07	5.05	0.00
	4	6, 1	28.57	65.62	42.24	2.17	8.67
	5	5	57.14	87.50	28.88	0.00	8.67

Table 11.5 Results of the evaluation criteria in Eqs. (11.3)–(11.8) for preference vectors [100, 0], [0, 100], and [50, 50], respectively, for Example 1

Metric	Run #1 [100, 0] (%)	Run #2 [0, 100] (%)	Run #3 [50, 50] (%)
1	80.15	81.23	86.28
2	19.85	18.78	13.72
3	19.50	15.17	17.34
4	58.33	58.33	86.66
5	37.50	37.50	37.50
6	36.28	39.11	37.29

communication occur when the algorithm minimizes bandwidth and sporadic communication, respectively. Running time of the shown example was less than 2 s.

11.6.2 Example 2: Profiling Results

The proposed algorithm is then used with the same example shown in [7], where the optimization criterion is to minimize the bandwidth only. Our proposed algorithm gives exactly the same results given by the mathematical optimizations, but it is more than 15× faster in computation time. Although having the same results from two different algorithms in one example never means that they will always give same results. Still, it gives confidence that our approach can give results, which are very near to the optimal solution, but in a way which is significantly faster than the mathematical approach given in [7], for the same number of SWCs , ECUs, and signals.

11.6.3 Example 3: Profiling Results

We here use a higher order example, where a random data is generated for a system consisting of 80 ECUs, 230 SWCs, and 1000 signals. The amount of time needed to parse the AUTOSAR XML files is 8 min, while the processing time of the algorithm is approximately 22 min. This example shows that the time response of the "Partition-and-Map" algorithm is still acceptable for high-dimensional systems.

11.7 Conclusion and Future Work

This chapter presents a framework for solving the problem of software component allocation using graph-theoretic techniques. The proposed model opens the venue to use a set of proven algorithms from graph theory in order to minimize multi-criteria problems.

The graph representation is simpler than the mathematical equivalent in terms of adding new optimization criteria to the model. The proposed "Partition-and-Map" algorithm is proven to minimize the given optimization criteria with a significant computation speedup compared to mathematical optimization techniques.

Further investigation is required for handling situations other than discussed here, like physical location of ECUs and wiring cost. Other directions for research include how to handle the multi-network situation where the designer is allowed to use a heterogeneous networking platform (Controller Area Networks (CANs), Local Interconnect Networks (LINs), FlexRay). The allocation algorithm is then needed to take into consideration which ECU is connected to which network type and the congestion on the nodes, which bridge between the different types of networks.

References

1. Abou-Rjeili A, Karypis G (2006) Multilevel algorithms for partitioning power-law graphs. In: Proceedings of IEEE international parallel and distributed processing symposium (IPDPS) 2006, Rhodes Island, Greece, pp 16–26. doi:10.1109/IPDPS.2006.1639360
2. AUTOSAR committee (2013) AUTOSAR: Technical overview. http://www.autosar.org/
3. Ercal F, Ramanujam J, Sadayappan P (1988) Task allocation onto a hypercube by recursive mincut bipartitioning. In: Proceedings of the conference on hypercube concurrent computers

and applications: architecture, software, computer systems, and general issues (SIGARCH) 1988, Pasadena, CA, USA, pp 210–220. doi:10.1145/62297.62323

4. Fiduccia CM, Mattheyses RM (1982) A linear-time heuristic for improving network partitions. In: Proceedings of the IEEE/ACM design automation conference (DAC) 1982, Las Vegas, CA, USA, pp 175–181. doi:10.1109/DAC.1982.1585498

5. Fummi F, Lovato G, Quaglia D, Stefanni F (2010) Modeling of communication infrastructure for design-space exploration. In: Proceedings of the forum on specification and design languages (FDL) 2010, pp 1–6. doi:10.1049/ic.2010.0135

6. Hardung B (2006) Optimisation of the allocation of functions in vehicle networks. PhD thesis, Technische Fakultät, Universität Erlangen-Nürnberg, Erlangen, Germany

7. Honnavara V (2008) Cost optimization by method of allocating software component units to electronic control units for model-driven designs. Master's thesis, North Carolina State University, Raleigh, NC, USA

8. Kernighan BW, Lin S (1970) An efficient heuristic procedure for partitioning graphs. Bell Syst Tech J 49:291–307. doi:10.1002/j.1538-7305.1970.tb01770.x

9. Kumar V, Karypis G, Schloege K (1999) A new algorithm for multi-objective graph partitioning. Technical report, University of Minnesota, Minneapolis

10. Lam Y, Coutinho J, Luk W (2008) Integrated hardware/software codesign for heterogeneous computing systems. In: Proceedings of the southern conference on programmable logic (SPL) 2008, San Carlos de Bariloche, Argentina, pp 217–220. doi:10.1109/SPL.2008.4547761

11. Leonardi F, Pinto A, Carloni L (2011) Synthesis of distributed execution platforms for cyber-physical systems with applications to high-performance buildings. In: Proceedings of the IEEE/ACM international conference on cyber-physical systems (ICCPS) 2011, pp 215–224. doi:10.1109/ICCPS.2011.23

12. Liu W, Gu Z, Xu J, Wu X, Ye Y (2011) Satisfiability modulo graph theory for task mapping and scheduling on multiprocessor systems. IEEE Trans Parallel Distrib Syst 22(8):1382–1389. doi:10.1109/TPDS.2010.204

13. Shuiping L, Xiaoxue W (2010) A global method for the limited K-partitioning of hypergraphs representing optimal design problems in complex machine systems. Int J Syst Cybern 39(6):980–989. doi:10.1108/03684921011046753

14. Wang LY, Lai YT (2001) Graph-theory-based simplex algorithm for VLSI layout spacing problems with multiple variable constraints. IEEE Trans Comput Aided Des Integr Circuits Syst 20(8):967–979. doi:10.1109/43.936378

15. Xun X, Jiangqing W (2008) An application of vertex partition for parallel test tasks scheduling in automatic test system. In: Proceedings of the international conference on computer science and software engineering (CSSE) 2008, Wuhan, China, vol 2, pp 723–726. doi:10.1109/CSSE.2008.1067

Chapter 12
Fine-Grained Adaptive Simulation

Marcus Eggenberger and Martin Radetzki

Abstract While gate level simulation precision is still required for certain tasks, its high simulation detail is rarely required throughout the total simulated duration. To overcome the low simulation performance that comes along with gate level simulation, we present an adaptive simulation approach allowing to choose between register transfer and gate level abstractions online during simulation. This is achieved by adaptive SystemC models encapsulating an RTL model and a synthesized gate level counterpart. Adaptive models multiplex between fixed abstraction models and efficiently perform the necessary state transfer, which is enabled by giving adaptive models limited access to the simulation kernel. Adaptivity works with almost any given RTL model and is established in an automated, seamless way. Simulation performance is further increased by the fine-grained selection of simulation precision on a submodule basis. The benefits of having spatial and temporal freedom to choose the abstraction level online during simulation have been confirmed in our evaluations where speedups of up to 150 times have been achieved.

Keywords Adaptive simulation · Register Transfer Level (RTL) · Gate Level (GL) · SystemC · VHDL · Mixed-language modeling · Multi-level simulation · Network on Chip (NoC) · Abstraction

12.1 Introduction

With technology nodes still continuing to decrease, the number of integrated transistors per die further grows. And while this allows to satisfy the every increasing need for more computational power, it comes at the price of high simulation times, calling for novel simulation techniques to improve simulation performance. One way to accelerate simulations is by raising the level of abstraction, and techniques like

M. Eggenberger (✉) · M. Radetzki
Embedded Systems Engineering Group, University of Stuttgart, Stuttgart, Germany
e-mail: marcus.eggenberger@informatik.uni-stuttgart.de

M. Radetzki
e-mail: martin.radetzki@informatik.uni-stuttgart.de

© Springer International Publishing Switzerland 2015
M.-M. Louërat and T. Maehne (eds.), *Languages, Design Methods,
and Tools for Electronic System Design*, Lecture Notes in Electrical Engineering 311,
DOI 10.1007/978-3-319-06317-1_12

system level modeling and Transaction-Level Modeling (TLM) have been successfully employed to increase simulation performance at the cost of reduced simulation accuracy [10]. However, while TLM is very helpful in the early design phases, its application becomes more and more limited in the later phases, where a higher degree of accuracy is needed for design verification. For example, TLM can be used to evaluate faults in a system context, but the precise nature and effect of actual faults is lost as faults must be modeled in an abstract way as well. On the extreme end, precise fault simulation in terms of structural fault injection requires gate level models, but the complexity of gate level models prohibits gate level simulations of full systems on chips. Moreover, the small feature sizes, we have reached in the nano scale area, make chips highly susceptible to faults, which requires integration of fault resilience methods into the designs and thorough system simulation under various fault conditions to evaluate such counter measures.

However, even for structural fault simulations, gate level simulation detail is not necessarily required through the whole simulation. For example when simulating a degrading system with permanent hardware defects occurring over time, the system can be simulated at a higher abstraction level until the first error occurs. And even from that point on, only the part of the system where the fault site is located must be simulated at gate level precision. When simulating transient faults, the time during which gate level precision is required is even less, as the simulation of a single cycle at gate level is enough to determine whether a transient fault will cause a bit flip in a register or not.

With the increasing chip complexity, we currently see a trend towards multi and many core architectures, as single core performance has not significantly improved in recent years. Since traditional bus-based communication architectures do not scale with the number of connected cores, Networks on Chip (NoCs) have been proposed as a new communication paradigm for such architectures to overcome these limitations [4]. Many features of large scale networks can be applied to NoCs as well; therefore, and due to their inherently redundant communication paths, NoCs allow tackling faults on many different network or abstraction levels. This makes NoCs an especially interesting candidate for a simulation approach, where high simulation detail is only provided when actually needed.

To achieve both low level fault simulation and fault evaluation in a system level context, we propose a novel adaptive simulation approach, which allows simulating individual subcomponents at gate level only when actually needed. For the remainder of the simulation, components are simulated at register transfer level to ensure a high simulation performance. While NoCs are an ideal candidate for adaptive simulation, our approach is not limited to the field of NoCs, and it can be applied to speedup simulations of almost any arbitrary Register Transfer Level (RTL) model, when gate level precision is not required all the time. For that reason, we not only demonstrate the applicability on the example of NoCs but also using a freely available microprocessor.

The rest of this chapter is organized as follows: Sect. 12.2 introduces relevant work in the field of adaptive simulation in general and fault simulation in particular. Section 12.3 identifies the targeted scope of this work and defines the problem statement. Section 12.4 details how adaptive simulation is enabled. Finally, Sect. 12.5 provides evaluation results and Sect. 12.6 concludes this paper.

12.2 Related Work

While this work does not directly cover fault simulation specifically, it aims to be the foundation for efficient fault evaluation at a system level, by providing means to change the abstraction level of a system at a subcomponent level at runtime. In this chapter, we cover existing work for both runtime adaptive simulation as well as fault simulations, either offering gate level precision or allowing fault evaluation in a system context.

12.2.1 Structural and System Level Fault Simulation

Detailed fault simulations, such as the evaluation of the performability of a degrading NoC switch [3] requires simulation at gate level precision. However, the evaluations are often limited to single module instances (i.e., a single NoC switch) as the complexity of gate level models prohibits examinations at a system level scope.

To evaluate the gate level faults in a system context without suffering performance penalties induced by full gate level simulations, Kochte et al. [9] integrate a gate level model, on which fault injection is performed, into a TLM-based system model. However, the gate level component(s) must be chosen statically in advance, rendering this method ineffective when faults in several subcomponents need to be simulated in one simulation run, as again the growing number of gate level components constitutes a prohibitive factor on simulation performance.

Fast system level fault evaluation is proposed in [2], which leverages the benefits of TLM. The necessary models are automatically generated on the basis of RTL models and generated TLM test patterns can be synthesized to RTL test patterns. However, being based on RTL models, the details of structural fault simulation are unavailable.

While providing dynamic switching between behavioral and gate level models, the sequential fault simulation [12] offers an analytical methodology to investigate fault propagation and cannot be used to evaluate system level methods for fault resilience. The dynamic switching automatically selects between behavior and gate level models, depending on whether a fault must be propagated or actually simulated. Thus the dynamic switching offers a different type of adaptivity than the one proposed in this work.

12.2.2 Adaptive Simulation

Adaptivity in terms of simulation accuracy has been proposed and employed in several different forms.

Aiming at software performance evaluation, SimOS [15] can perform simulations at different level of precision by switching hardware models at runtime. However, aiming at software evaluation SimOS does not offer a general adaptive simulation concept.

Hines and Borriello [8] propose an adaptive communication model, where communications can be either simulated at full detail by simulating all steps of the communication stack and using detailed interconnection models, or by using shortcuts and end to end communications at different levels of the communication stack.

In the field of transaction level modeling, adaptive simulation has been employed by using differently detailed timing annotations to trade off simulation performance for timing accuracy [13, 16]. Specialized on timing, this form of adaptivity is not suitable for fault simulations.

Another adaptive simulation methodology for transaction level models was introduced in [1], which multiplexes fixed abstraction models. Changing the level of abstraction is only possible at transaction boundaries. This work differs as we target adaptivity between register transfer and gate level models and enable abstraction changes at any given simulation cycle.

To the best of our knowledge, this is the first adaptive simulation concept that offers switching between gate and register transfer abstraction level at runtime on a subcomponent granularity. We also offer a generic solution as switching abstraction levels is not only working for combinational but also for sequential models, as we provide means to automatically transfer the state from one abstraction level to another.

12.3 Preliminaries and Problem Statement

In this work we focus on the adaptive simulation of RTL models and their synthesized gate level counterparts, where adaptivity is enabled in an automated, generic way. Both RTL and gate level models are used during the simulation, but at any given time, only one instance of abstraction level is active. While such adaptivity can be achieved seemingly easily by multiplexing models of the individual abstraction levels, there are several challenges that must be taken care of.

With the exception of purely combinational models, models are associated with a state, which changes during simulation. Since in adaptive simulation there is always only one active model, the state of the model must be transferred to the newly activated model whenever switching abstraction levels. This state transfer must be done in an automated, generic way to avoid modeling overhead and to reduce the possibility of implementation errors. Furthermore, it is desirable that the degree of expressiveness of the modeling language is constrained as little as possible, i.e., switching of abstraction level, including state transfer, should be possible for almost any given model. Additionally, since the total simulated model is larger in size than any individual model, it must be ensured that the increased model complexity does not affect simulation performance noticeably, i.e., the inactive model must be shut down in such a way that it does not reduce simulation performance.

Solving the aforementioned challenges lays down certain requirements on the simulator. To transfer the state from one model to another, it is not only necessary to have insight into arbitrary data structures of a model, but also to have the possibility to manipulate the current state of the simulation by altering the current state of individual models. In this work we use Mentor Graphics ModelSim, which meets the requirements with its Foreign Language Interface (FLI), a C-Library that can be used "to traverse the hierarchy of an HDL design, get information about and set the values of VHDL objects in the design, get information about a simulation, and control (to some extent) a simulation run" [11].

12.4 Adaptive Modules

The core concept of our adaptive simulation is an advanced form of mixed simulation, not only combining SystemC and VHDL models, but also by coupling SystemC models and ModelSim's simulation kernel in a unique way.

The centerpieces of the adaptive simulation are special SystemC models, called *Adaptive Modules*, encapsulating RTL and gate level VHDL models. Adaptive Modules perform input and output (de)multiplexing of the fixed abstraction level models and provide the necessary means to read and write the state of these models. The structure of an Adaptive Module is shown in Fig. 12.1. A complete simulation model can consist of many different Adaptive Modules and they are supported on arbitrary levels of the model hierarchy. This enables fine-grained control over the abstraction levels, while at the same time delivering a high simulation performance, as only the parts actually requiring high simulation accuracy are simulated using a low abstraction level.

12.4.1 Multiplexing Fixed Abstraction Level Models

The basis of an Adaptive Module provides a synthesizable RTL model, which is given in the hardware description language VHDL. For this model, a gate level counterpart is generated using hardware synthesis. In this work, we used Synopsys Design Compiler (DC) to synthesize the gate level model, which is then exported as a VHDL model. In combination with cell models of the targeted vendor library, this model can be used for a timing accurate simulation. Having both abstraction levels incorporated into one model allows selecting between fast simulation performance and high simulation precision at simulation time.

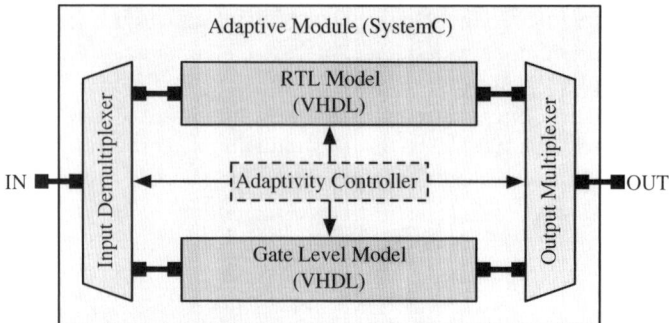

Fig. 12.1 Adaptive module structure

Whereas the fixed abstraction level models are provided in VHDL, the wrapping Adaptive Module is designed as a SystemC module. While multiplexing between the different submodules is possible in VHDL, it is the mixed simulation model that makes efficient adaptive simulation possible. Using SystemC, and thus C++, gives access to the full scope of ModelSim's Foreign Language Interface (FLI) by linking against the respective C-Library. This, in turn, provides Adaptive Modules with introspective capabilities giving insight into its fixed abstraction submodules and enabling the manipulation of their states.

To transfer the state between two similar models, it is required to know how the state is represented in each model and how to transform between the individual representations. In both RTL and gate level models, the state is stored in registers. While in RTL designs, registers are modeled using VHDL signals, possibly of aggregate data types, gate level models use flip-flop instances. Since a flip-flop only stores a single bit, multiple flip-flops might be necessary to store the content of a single RTL register. Therefore, a bijective mapping from RTL registers to gate level flip-flops is required to enable state transfer. During the elaboration phase of the simulation, this mapping is automatically generated by each Adaptive Module. Building the mapping is possible from within a SystemC model by using FLI functions for enumerating subregions and signals within a given region.

Creating the register map consists of two steps: First, an intermediate mapping between the individual subregions of the fixed abstraction models is built. This map of matching regions is necessary due to inconsistent region names between autogenerated regions in ModelSim and Synopsys DC and because of flattened region levels, which is especially the case for regions created using VHDL's `generate` statement. In the second step the actual mapping between RTL registers and corresponding gate level registers is established. For each matching pair of items in the region map, the gate level region is scanned for flip-flops. Here, flip-flops are instances of gate level models, which can be identified by their entity and architecture names, as given in the vendor library. Synopsys DC derives names for flip-flop instances based on the names of their corresponding RTL signals. E.g., an RTL signal named `value` of unconstrained integer type, would be synthesized into flip-flop instances named value_reg_0_inst to value_reg_31_inst. We use this naming convention to find the matching RTL signal and to identifiy the corresponding bit position inside the RTL signal in the case of array, scalar (integer) and enumeration types.

Note that for synthesizing the gate level model, the Synopsys DC command ungroup must not be used as it destroys the model's hierarchy and with that removes information necessary to establish the register mapping.

Input and output (de)multiplexers are used to select between the individual, fixed abstraction level models. Output multiplexers forward the values of the currently active model to the output ports of the Adaptive Module and suppress all events from the currently unused model. For this purpose, output multiplexers contain special, suspendable signals, derived from sc_signal. When disabled, these signals prohibit event generation when being written to. While the output multiplexers are needed for functional reasons, the input demultiplexers enable fast simulation performance. The input demultiplexer only forwards new values to the currently active module, and for the inactive model, the input ports are kept at the last active value, which ensures that no new events are generated inside the model and thus that the unused model is effectively deactivated.

For the input and output (de)multiplexers, using SystemC is also beneficial as it makes them not only more efficient, but also simpler. For each input and output port, one demultiplexer or multiplexer is needed, and the data type of the port can be arbitrary and is not known in advance. While in VHDL, that would require one (de)multiplexer definition (not only instance) per data type, using SystemC allows a generic design by using template classes for the (de)multiplexers having the data type of the port as a template parameter. For example, a VHDL outport of type std_logic corresponds to a SystemC sc_out<sc_logic>, which, in the Adaptive Module, corresponds to a OutputMultiplexer<sc_logic, 2>.[1]

12.4.2 3-Phase Switchover

Switching the abstraction level of an Adaptive Module consists of three distinct phases, which must be performed in a strict order to ensure a correct simulation outcome. In order, these phases are:

1. Switch the input demultiplexers to forward values to the newly activated model and to inhibit updates for the old one.
2. Transfer the state of the previously active model to the new model.
3. Switch the output multiplexers to forward values from the new model and to ignore updates from the old model.

When switching adaptivity without strictly adhering to this order, the external view of the Adaptive Module can deviate from that of a fixed abstraction level model. That is, signal toggles, caused by the switchover, can propagate to an output port before

---

[1] The value of 2 determines the number of multiplexer inputs.

reaching a stable state, thus leading to an erroneous behavior of the model. The reason for the switchover order is as follows: Fairly obviously, switching the output multiplexers must be performed as the last step. This ensures that until the newly activated model has reached a stable state, the still correct values of the previously active model are output and no glitch resulting in input switching or state transfer will be visible at the outputs of the Adaptive Module.

State transfer must be performed after switching the inputs to ensure that the state is not altered after transfer. Switching the inputs will cause new input values to be applied to the model, which in turn causes VHDL processes to run. If the inputs were switched after state transfer, this could cause an erroneous state. For example, imagine a simple counter model, incrementing its value with each clock cycle; if the counter has a value of N at the time of switchover and the clock input is connected after the state transfer, this can result in an unwanted increment of the counter, effectively setting a value of $N+1$. Therefore, the state must be transferred after input switching, and combined with necessity to switch outputs last, yields the aforementioned order.

12.4.3 State Transfer

Transferring the state between the models is efficiently done by copying the current register values from one model to the other, using the register map created during elaboration (cf. to Sect. 12.4.1). The process illustrated in Fig. 12.2, where the flow for transfer from RTL to gate level follows top-down direction and vice versa for gate

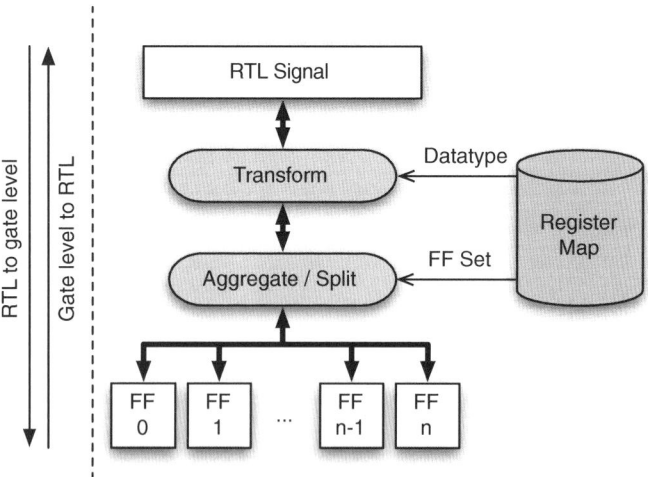

Fig. 12.2 State transfer between RTL and GL models. State transfer consists of a transformation and aggregation or splitting. The necessary meta information is provided by the register map, established during elaboration

level to RTL. When switching from RTL to gate level abstraction, the RTL values are first transformed into bitvector representations before they are split into the individual bit values, ready to be assigned to the individual flip-flops. In the other direction, gate level to RTL, flip-flop values are first aggregated to a bitvector, which is then transformed into the RTL data type. The necessary meta informtion for this process (data type, corresponding set of flip-flops) is readily available in the register map enabling an efficient state transfer.

Due to the similar data types employed in RTL and gate level modeling, the transform stage is optional and can be skipped in many cases. If the RTL value is based on a data type used for digitally encoded values, the transform stage is bypassed and the values are directly copied to the flip-flops. This is the case for the data types `std_logic`, `std_logic_vector`, `signed`,[2] `unsigned`,[2] and any of their subtypes.

For scalar and enumeration types, the transformation is mandatory. In gate level, scalar data types are encoded using two's complement for `integer` type and simple binary encoding for the types `natural` and `positive`, which do not include negative values. Unfortunately, the ModelSim FLI does not differentiate between those individual scalar types, which makes it impossible to determine whether a value is encoded using two's complement or not. For example, in gate level, a value of 1111_2 could either represent a value of -1, if the corresponding RTL type is `integer`, or 15, if the RTL type is `natural` or `positive`. To overcome this limitation of the FLI, we perform transformation of scalar types by applying bit manipulation operations on the C data type, representing the RTL value. For example, when transforming from RTL to gate level, a bit mask is shifted and applied to the RTL value, to detect which bits have been set. The transformation from gate level to RTL is handled similarly.

To transfer the value of enumeration types, which are commonly used for state machines, the encoding of the individual states must be known. Fortunately, Synopsys DC encodes states linearly according to the order they have been declared in. Also, Synopsys DC does only little optimizations to the state encoding: there is no reordering of states, no state reduction due to equivalence, and states that can never be reached are only removed to save flip-flops if they are at the end of the state declaration list. Example, assume a type `state_t` having states `s0`, `s1`, `s2`; if only `s0` and `s1` are used, only one flip-flop is used to encode the state, but two flip-flops are synthesized even if only `s0` and `s2` are used. Due to the linear encoding and the minimal optimizations, a fixed bijective relation between abstract RTL values and bit-encoded gate level values exists, which enables state transfer of enumeration types in a general way.

---

[2] Like `std_logic_vector`, `signed` and `unsigned` types from package `ieee.numeric_std` are arrays of `std_logic`. However, they have added numerical interpretation. `signed` and `unsigned` types are not to be confused with scalar types `integer`, `natural` or `positive` types!

12.4.4 Controlling the Level of Abstraction

While the Adaptive Modules provide the means for input and output switching as well as state transfer, the actual execution of the 3-phase switchover protocol, described in Sect. 12.4.2, is not carried out by themselves. Between every two steps of the 3-phase switchover, we have to ensure that all events generated by a previous phase have been processed before continuing with the next phase. These pending events are processed by a centralized instance, called Adaptive Module Manager (AMM). This not only allows efficient abstraction switching of multiple Adaptive Modules at the same time but also ensures that the proper method to process pending events is selected and the same for all Adaptive Modules. The latter is necessary as our adaptive simulation approach supports two types of switching the abstraction level. These are *interactive switchover* and *on-the-fly switchover* and are described in detail in the remainder of this section.

12.4.4.1 Interactive Switchover

The purpose of interactive switchover is to allow users controlling the abstraction level of the individual Adaptive Models using ModelSim's Command Line Interface (CLI) and to enable abstraction switching in scripted simulation runs.

The ModelSim FLI allows exporting custom C-functions to ModelSim's CLI augmenting the simulation with new functionality. However, such C-functions do not have any knowledge of C++ objects and therefore cannot call methods of Adaptive Modules, which, in turn, prohibits accessing Adaptive Modules via the CLI. To overcome this limitation, the Adaptive Module Manager is implemented as a singleton [7], which enables global access to its instance, including plain C-functions exported to the CLI. And since all Adaptive Modules automatically register with the manager, the AMM can forward calls to individual Adaptive Modules. For the purpose of abstraction switching, the manager provides a static method `SwitchAdaptiveModule(...)`, which is directly exported to the CLI. From there, it can be called as `SwitchAdaptiveModule <AM_NAME> <ABSTR_LVL>` having the name of an Adpative Module and the desired abstraction level as parameters. The effective call sequence when accessing an Adaptive Module via the CLI is shown in Fig. 12.3. Note that the FLI acts as a bridge connecting simulation control and the simulation model. This, in combination with the singleton AMM, exposes control over the individual Adaptive Modules to the CLI, which otherwise would not be possible.

When `SwitchAdaptiveModule` is called, all currently pending events must be processed before starting with the first phase of the actual switchover. Since CLI functions can only be called when the simulation is currently halted or paused, processing the pending events means advancing the simulation accordingly. Unfortunately, the FLI does not provide any function to run the simulation. However, the FLI can execute CLI commands, and we use this feature to call `run -all`, which

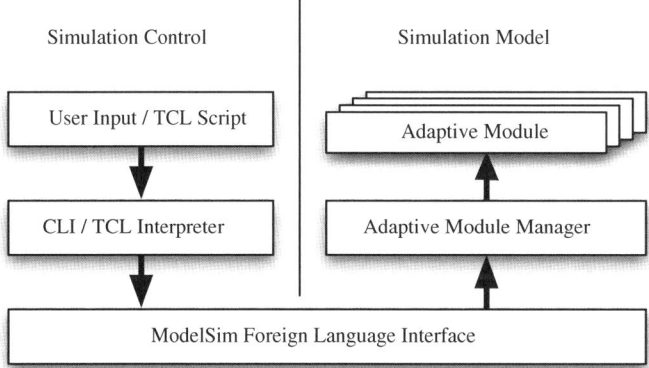

Fig. 12.3 Call sequence for interactive control over adaptive modules

advances the simulation until all currently pending events have been processed. After all events have been processed, the AMM sequentially executes the three phases, interleaved by calls to `run -all` to ensure that all newly generated events have been processed. To efficiently change the abstraction level of several Adaptive Modules at the same time, the AMM keeps track of the phase an Adaptive Module is currently in. This allows different Adaptive Modules to be in different switchover phases at the same time, which reduces the number of calls to `run -all` and speeds up abstraction switch of multiple modules.

While interactive switchover is useful for manual simulation runs and evaluation, ModelSim also allows scripted execution of simulations in terms of `.do`-files using the interpreted language TCL. All CLI commands, including custom commands exported using the FLI, can be used in TCL scripts to control the flow of a simulation. This enables an automated simulation run with script-controlled adaptivity. With that, interactive switchover can be used to perform fault simulation at gate level. Here, Stuck-At faults are injected using the CLI command `force`, which overrides any value applied by a regular driver of the signal with a fixed value. An exemplary script for the simulation of a transient fault is shown in Listing 12.1. First the system undergoes a warmup phase at register transfer level. Then, to simulate the transient fault the abstraction level of a submodule is changed to gate level, which allows fault injection by forcing a specific signal to logical 1. The simulation continues to evaluate the effect of the fault before removing the injected fault and bringing the submodule back to RTL.

12.4.4.2 On-The-Fly Switchover

On-the-fly switchover enables models to programmatically change the abstraction level online during a running simulation with triggers based on observered or timed events. On-the-fly switchover increases the range of applications for adaptive

Listing 12.1 Example use case: TCL script for transient fault simulation

```
1   # warm up system at RTL
2   run 10 us
3   # switch /dut/inst to gate level
4   switchAdaptiveModule /dut/inst GATELVL
5   # inject fault
6   force /dut/inst/signal -freeze 1
7   # perform fault simulation
8   run 20 ns
9   # remove fault
10  noforce /dut/inst/signal
11  # return /dut/inst to RTL
12  switchAdaptiveModule /dut/inst RTLLVL
13  # continue with simulation
14  run 10 us
```

simulation, as interactive switchover can only be performed when the simulation is not running. However, because any observerable condition can be used to trigger abstraction switching, defining and testing the conditions is highly model-specific and therefore cannot be done in a generic way. Thus, triggering automatic changes of abstraction level requires user implemented functionality to check for the occurence of the conditions of interest, e.g., test for a specific input or output bit pattern.

If the condition of interest is observable at the input or output ports, a simple SC_METHOD, sensitive to the respective ports, can be added to the Adaptive Module. The method has to check for the occurence of the condition and, in the case of a match, call the Adaptive Module Manager's SwitchOnTheFly method to trigger an online change of abstraction level. If the condition is not observable at the input or output ports, the checking method can be implemented as an SC_THREAD, which periodically checks the internals of the model for the occurence of the condition. Since the SystemC model has no immediate insight into the fixed abstraction models, the checking method has to rely on FLI functions, which provides the Adaptive Module with introspective capabilites. This makes it possible to test for any arbitrary condition inside both RTL and gate level models. Again, when the triggering condition is met, the SC_THREAD calls the AMM's SwitchOnTheFly method taking care of the switchover.

Since on-the-fly switchover happens while the simulation is running, pending events cannot be processed by simply calling run -all, as it is handled in the interactive case. But since the simulation is running anyways, the AMM simply queues the Adaptive Module for switchover and waits until all events for the current simulation time have been processed before it starts executing the individual switchover phases. Similarly to interactive switchover, the succeeding phases must not be executed until all pending events have been processed, and thus the AMM waits for these events to be processed.

Unfortunately, neither the FLI nor any CLI command offer the possibility to detect whether there are pending events or not. Therefore, the only way to wait for events to be processed is by waiting a fixed amount of delta cycles. This timeout is highly model dependent, and chosing a timeout value that is too short can lead to an erroneous switchover possibly voiding the simulation outcome. On the other hand, too long timeouts lead to unneeded additional delta cycles. Fortunately, experiments showed, that this overhead does not impose a significant performance penalty, which is most likely because there are almost no actions to be performed during these unnecessary delta cycles.

While online adaptivity changes come at the price of additional implementation efforts, they are also a powerful tool to get detailed insight into a complex model's behavior without suffering the low simulation performance of a full gate level simulation. For example, on-the-fly switchover allows conditional fault injection to observe the behavior under given faults only for specific packets. Another possible use case is tracing a packet at gate level through a full NoC model, where switches not currently transporting the packet are simulated at RTL ensuring a high simulation performance. However, the field of application is not limited to these examples, and custom implemented trigger functions can be used to speed up nearly any simulation scenario, where gate level simulation is required only in certain situations.

12.4.5 RTL Modeling Requirements and Simulation Limitations

By and large, we achieved our goal of enabling adaptive simulation for any arbitrary given RTL model. However, RTL models, where the state is stored in a variable of a VHDL process, prohibit adaptive simulation. If other signals or outputs of the model depend on this variable, it is necessary to run the process, to which the variable belongs to, in order to propagate the changed value of the variable. While it is generally possible to force a process run via the FLI, such state carrying variable can only occur in sequential processes, which have a control flow that possibly prohibits a proper process run. In VHDL, sequential processes active on the rising edge of a clock signal `clk` are usually modeled similiar to the example shown in Listing 12.2. If the switchover is performed at a time when `clk = 0`, the if clause will not be satisfied and the branch will not be taken. On the other hand, if a model is deactivated and activated again, when at both times `clk = 1`, then the condition of the if clause again will not be satisfied since the old and current value of `clk` are the same and thus `clk'event` will evaluate to `false`. As a result, the part of the process responsible for handling the variable will never be executed, and thus the depending signals and outputs will not be properly updated.

Instead, it is recommended to use a clean design methodology, such as the *two-process* design method [6], where combinational logic and sequential logic are split into separate processes. While such design methods are generally recommended for the design of synthesizable models, such methods also ensure that the state is never stored in VHDL variables and help identifying registers at design time.

Listing 12.2 Sequential Process in VHDL

```
1  process (clk) begin
2    if clk'event and clk = '1' then
3      ...
4    end if;
5  end process;
```

There is also a minor limitation when using timing-accurate gate level models. In such models, it is not possible to switch from gate level to RTL abstraction immediately at a simulation time of a positive clock edge. On positive clock edges, registers will latch the value currently applied to their inputs. However, a timing accurate simulation respects setup and hold times of registers as well as wire delays. Therefore, as opposed to an RTL model, a gate level register will not output the new value of the next cycle until the setup time of the register has passed. As long as only registers triggering on the positive clock edge are used, a safe time to switch from gate level to RTL is at the negative clock edge. However, this limitation can be simply overcome by delaying the switchover until the register's output has settled for the new value. Note that it is still possible to switch abstraction levels in any given cycle.

12.4.6 Building an Adaptive Module

Building an Adaptive Module is an automated task, for which we wrote a small program in Python. Mixed-language simulation with VHDL entities inside SystemC modules requires special SystemC stub modules matching the interface of the VHDL models. These stub modules can be generated by ModelSim using the scgenmod command. We use the output of this command to automatically build an Adaptive Module. The generated SystemC module derives from a base class providing the state transfer logic, including register map generation, and it instantiates all necessary input and output (de)multiplexers. If necessary, the generated Adaptive Module can be extended according to individual needs, e.g., by implementing methods to perform on-the-fly switchover (cf. to Sect. 12.4.4.2).

12.5 Evaluation

Evaluation of our adaptive simulation approach was carried out in two ways: First, we evaluated the general applicability of our simulation model to a freely available microprocessor model; and second, we used a NoC model, implemented by ourselves, to evaluate performance benefits of adaptive simulation on a submodule level.

12.5.1 General Applicability of Adaptive Simulation

In Sect. 12.3, we identified the goal to have a generic approach for adaptive simulation, which can be applied to any existing model. To back this claim, we applied the methods developed in this work to the freely available microprocessor model Plasma [14], a System on a Chip (SoC) using a MIPS I microcontroller, which has been published in the public domain as an opencores project. However, we had to make two minor modifications to the Plasma SoC. These modifications were necessary to make Plasma properly synthesizable with Synopsys DC, as it was designed for usage with Field-Programmable Gate Arrays (FPGAs). For on-chip memory, the SoC instantiated hard blocks of random access memory, available in FPGAs. Obviously, these FPGA-specific memory blocks are unavailable in general synthesis tools, and we had to replace them with a synthesizable memory model. The second change was necessary because the register file was modeled in such a way that it relied on signal initialization, which is not supported by Synopsys DC, instead of initializing the register file on reset. Note that while we made changes, these changes were only necessary to make the SoC synthesizable and to make the synthesized model match the RTL model. The changes were not caused due to a lack of support by our adaptive simulation.

The gate level model of Plasma was synthesized for the `LSI_10K` target library. Based on the SystemC stub, generated using ModelSim's `scgenmod`, the corresponding Adaptive Module of Plasma was generated using the Python program described in Sect. 12.4.6. The Adaptive Module was then instantiated in the test bench that comes with the Plasma SoC. We also instantiated a fixed abstraction, pure RTL model of Plasma to check the correctness of the Adaptive Module. The simulations were carried out using ModelSim in version 6.6d, running on a simulation platform featuring an Intel® Core™ 2 Quad processor and 8 GB of main memory. We ran the simulation for 50 μs and changed the abstraction level of the Adaptive Module at various points in time. The outputs of the two models were compared on every clock edge. At no time, there was a difference in the output, which proves the correctness of the Adaptive Module and thus the general applicability of our adaptive simulation approach.

We also carried out initial performance evaluation of our adaptive simulation using Plasma. Again, the test bench provided by Plasma was used, this time without the overhead of checking the outputs against a reference model, i.e., only a single Plasma instance was used. We used the CPU time spent by the simulation kernel as performance metric. The results are summarized in Table 12.1. First, a reference baseline is established using two fixed abstraction, pure VHDL models. We then repeated the simulation using the adaptive model, setting the abstraction level to RTL or GL at the start of the simulation and no abstraction switching to measure the overhead of adaptive simulation. In the pure RTL case, the adaptive simulation imposes a significant overhead, having a CPU time of 1.623 s compared to 0.02 s in the pure VHDL case. However, further investigations showed, that the overhead in RTL does not depend on the simulation duration, as running the same scenario for 500 μs

Table 12.1 CPU time required for a 50 μs simulation of Plasma [14]

Model type	Abstraction level	CPU time (s)
Fixed RTL Model	—	0.02
Fixed GL Model	—	606.04
Adaptive Model	100 % RTL	1.62
Adaptive Model	100 % GL	616.31
Adaptive Model	20 % RTL, 80 % GL	467.06

only increased the CPU time to 1.9 s. Moreover, in the pure gate level scenario the overhead is negligible (about 1 %), where the CPU time increased from 606 to 616 s. Since the idea of adaptive simulation is to use—at least to some degree—the low gate level abstraction, the overhead pays off even when the RTL abstraction is only used for a short time.

For an example of adaptive simulation, we first ran the simulation for 10 μs at RTL abstraction level, and then simulated the remaining 40 μs at gate level. In this case, the consumed CPU time was only 467 s, which is about 77 % of a full gate level simulation (606 s). The super-linear CPU time reduction[3] is possible because the CPU time required to simulate a time step at gate level depends on the current switching activity, which can change throughout a simulation run. This supports that adaptive simulation is also beneficial if it is only used to warm up a simulation model, e.g., until a boot up sequence has finished.

12.5.2 Performance Evaluation

To evaluate the performance of our adaptive simulation, especially on a sub-module level, we designed an adaptive NoC model, based on a self-designed VHDL switch. The model consists of 16 adaptive switches aranged in a 4 × 4 mesh topology. For the switches, we used a basic model, featuring XY-routing and wormhole switching [5]. Again, the corresponding gate level model was synthesized for the `LSI_10K` target library using Synopsys DC. Figure 12.4 shows the general structures of the adaptive NoC model (left) and the adaptive switches (right). In addition to the four ports shown (left, right, top, bottom), every switch has an additional local port connected to a Processing Element (PE), which is responsible for generating and receiving test data. The PEs are generating synthetic network traffic by sending a fixed message to a predefined destination PE repeatedly. Destination PEs are selected in a transpose pattern.

The simulation environment is the same as in the previous section, and performance metric is again CPU time spent by the simulation kernel. All simulations were run for 2,500 simulated cycles.

---

[3] 77 % CPU time compared to 80 % gate level simulation time.

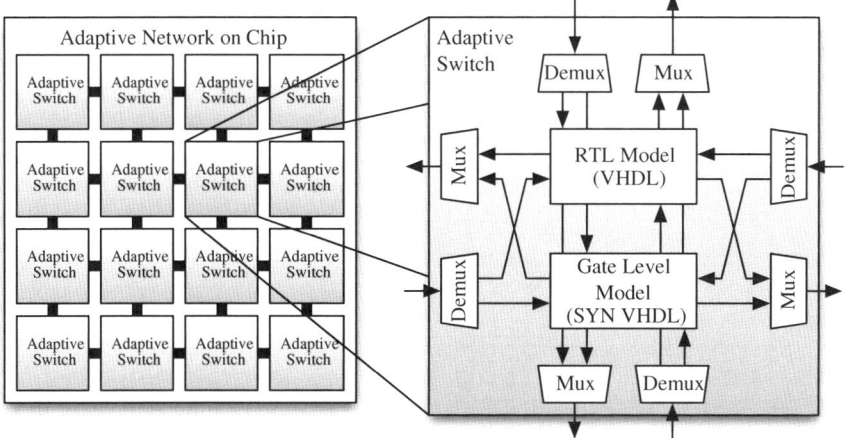

Fig. 12.4 Adaptive NoC model of 4 × 4 mesh topology and structure of an adaptive switch

First, we evaluated the performance, when only a part of the model is simulated at gate level. For that purpose, we performed simulations with different number of switches simulated at gate level. Each switch was fixed at either gate level or RTL throughout the whole simulation, and the abstraction level was never changed during a simulation run. Simulations were repeated with $gl = 0$ to 16 switches at gate level, and the results are shown in Fig. 12.5. As expected, the simulation duration increases nearly linearly with the number of switches simulated at gate level. This shows that adaptive simulation is helpful even if only a part of a model needs to be simulated at lower abstraction levels. While this can also be achieved by using fixed abstraction instances of RTL and gate level models, the adaptive simulation has the advantage that the models to be simulated at gate level can be selected at runtime and there is no need to change the simulation model. For example, this can be helpful when evaluating fault tolerance methods for different fault scenarios. Based on these measurements, we can derive the average simulation times for a single switch at RTL and gate level as approximately 0.3 and 55 s respectively. The large ratio of these simulation times confirm that the gate level models are properly shut down when simulating at RTL and that overhead of adaptive simulation is negligible.

We also evaluated how adaptive simulation can reduce simulation times when gate level precision is only required for a short period of time, but repeatedly throughout the simulation run and for different submodules. For this purpose, we evaluated the performance of a transient fault simulation. While the general simulation was performed at RTL abstraction, the simulation of transient faults requires gate level precision for a short time. Therefore, we periodically changed the abstraction level of one switch to gate level precision, and after one cycle switched back to RTL. This time is sufficient to evaluate whether a transient fault affecting a certain wire will manifest in a flipped register bit or not. We adaptively changed precision to gate level every n cycles and repeated simulations with doubling n between simulation

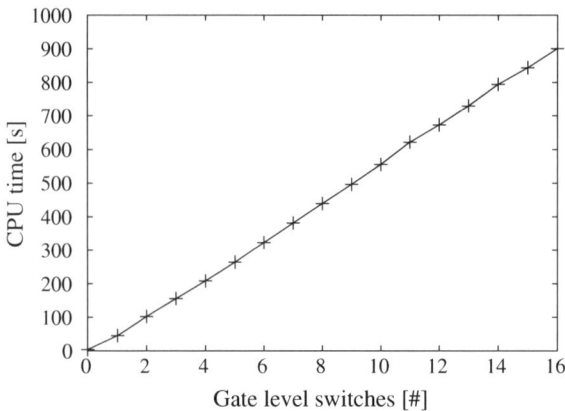

Fig. 12.5 Simulation performance for fixed number of NoC switches at gate level. No adaptivity

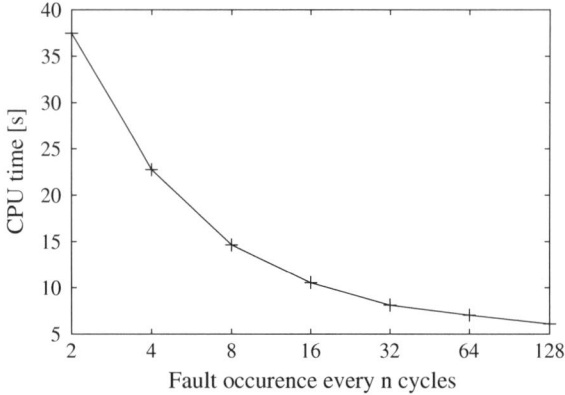

Fig. 12.6 Simulation performance for transient fault simulation. One NoC switch is changed to gate level precision every *n* cycles for a duration of 1 cycle

runs, *n* ranging from 2 to 128. The performance of the transient fault simulation is shown in Fig. 12.6. It can be clearly seen that adaptive simulation pays off even for very high transient fault probabilities. Especially in the case of the lowest fault probability ($n = 128$), the consumed CPU time (6.1 s) was only about 50 % higher than the CPU time of a pure RTL simulation (4.1 s). Note that without adaptive simulation, a full gate level simulation would be required, which has a CPU time of 900 s (cf. to Fig. 12.5). This yields a speedup of 150 over traditional, fixed abstraction simulation.

12.6 Conclusion and Future Work

We presented a novel methodology for adaptive simulation allowing to change the simulation accuracy between register transfer level and gate level online during the simulation. The fine-grained adaptivity at submodule level makes it possible to simulate only the parts actually needed at gate level while keeping the rest of the model at register transfer level to ensure a high simulation performance.

For this, we introduced Adaptive Modules taking care of abstraction switching in an automated way. The necessary state transfer is enabled by an automatically generated mapping between RTL registers and synthesized flip-flops. To generate this mapping and to copy the state from one model to another, we incorporated usage of ModelSim's Foreign Language Interface into the adaptive SystemC wrapper models in a novel way giving limited access to ModelSim's simulation kernel.

The general applicability was verified using Plasma [14], a freely available microprocessor. And our evaluations show the benefits of adaptive simulation, whenever gate level precision is required only for a part time of the simulated duration. Speedups of a factor of 150 compared to full gate level simulation highlight the applicability of our approach especially in the context of transient fault simulation, where arbitrary parts of the simulation model are required at gate level repeatedly for a short duration each time.

For future work, we intend to incorporate additional abstraction levels into the Adaptive Modules including TLM-based models.

Acknowledgments This work has been supported by the German Research Foundation (Deutsche Forschungsgemeinschaft, DFG) under grant Ra 1889/4-1.

References

1. Beltrame G, Sciuto D, Silvano C (2007) Multi-accuracy power and performance transaction-level modeling. IEEE Trans Comput Aided Des Integr Circuits Syst 26(10):1830–1842
2. Bombieri N, Fummi F, Guarnieri V (2011) Accelerating RTL fault simulation through RTL-to-TLM abstraction. In: Proceedings of the European test symposium (ETS) 2011, pp 117–122
3. Dalirsani A, Holst S, Elm M, Wunderlich H (2011) Structural test for graceful degradation of NoC switches. In: Proceedings of the European test symposium (ETS) 2011, Trondheim, pp 183–188
4. Dally W, Towles B (2001) Route packets, not wires: on-chip interconnection networks. In: Proceedings of the IEEE/ACM design automation conference (DAC) 1999, Las Vegas, pp 684–689
5. Flich J, Skeie T, Mejia A, Lysne O, Lopez P, Robles A, Duato J, Koibuchi M, Rokicki T, Sancho J (2012) A survey and evaluation of topology-agnostic deterministic routing algorithms. IEEE Trans Parallel Distrib Syst 23:405–425
6. Gaisler J (2004) A structured VHDL design method. http://www.gaisler.com/doc/vhdl2proc.pdf
7. Gamma E, Helm R, Johnson R, Vlissides J (1994) Design patterns: elements of reusable object-oriented software. Addison-Wesley, Boston

8. Hines K, Borriello G (1997) Dynamic communication models in embedded system co-simulation. In: Proceedings of the IEEE/ACM design automation conference (DAC) 1997, Anaheim, pp 395–400
9. Kochte M, Zoellin C, Baranowski R, Imhof M, Wunderlich H, Hatami N, Di Carlo S, Prinetto P (2010) Efficient simulation of structural faults for the reliability evaluation at system-level. In: Proceedings of the Asian test symposium (ATS) 2010, pp 3–8
10. Kohler A, Radetzki M (2009) A SystemC TLM2 model of communication in wormhole switched networks-on-chip. In: Proceedings of the forum on specification and design languages (FDL) 2009, Sophia Antipolis, pp 1–4
11. Mentor Graphics® Corp (2010) ModelSim® SE Foreign Language Interface Manual. Version 6:6d
12. Navabi Z, Mirkhani S, Lavasani M, Lombardi F (2004) Using RT level component descriptions for single stuck-at hierarchical fault simulation. J Electron Test 20(6):575–589
13. Radetzki M, Salimi Khaligh R (2008) Accuracy-adaptive simulation of transaction level models. In: Proceedings of the design, automation and test in Europe conference (DATE) 2008, Munich, pp 788–791
14. Rhoads S (2013) Plasma-most MIPS I™ opcodes. http://opencores.org/project,plasma
15. Rosenblum M, Herrod S, Witchel E, Gupta A (1995) Complete computer system simulation: the SimOS approach. Parallel Distrib Technol: Syst Appl 3:34–43
16. Salimi Khaligh R, Radetzki M (2010) Modeling constructs and kernel for parallel simulation of accuracy adaptive TLMs. In: Proceedings of the design, automation and test in Europe conference (DATE) 2010, Dresden, pp 1183–1188

Part V
Model-Driven Engineering
for Embedded System Design

Chapter 13
Model-Based Design of Real Time Embedded Application Reconfiguration

Mouna Ben Said, Yessine Hadj Kacem, Nader Ben Amor, Mickaël Kerboeuf
and Mohamed Abid

Abstract Maximizing the system output quality under resource constraints presents an inherent challenge in the design of RTES. To deal with this issue, scaling the application quality level through algorithmic or parameters tuning is an interesting adaptation mechanism since it permits to handle the complexity of modern embedded applications. Unfortunately, this adaptation mechanism is still underexplored by existing model-based design approaches. It is also not supported by the UML MARTE profile. Therefore, we propose in this chapter a model-based design of application reconfiguration using the MARTE standard. We define an additional package extending the Software Resource Modeling sub-profile. Then, in order to promote reusability of our proposed extension and facilitate its use by non-experts, we exploited it in the definition of a design pattern for an adaptation RTES decision making process.

Keywords Model driven engineering · UML/MARTE profile · MARTE extension · Real-Time Embedded System (RTES) · Monitor Analyze Plan Execute (MAPE) adaptation loop · Fine-grain adaptation · Application reconfiguration · Quality level · Advanced Video Coding (AVC) video encoder · Design pattern · Model reuse · Decision making

M.B Said (✉) · Y.H Kacem · N.B Amor · M. Abid
CES Laboratory, University of Sfax, ENIS, Soukra km 3, 5 B.P.: 1173-3000, Sfax, Tunisia
e-mail: mouna.ben-said@ceslab.org

Y.H Kacem
e-mail: yessine.hadjkacem@ceslab.org

N.B Amor
e-mail: nader.ben-amor@ceslab.org

M. Abid
e-mail: mohamed.abid@ceslab.org

M. Kerboeuf
University of Brest, MOCS Team, Lab-STICC, Brest, France
e-mail: kerboeuf@univ-brest.fr

© Springer International Publishing Switzerland 2015   245
M.-M. Louërat and T. Maehne (eds.), *Languages, Design Methods,*
and Tools for Electronic System Design, Lecture Notes in Electrical Engineering 311,
DOI 10.1007/978-3-319-06317-1_13

13.1 Introduction

A self-adaptive system is a system that is able to change its structure or behavior at run-time in response to the execution context variations and according to adaptation engine decisions [14]. The design of adaptive embedded systems presents many challenges due to the complexity of the problem it handles. A common basic challenge is optimizing system non-functional properties (e.g., maximizing output quality) while meeting internal and external constraints (e.g., real-time constraint). For example, a high quality of service may require a high utilization of system resources, such as CPU cycles and memory space, and implies high energy consumption. Numerous approaches for the development of self-adaptive Real-Time Embedded System (RTES) have been proposed in the literature. They tackle various layers of the system, ranging from the application to the hardware, with different adaptation scope granularities and different performances [19, 20]. The software adaptation is fast but performed locally, while the hardware adaptation is complex but has more global effect. Thus, adaptation needs to be integrated at different system levels, and with different granularities (fine and coarse) to yield considerable benefits in resource utilization.

Due to the hardness of self-adaptive systems development at a low abstraction level, designers have resorted to high level design methods [23] based on the Model-Driven Engineering (MDE) paradigm. Using MDE [22] with the Unified Modeling Language (UML) is becoming a promising solution to decrease the complexity of RTES design via UML profiles. The recent profile Modeling and Analysis of Real-Time and Embedded systems (MARTE) [13] comes to provide a rich terminology for the specification and analysis of RTES. High level existing design methods of adaptive RTES are restricted to the modeling of coarse-grain adaptation techniques which bring modification to the whole system configuration. A typical modification is an allocation scenario of the software part on the execution resources. However, modern applications, typically multimedia ones, are increasingly complex and require more and more high computational capacity that may exceed existing systems capacity. Fine-grain application adaptation permits to manage this complexity by scaling the applications output quality according to context variations and resources availability. It is therefore an important capability of embedded systems that has been proven to be beneficial but is unfortunately still under-explored. Indeed, the MARTE standard offers features to model global adaptation but does not offer explicit semantics to support local application adaptation. Moreover, there is still a lack of reusable designs that are sufficiently generic to fit different systems and permit to fasten the designer task.

The main contribution of this work consists in extending the MARTE standard in order to cover the lack of fine-grain application reconfiguration support and permit the modeling of more complex adaptation approaches through hierarchical solutions yielding both local and global adaptations. We also exploit our extension in the definition of a generic model, taking the form of a design pattern [8], for the design of an adaptation decision process for real-time and embedded systems. The proposed pattern enables experts of adaptive RTES to get fast and reusable designs of fine-grain reconfigurable systems. It also gives them the liberty to insert additional details that are specific to their case studies.

The remainder of this paper is organized as follows. In Sect. 13.2, we give some useful definitions and details concerning the application adaptation. In Sect. 13.3, we briefly review some related works that address MDE-based software adaptation design for real time embedded systems. We present in Sect. 13.4 a targeted overview of the UML MARTE profile, where we outline its capabilities and limitations regarding RTES reconfiguration modeling. Section 13.5 describes our proposed extension of MARTE and presents a case study illustrating the use and importance of this extension. In Sect. 13.6, we show how the proposed reconfiguration package is used for the development of an adaptation decision design pattern. Finally, we conclude the paper and present future works in Sect. 13.7.

13.2 Self-Adaptation in RTES

In order to ease the understanding of the contributions presented in this chapter and keep the paper self-contained, the underlying concepts of self-adaptive systems are briefly reviewed in the following subsections.

13.2.1 The MAPE Adaptation Loop

Self-adaptation can be conceived in different ways depending on various aspects such as target platform, application domain, adaptation goals, users' requirements, system constraints, context changes, adaptation mechanisms, targeted system layers, adaptation scope, and many others [1]. However, there is a common structure of the adaptation mechanism that a self-adaptive system embodies. It is an adaptation loop referred to as the Monitor, Analyze, Plan, Execute (MAPE) loop [10]. It is composed of sensors, effectors, and four basic modules, which are the monitoring, analyzing, planning (or deciding), and executing (or acting). Figure 13.1 presents a bloc diagram illustrating the global structure of a self-adaptive system.

The entities forming the MAPE loop are described as follows:

- *Sensors* collect data about the status of the system and its environment.
- The *Monitoring module* processes the collected data to decide about relevant changes and then trigger change events.
- The *Analyzing process* examines the received events to detect if an adaptation is required. It can also identify the source of the change. Monitoring and Analyzing processes stand for all forms of observation and evaluation of systems' execution such as performance monitoring, safety inspection and constraint verification [14].

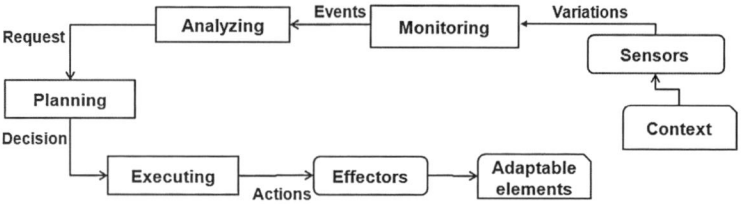

Fig. 13.1 Global structure of a self-adaptive system

- The *Planning process* generates an adaptation decision which specifies what elements to change and how to change them in order to best meet system requirements. Two common approaches are used in the literature to construct Decision makers: rule-based approaches and intelligent approaches. The second approach does not fit the real-time and embedded systems domain because of its requirements in terms of computing time. Adaptation decision mechanisms can be classified in two categories [12]: Parameters tuning and compositional. The former reconfigures the application through parameters' values or algorithms modification to scale the application quality level. It is an application-specific adaptation, called fine-grain adaptation that is performed locally on the application. However, the latter reconfigures the system by modifying its software/hardware allocation or changing its structural components to improve its outcome. It is an application-independent adaptation, called coarse-grain adaptation that is applied globally for the whole system in case of reconfigurable architectures (like FPGA-based architectures).

  In this work, we are interested in the fine-grain application reconfiguration mechanism which is described in the next subsection.
- The *Executing module* applies the decision to the system. It maps actions to effectors' interfaces.
- An *effector* is related to an adaptable system element and is responsible for applying adaptation actions to it.

13.2.2 Application of Reconfiguration-Based Decision Making

The fine grain application adaptation mechanism consists in dynamically reconfiguring the running applications to scale their output quality levels in order to meet resource allocations. The reconfiguration consists in algorithmic or parameters modification. To do that, each reconfigurable application has a set of modes (i.e., algorithmic versions), each yielding a different quality level. Each algorithm or parameters combination provides, for the same task, different values of non-functional properties, such as execution time, and different output quality such as video quality. The set of non-functional properties and the corresponding output quality define an application quality level that we also call Q-level. Q-levels range from <highest quality/highest complexity> to <lowest quality/lowest complexity>. The more we dispose of algorithmic versions, the more we can adapt efficiently. This adaptation decision mechanism fits very well many multimedia applications, such as video codecs (H264, MPEG), 3D object synthesis and cryptography application, which are highly configurable.

13.3 Related Work

Due to the increased complexity of embedded applications, researchers have resorted to high abstraction modeling methods to decrease the design complexity of embedded systems. In the present research, we are interested in application adaptation modeling for real-time embedded systems using UML/MARTE profile. Multiple works in the literature have dealt with embedded systems modeling and have shown the benefits of using the model-based approach and the UML/MARTE standard [5]. However, adaptive embedded systems modeling is a recent research area.

Most of existing works on adaptation modeling [3, 11, 17, 26, 27] were interested in the modeling of reconfigurable systems and especially in dynamic and partial reconfiguration. For example, in Rafiq Quadri et al. [17], a SoC co-design approach for reconfigurable systems modeling is developed under the GASPARD [7] framework. The reconfigurable system is modeled using a mode automata composed of a Mode Switch Component and a State Graph. In this work, the software application is transformed from high abstraction model into a hardware accelerator, which is then considered as a reconfigurable region with several implementations. A model-based approach for software reconfiguration in Distributed Real-time Embedded (DRE) systems is also proposed in Krichen et al. [11]. The authors propose a solution for global adaptation of a DRE system by reconfiguring it using a non-predefined set of configurations, which are dynamically captured using mode structure concept. They present a MARTE and AADL inspired meta-model, which is a combination of model and component paradigms. A configuration is described by a set of structured components, connections between them, their configuration, and their allocation on the execution supports. This work adds valuable extension to the support of software adaptation in embedded systems using UML, MARTE, and AADL. However, a configuration only describes the software part allocation on execution supports. Additionally, the authors used MARTE reasoning in dealing with the reconfiguration issue, but they do not reuse or make a link between the proposed meta-model and the useful reconfiguration-related semantics already defined in the standard.

The common downside of the cited approaches is that they are limited to global adaptation. Coarse grain adaptation requires, at every event occurrence, to reconfigure the whole system by changing its operational mode, which is time and power consuming. Additionally, modes are often defined as black boxes with a simple name and an associated allocation scenario. There is no detail about the actual effect of the mode on resource consumption and output quality. These are important facts that have to be taken into account in modes modeling. To the best of our knowledge, fine grain adaptation modeling on the application level using the MARTE standard is not tackled in literature. The notion of application Q-level is also absent. Moreover,

MARTE does not offer explicit semantics to support local application adaptation. It only offers features to model global adaptation. What makes the present work different is that we extend the MARTE standard to permit fine grain adaptation modeling by specifying an adaptive application with per-application Q-levels and configuration.

All the previously described approaches are beneficial since they facilitate and fasten the development of adaptive systems. However, they present some weaknesses. They are not sufficiently generic since they tackle specific adaptation problems, which consequently compromises their reusability as well as their ability to adapt to new system requirements and constraints. The development of design patterns is a promising alternative approach to deal with the above problems. A design pattern gives a higher abstraction view of a commonly recurring problem, thus promoting the reusability and extensibility of the design.

Research works tackling pattern-based adaptation are limited. We classify them into two classes: structural patterns which deal with the structure of adaptive systems and behavioral patterns which rather focus on the internals of adaptation modules. Concerning the structural patterns, Gomaa and Hashimoto[9] proposed a dynamic self-adaptation pattern for distributed transaction management in Service-Oriented Applications (SOAs). SOA coordination patterns are used to deal with the coordination of distributed transactions. In Weyns et al.[28], the authors proposed patterns to decentralize multiple adaptation loops in large and complex self-adaptive systems. In Puviani et al.[16], a taxonomy was proposed for self-adaptation patterns at both component and ensemble levels. As for the behavioral patterns, Gamma et al. [8] proposed design patterns to specify the behavior of dynamically reconfiguring software architectures. Schmidt et al.[21] proposed a set of patterns that can be used for the development of adaptive middleware such as the virtual component pattern and the component configurator pattern [4]. Ramirez and Cheng[18] proposed a set of patterns aiming at adapting distributed networked systems. They classified them into three principle categories: monitoring, decision-making and reconfiguration activities. These patterns are useful for the development of adaptive systems in different domains. However, they are most appropriate for distributed systems and do not fit the real-time and embedded systems domain since they do not deal with RTES constraints.

13.4 MARTE Capabilities for Software Adaptation Modeling

This section presents an overview of the UML/MARTE profile on which the present work is based, outlining the concerns related to software adaptation issues in RTES modeling. MARTE consists of three major packages. MARTE Foundations Package represents the foundational concepts for RTES design. It allows the specification of basic real time concepts such as non-functional properties (NFPs), time constraints and useful resources. The other two packages are refined from the first one. The second package named MARTE Design Model is dedicated for a detailed hardware and software description. The third package, the MARTE Analysis Model package, offers annotations for generic basis of quantitative performance and schedulability analysis.

In the MARTE Design Model package, the Software Resource Modelings (SRMs) and the Hardware Resource Modelings (HRMs) sub profiles present a specialization of the Generic Resource Modeling (GRM). SRM is also intended to describe multi tasking software platforms such as Real Time Operating System (RTOS). In the MARTE Foundations package, the Core Elements profile defines general basis for the specification of most elements of the other MARTE packages. It is split in two sub-packages. The Foundations package describes the basic elements for model-based design of RTES such as the classifier root concept. The Causality package permits the description of any dynamic model of RTES through behavioral modeling and run-time semantics definition. A behavior describes the dynamic of a system or its elements at run-time. In our context, i.e. reconfigurable RTES, the main concepts related to the system behavior are described through the Common Behavior package of Causality, and more specifically the modal behavior model.

The modal behavior relates to the notion of operational mode and configuration which represent an operational state of a system that may be changed at run time to permit dynamic system reconfiguration. The system is modeled by a set of modes stereotyped ≪Mode≫ and each characterized by a configuration, and transitions between modes stereotyped ≪ModeTransition≫. State machines are used to describe the dynamics of system modes. The system configuration under a given mode is described by a composite structure. The MARTE standard represents a common conceptual basis for coarse-grain reconfiguration behavior of RTES, however, it lacks explicit support of fine-grain adaptation modeling, in particular the application modal behavior. It gives an abstract definition of system mode which is simply modeled using a name and a configuration illustrating an allocation scenario. Thus, it does not offer support for application Q-level modeling which requires details about the implied resource consumption and output quality of an application mode. In the rest of this chapter, we propose an extension of MARTE for the modeling of fine-grain adaptation applied to the application level.

13.5 The Proposed Extension

Since it relates to the software part of the system, our extension is integrated in the Software Resource Modeling (SRM) sub-profile as an additional package named SW_Reconfiguration. The overall structure of the extended profile is illustrated in Fig. 13.2.

The existing SRM sub-profile is composed of four packages. The package SW_ResourceCore provides the basic software resource semantics. The package SW_Concurrency specifies the execution context of concurrent entities. It defines, in particular, ≪SwSchedulableResources≫, which are CPU competing resources brokered by a software scheduler which decides on the order and timing of their

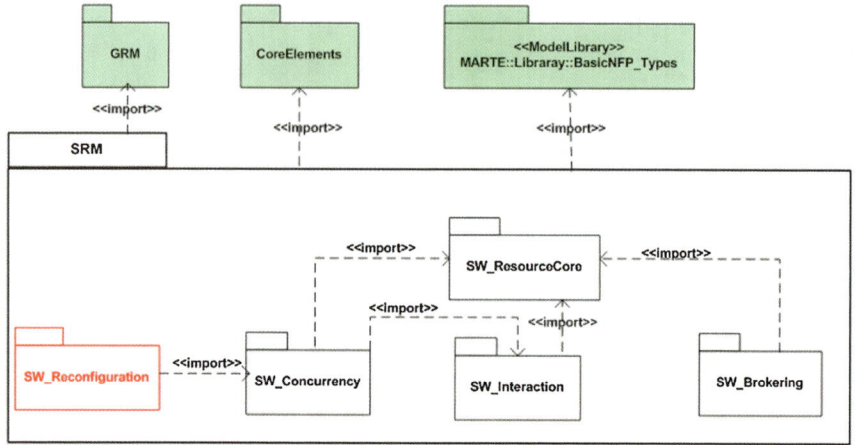

Fig. 13.2 Structure of the extended SRM modeling framework

execution. Traditionally, software applications in an OS-based multitask system are stereotyped ≪SwSchedulableResource≫. The package SW_Interaction defines two interaction modes between concurrent resources through communication and synchronization resources. The package SW_Brokering focuses on the management of software and hardware resources.

In next subsections, we give a brief description of the new reconfiguration package. Then we detail the modeling of the reconfiguration behavior key features. We illustrate our extension through a case study.

13.5.1 The SW_Reconfiguration Package

The structure of the SW_Reconfiguration package is shown in Fig. 13.3. The metamodel specifies the structure and behavior of a reconfigurable software resource stereotyped ≪SwAdaptiveResource≫. The reconfiguration behavior of a software adaptive resource (software resource is assumed to be called application for the rest of this paper) consists in a switch between a set of operational modes according to mode transitions. The adaptive application is associated to a software adaptor which controls this behavior. In order to enable MARTE to cohabit both coarse and fine grain adaptation, we kept unchanged the existing ≪Mode≫, ≪ModeBehavior≫ and ≪Configuration≫ stereotypes of CoreElements::Causality::CommonBehavior package, while we define three new stereotypes to model per-application modes: ≪ElementaryMode≫, ≪ElementaryModeTransition≫ and ≪SwAdaptor≫. The association between ≪ElementaryMode≫ and ≪Mode≫ means that an elementary mode may be part of one or several system mode. A whole system mode may be composed of a set of elementary application modes.

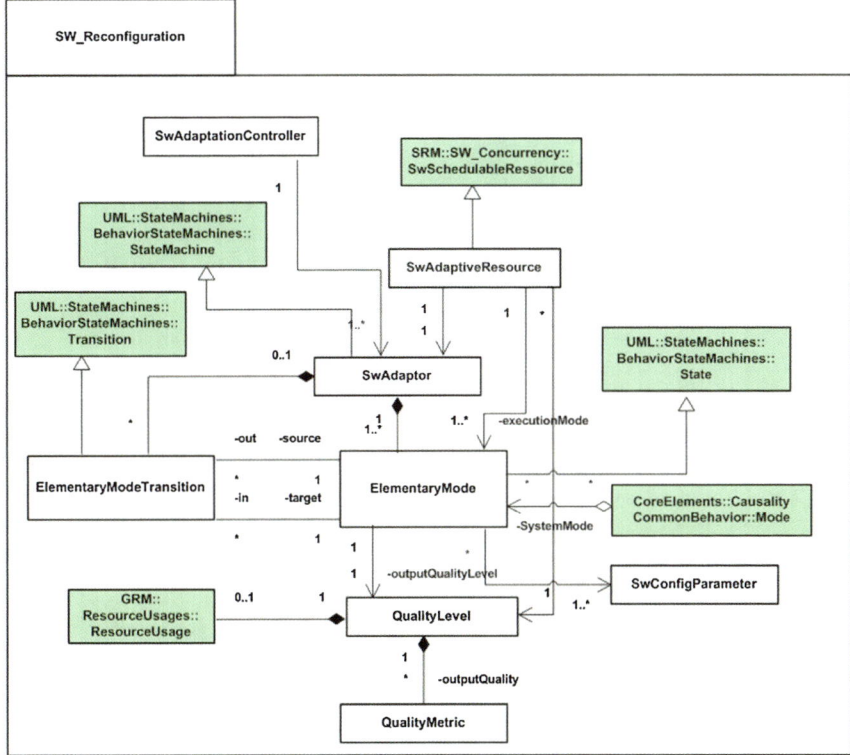

Fig. 13.3 The SW_Reconfiguration package overview

The ≪SwAdaptor≫ is composed of one or several ≪ElementaryMode≫ connected by ≪ElementaryModeTransition≫. It extends the UML StateMachine metaclass. An elementary mode is characterized by a combination of the application configuration parameters and a given output quality level indicating the implied resources usage amounts and the output quality.

The purpose and content of each of these new stereotypes are described in subsequent sub-sections.

13.5.2 The Software Adaptive Resource Modeling

We describe in this section both structural and behavioral views of a software adaptive resource. The former is defined by the ≪SwAdaptiveResource≫ stereotype. The latter is represented by the ≪SwAdaptor≫ stereotype.

Fig. 13.4 The software
adaptive resource structural
view

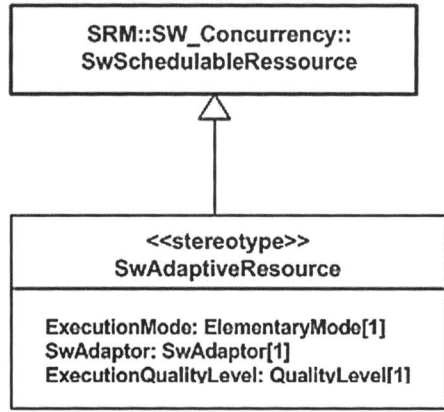

13.5.2.1 Structural View

Since we consider a context of multitask systems where competing tasks execution is managed by a scheduler, a ≪SwAdaptiveResource≫ is then both schedulable and concurrent. Hence, it is a specialization of the ≪SwSchedulableResource≫ of the SW_Concurrency package. An import association between this and the new package is added to the SRM profile structure (Fig. 13.2). The ≪SwAdaptiveResource≫ structure is illustrated in Fig. 13.4. An adaptive application is characterized by an execution mode representing its current operational elementary mode. It has a swAdaptor defining the dynamics of its modes. It saves its actual execution quality level in a ≪QualityLevel≫ attribute. This information is to be used in further extension for an evaluation purpose of the reconfiguration behavior.

13.5.2.2 Behavioral View

The behavior of adaptive application is controlled by a ≪SwAdaptor≫ entity. It is represented by a UML state machine in order to clearly describe the different application modes stereotyped ≪ElementaryMode≫ and the switching transitions stereotyped ≪ElementaryModeTransition≫. The software adaptor structure is described in Fig. 13.5. ≪ElementaryMode≫ extends the UML State meta-class and ≪ElementaryModeTransition≫ extends the UML Transition meta-class. The elementary modes are mutually exclusive, i.e., only one mode is active at a given instant.

When an event occurs, the adaptor decides of the convenient mode transition to be triggered to switch to the next execution mode. An important issue concerning mode switch needs to be revised here; in the context of embedded systems, which have a fluctuant environment, transitions from one mode to another are not always straightforward. Modes may not be related to each other since the transition from

Fig. 13.5 The per-application
software adaptor structure

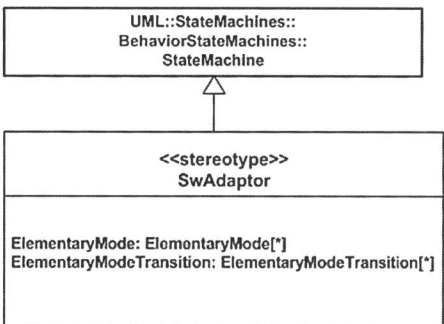

the current mode to the next one is not static. It varies according to the environment variation. A typical example is the network BandWidth (BW) variation, which does not follow a specific law. At instant i, $BW = maxBW$, at instant $i + 1$, there is no network coverage, at instant $i + 2$, $BW = maxBW/2$. Therefore, each transition of our state machine is triggered by an Event-Condition-Action tuple as follows: Event(arg1,...)[Condition]/Action

Transitions between modes are triggered by a resource constraint event, which indicates that a variation of one or several system resources has occurred and is accompanied by a set of newly required resource amounts for the application execution. A condition is a comparison between the required amounts of constrained resources brought by the event and the corresponding resource usage implied by the target mode. Both condition elements are variable in order to permit run-time updates of the NFP values when necessary. The selected elementary mode is the one, which resource usage satisfies the application resource requirements while maximizing its output quality.

13.5.3 Application Modes Modeling

An elementary mode represents a Quality level implied by a certain combination of algorithmic parameters values. Thus, an ≪ElementaryMode≫ stereotype is essentially characterized by one or more ≪SwConfigParameter≫ and a corresponding ≪QualityLevel≫. Figure 13.6 illustrates an ≪ElementaryMode≫ structure. A configuration parameter has a couple of attributes indicating its name and value. A quality level represents the application output quality and the resources usage implied by the configuration parameters combination. Therefore, it needs to be characterized by:

- A set of non functional properties representing the consumed amounts of resources, such as the worst case execution time taken from a computing resource, the memory and energy consumption and the number of generated bytes to be transferred through a network. The ≪ResourceUsage≫ stereotype is finely suitable here. We therefore define a ≪ResourceUsage≫ attribute for the quality level. We have to

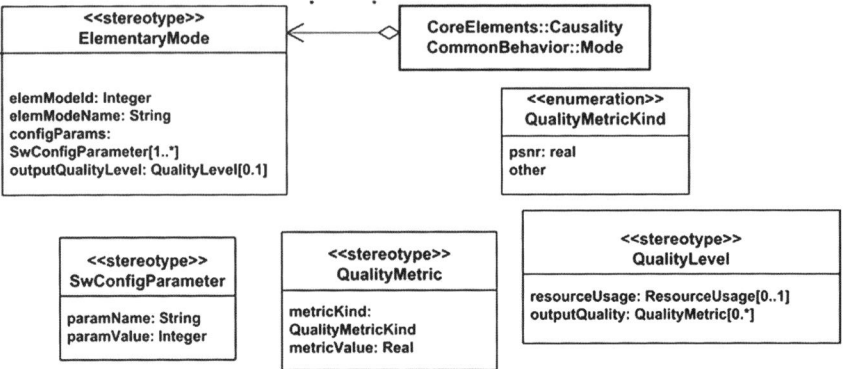

Fig. 13.6 The elementary mode modeling

mention here that for rigorous quality levels definition, the resource usages implied by a quality level must be defined by intervals of values instead of one value. In fact, in the context of data flow systems, in particular multimedia systems, the resource consumption of an application varies with the input data. A typical example is the different CPU times consumed by a video encoder when treating a slow motion news video and a motion-based action film. Intervals of NFPs may be expressed as follows:

ResourceUsage NFP = (value, unit, max),(value, unit, min)

Example: execTime = (40, ms, max),(20, ms, min)

- One or several quality metrics stereotyped ≪QualityMetric≫ to evaluate the offered output quality of the mode. A quality metric is defined by a metric kind, such as the PSNR for video quality, and a metric value quantifying the quality.

13.5.4 Modeling of the Adaptation Controller

When a variation of the system context occurs a variation event is launched with information about new context data. An adaptation controller stereotyped ≪SwAdaptationController≫ analyses the input context data (resource constraints), re-allocates resource budgets for applications and notifies application adaptors of the newly allocated per-application resource budgets. It has therefore two input ports for context variation events and context data and one output port for the list of allocated resource amounts, as illustrated in Fig. 13.7.

The control must be applied periodically. Therefore we define a period attribute for the ≪SwAdaptationController≫. Details about the internal functioning of this component will be presented in next works.

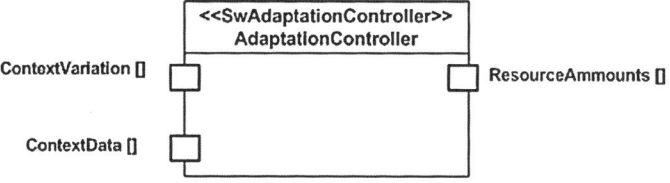

Fig. 13.7   The adaptation controller modeling

13.5.5 Case Study: An Adaptive H264/AVC Video Encoder

We present here a case study to illustrate our extension for the modeling of fine grain software adaptation. The case study relates to a highly reconfigurable multimedia application, the H264/AVC video encoder. The standard H.264 or MPEG-Four part Ten Advanced Video Coding (AVC) is one of the most important developments in video coding in the last few years that has been defined jointly by the Joint Video Team (JVT) of the ITU and the ISO/IEC. It has been proven in Wiegand et al. [29] to be the best video encoder compared to previous standards. H.264/AVC-compliant encoders achieve essentially the same reproduction quality as encoders that are compliant with the previous standards (the quality is measured by the Peak Signal-to-Noise Ratio (PSNR) value and subjective testing) while typically requiring 60 % or less of the bit rate. The H264 encoder is characterized by a big set of configuration parameters, each has a number of possible values. As examples of parameters, we cite the Quantization Parameter (QP) (Allowable values are from 0 to 51), the period of Intra-coded frames, the Motion Estimation (ME) search algorithm (−1 for Full Search, 0 for Fast Full Search, 1 for UMHexagon Search). More details can be found in Ostermann et al. [15], Tourapis et al. [24]. We are interested in this case study in the respect of the real-time constraint.

Figure 13.8 defines a state machine representing a ≪SwAdaptor≫ that is associated to the H264_encoder application. It is composed of three elementary modes, highVideoQ, mediumVideoQ and lowVideoQ. These modes indicate three levels of video quality from the highest to the lowest quality. Since our modes are not related between them, we define a particular state named CurrentMode to be always the entry point of the adaptor. To do this, we define direct transitions from all modes to the current mode with a simple action demanding the update of the current mode with the selected one.

Since we are concerned with the temporal constraints, transitions are triggered by a TaskEntry event which indicates an increase of the total CPU load causing the exceed of the CPU computing capacity and leading to deadlines misses. Upon receiving the resource requirements, the H264 adaptor selects the transition whose condition is validated, i.e., the target elementary mode satisfies the required resource amounts. The corresponding mode is then activated to be the application execution mode.

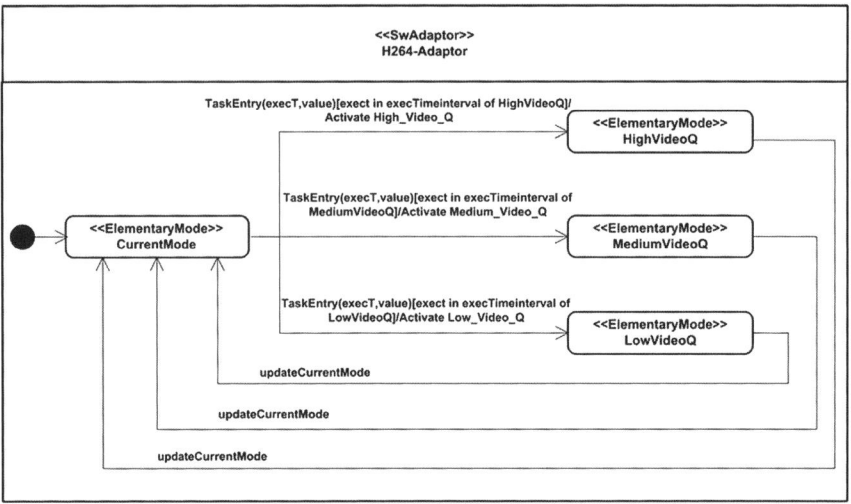

Fig. 13.8 The software adaptor of the H264 encoder case study

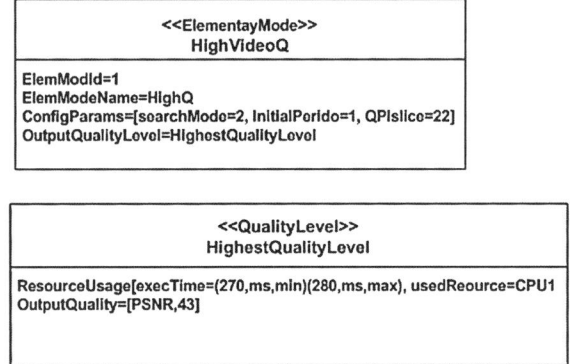

Fig. 13.9 Modeling of the H264 encoder HighVideoQ elementary mode

Figure 13.9 illustrates the structure of the HighVideoQ elementary mode. This mode is defined by a combination of values of four H264-specific ≪SwConfigParameter≫, which are SearchMode, IntraPeriod and QPISlice. Each parameter is defined by a couple of name and value. Since we only consider the timing constraint, we are content with defining the range of CPU time usage implied by the mode. The output quality is quantified using the PSNR quality metric.

13.6 Using the Proposed Extension in a Decision Making Pattern

As we have already mentioned, we focus in this chapter on the decision making process of the adaptation loop. We propose a pattern for adaptation decision mechanisms according to the [2] pattern template. However, we only give details of the five fundamental fields which are the pattern name, intent, context, problem and solution.

Different classifications of decision strategies have been proposed in the literature. In Andersson et al.[1], McKinley et al.[12], Salehie and Tahvildari[20], the authors proposed a classification in two types according to the level of granularity and complexity of the change: Parametric and structural strategies. The parametric strategy modifies parameters of system components which have effect on the system behavior. It is a low-cost fine grain adaptation applied locally on system elements. However, structural strategy modifies the system structure such as components allocation change and their activation/inactivation. It is a high-cost coarse grain adaptation that involves the entire system. A self-adaptive system may have a parametric or structural adaptation strategy, or a combination of these. This latter case is called hierarchical adaptation that has been tackled by a number of research works like Diguet et al.[6], Vardhan et al.[25] and has been proven to be effective:

Name: RTE Decision Maker
Problem: The problem treated by this pattern is to decide what artifact in an RTES to adapt and how to adapt it to meet a set of requirements and constraints.
Intent: When an adaptation decision is required, the *RTE Decision Maker* pattern decides what system elements to change and how to meet requirements and constraints. This pattern defines the adaptation strategy to apply. It can be based on parameters tuning of system's changeable elements, the modification of system's structure or a hierarchical adaptation coordinating both strategies.
Context: This pattern is used when a real time embedded system exhibits new constraints or requirements due to change in its execution context.
Motivation: When modeling self-adaptive RTES, designers need to specify the adaptation strategy to use to calculate the adaptation decision. The *RTE Decision Maker* pattern permits to model three types of adaptation strategies.

Solution:

   Structural view: We designed the *RTE Decision Maker pattern* at a high abstraction level so that it is enough simple and generic to permit the design of a hierarchical adaptation decision-making by considering two different adaptation strategies at once: the parametric fine-grain and the structural coarse-grain strategies.

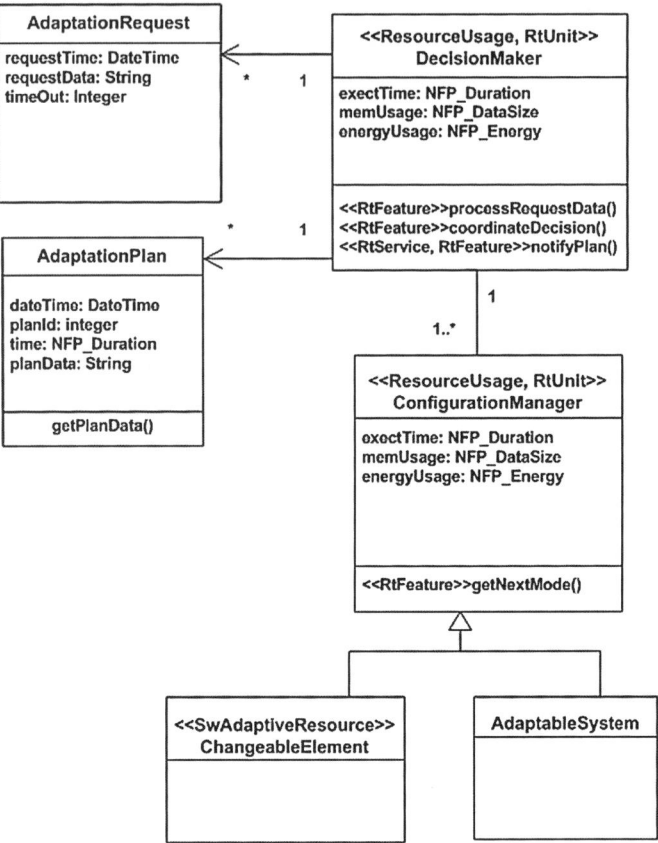

Fig. 13.10 Structural view of the RTE decision maker pattern

Participants: The structure of this pattern, depicted in Fig. 13.10, is com-
                     posed of two basic classes: a *DecisionMaker* (DM) and a *Con-
                     figurationManager*:

- *DecisionMaker* is the principal class of the pattern. It is responsible for generating the adaptation decision that best meets the received *adaptation request*
  requirements. It initiates a hierarchical decision making by cooperating with
  fine and coarse-grain configuration managers. It asks for partial reconfiguration decisions. Then it coordinates between them through its *coordinate()*
  method which generates the final decision that is encapsulated in an *Adap-
  tationPlan* and sent through the *notifyPlan()* method to an acting process.
  The DecisionMaker is therefore an active class stereotyped ≪RtUnit≫. An
  illustrative example of the hierarchical adaptation decision making is the
  GRACE platform proposed in Vardhan et al.[25]. Its authors proposed a
  hierarchical adaptation approach performing expensive global adaptations

occasionally at large system changes (e.g., application entry or exit) and low-cost limited-scope per-application adaptations frequently at the start of every frame. Since we are in the context of RTES, which behavior needs to be predictable, the overhead of adaptation activities has to be taken into account. It is presented by the Worst-Case Execution Time (WCET) and required resources of the adaptation process. We use the ≪ResourceUsage≫ stereotype to capture the deciding cost.

- *ConfigurationManager* is the generalization of *AdaptableSystem* and *ChangeableElement*. It is an ≪RtUnit≫ responsible for the management of adaptable elements configurations. It delivers, when required, the next mode that best responds to received requirements using its *getNextMode()* method. It is stereotyped ≪ResourceUsage≫ to capture its adaptation cost.
- *AdaptableSystem* represents the system to adapt as a whole. Its behavior is specified using a UML State Machine which manages global system reconfiguration through structural modifications.
- *ChangeableElement* represents an element of the adaptable system that is amenable to change. It is stereotyped ≪SwAdaptiveResource≫. Similarly to the *AdaptableSystem*, it owns a state machine managing fine-grain reconfiguration of a *ChangeableElement*, which is based on simple parameters modification.

Behavioral view: The behavior of the Decision Maker pattern is modeled using a sequence diagram accompanied with state machines of the adaptable system and changeable system elements. In fact, the configuration selection is modeled by state machines composed of a set of modes and transitions between them. The triggering of an event ensures the transition from one mode to another. To model global adaptation decision, we use MARTE capabilities for reconfigurable systems modeling defined by the modal behavior model of the CommonBehavior package. A state machine, stereotyped ≪ModeBehavior≫ is used to model the dynamics of the adaptable system configurations. It is composed of a set of mutually exclusive modes, stereotyped ≪Mode≫, each characterized by a configuration, and transitions between modes, stereotyped ≪ModeTransition≫. As for the design of local adaptation decision, we use our proposed extension. The behavior of a changeable element is controlled by a state machine stereotyped ≪SwAdaptor≫, which is composed of a set of configurations stereotyped ≪ElementaryMode≫ and switching transitions stereotyped ≪ElementaryModeTransition≫. An elementary mode represents a quality level of the changeable element. It is characterized by a combination of configuration parameters and its implied output quality and resources usage. These characteristics are to be compared to change requirements and constraints in order to select the best next mode.

When the DecisionMaker receives a valid adaptation request, it processes the data that it captures to determine requirements and constraints to take into consideration. Then it decides whether a local, structural or both decision strategies are necessary and which elements of the system to change. If local adaptation of a *ChangeableElement* is required, the DecisionMaker triggers mode switch of this element by invoking the *getNextMode()* method, giving them requirements and constraints to respect. The ≪ElementaryTransition≫ that best meets constraints is then activated and the destination ≪ElementaryMode≫ is selected and returned back to the DecisionMaker. the same scenario is applied to the adaptable system which returns back the next ≪Mode≫ of the whole system. Having received destination modes decisions, the DecisionMaker coordinates between them, if needed, generates the final adaptation plan, notifies the Actor.

13.7 Conclusion

A self-adaptive system is structured using an adaptation loop, referred to as the MAPE loop which is composed of four adaptation processes: monitoring, analyzing, deciding and acting, accompanied with sensors and effectors. In this chapter, we were interested in the deciding process. We dealt with the high abstraction level modeling of reconfigurable RTES using the UML/MARTE profile. The reconfiguration behavior consists in mode switch. It can be either coarse-grain adaptation which is applied globally for the whole system, or fine-grain that is locally performed for one application.

An overview of the current MARTE capabilities for the specification of reconfigurable RTES has shown that the current version of the standard is inadequate for a detailed description of per-application reconfiguration. A main contribution of this work is that it supplies the MARTE standard with new semantics to model application fine-grain reconfiguration. We illustrated the use and importance of our extension through a case study of a reconfigurable H264/AVC video encoder.

At the aim of promoting reusability and ease of use of our proposed extension for MDE and MARTE non-experts, we exploited it in the definition of a design pattern for an adaptation RTE decision making process. Additionally, the DecisionMaker pattern permits to handle concurrency and real-time features relative to the adaptation operations, which are key issues in RTES design.

We plan in future works to define generic models for the remaining modules of the adaptation loop. Indeed, we aim at proposing an MDE-based approach for the automatic generation of complex self-adaptive RTES.

References

1. Andersson J, Lemos R, Malek S, Weyns D (2009) Modeling dimensions of self-adaptive software systems. In: Cheng BH, Lemos R, Giese H, Inverardi P, Magee J (eds) Software engineering for self-adaptive systems. Springer, Berlin, pp 27–47. doi:10.1007/978-3-642-02161-9_2
2. Buschmann F, Meunier R, Rohnert H, Sommerlad P, Stal M (1996) Pattern-oriented software architecture: a system of patterns. Wiley Inc, New York
3. Cherif S, Trabelsi C, Meftali S, Dekeyser JL (2011) High level design of adaptive distributed controller for partial dynamic reconfiguration in FPGA. In: Proceeding of the conference on design and architectures for signal and image processing (DASIP) 2011, pp 308–315. doi:10.1109/DASIP.2011.6136896
4. Corsaro A, Schmidt DC, Klefstad R, ORyan C (2002) Virtual component—a design pattern for memory-constrained embedded applications. In: Proceedings of the conference on pattern language of programs (PLoP)2002
5. Dekeyser JL, Boulet P, Marquet P, Meftali S (2005) Model driven engineering for SoC co-design. In: Proceedings of the international IEEE-NEWCAS conference (NEWCAS) 2005, Quebec. http://hal.inria.fr/inria-00565172
6. Diguet JP, Eustache Y, Gogniat G (2011) Closed-loop-based self-adaptive hardware/software-embedded systems: design methodology and smart cam case study. ACM Trans Embed Comput Syst 10(3):1–28. doi:10.1145/1952522.1952531
7. Gamatié A, Le Beux S, Piel É, Etien A, Ben Atitallah R, Marquet P, Dekeyser JL (2008) A model driven design framework for high performance embedded systems. Research Report RR-6614, INRIA, http://hal.inria.fr/inria-00311115/en/
8. Gamma E, Helm R, Johnson R, Vlissides J (1995) Design patterns: elements of reusable object-oriented software. Addison-Wesley Longman Publishing Co., Inc, Boston
9. Gomaa H, Hashimoto K (2012) Dynamic self-adaptation for distributed service-oriented transactions. In: Proceeedings of the ICSE workshop on software engineering for adaptive and self-managing systems (SEAMS) 2012, pp 11–20. doi:10.1109/SEAMS.2012.6224386
10. Kephart J, Chess D (2003) The vision of autonomic computing. Computer 36(1):41–50. doi:10.1109/MC.2003.1160055
11. Krichen F, Hamid B, Zalila B, Jmaiel M (2011) Towards a model-based approach for reconfigurable DRE systems. In: Proceedings of the european conference on software architecture (ECSA) 2011, pp 295–302. doi:10.1007/978-3-642-23798-0_32
12. McKinley PK, Sadjadi SM, Kasten EP, Cheng BHC (2004) Composing adaptive software. Computer 37(7):56–64. doi:10.1109/MC.2004.48
13. OMG (2011) A UML profile for MARTE: Modeling and Analysis of Real-Time and Embedded Systems, ptc/2011-06-02. Object Management Group (OMG), standard
14. Oreizy P, Gorlick MM, Taylor RN, Heimbigner D, Johnson G, Medvidovic N, Quilici A, Rosenblum DS, Wolf AL (1999) An architecture-based approach to self-adaptive software. IEEE Intell Syst 14(3):54–62. doi:10.1109/5254.769885
15. Ostermann J, Bormans J, List P, Marpe D, Narroschke M, Pereira F, Stockhammer T, Wedi T (2004) Video coding with H.264/AVC: tools, performance, and complexity. IEEE Circuits Syst Mag 4(1):7–28. doi:10.1109/mcas.2004.1286980
16. Puviani M, Cabri G, Zambonelli F (2013) A taxonomy of architectural patterns for self-adaptive systems. In: Proceedings of the international C* conference on computer science and software engineering (C3S2E) 2013, ACM, New York, pp 77–85. doi:10.1145/2494444.2494470
17. Rafiq Quadri I, Meftali S, Dekeyser JL (2009) A model based design flow for dynamic reconfigurable FPGAs. Int J Reconfigurable Comput
18. Ramirez AJ, Cheng BHC (2010) Design patterns for developing dynamically adaptive systems. In: Proceedings of the ICSE workshop on software engineering for adaptive and self-managing systems (SEAMS) 2010, ACM, New York, pp 49–58. doi:10.1145/1808984.1808990

19. Said MB, Kacem YH, Amor NB, Abid M (2013) High level design of adaptive real-time embedded systems: A survey. In: Proceeedings of the international conference on model-driven engineering and software development (MODELSWARD) 2013, pp 341–350
20. Salehie M, Tahvildari L (2009) Self-adaptive software: landscape and research challenges. ACM Trans Auton Adapt Syst (TAAS) 4(2):1–42. doi:10.1145/1516533.1516538
21. Schmidt D, Stal M, Rohnert H, Buschmann F (2000) Pattern-oriented Softw Architect., Patterns for concurrent and networked objects. Vol 2 Wiley, New York
22. Schmidt DC (2006) Model-driven engineering. IEEE Computer 39(2). http://www.truststc.org/pubs/30.html
23. Schätz B, Pretschner A, Huber F, Philipps J (2002) Model-based development of embedded systems. In: Proceedings of the advances in object-oriented information systems workshop (OOIS) 2002, Springer, pp 298–312
24. Tourapis AM, Leontaris A, Sühring K, Sullivan G (2009) H.264/14496-10 AVC Reference Software Manual
25. Vardhan V, Yuan W, III AFH, Adve SV, Kravets R, Nahrstedt K, Sachs DG, Jones DL (2009) GRACE-2: integrating fine-grained application adaptation with global adaptation for saving energy. Int J Embed Syst 4(2):152–169. doi:10.1504/IJES.2009.02793
26. Vidal J, de Lamotte F, Gogniat G, Diguet JP, Soulard P (2010) UML design for dynamically reconfigurable multiprocessor embedded systems. Proceedings of the conference on design, automation and test in Europe (DATE) 2010, European design and automation association, 3001 leuven. Belgium, pp 1195–1200
27. Vidal J, de Lamotte F, Gogniat G, Diguet JP, Guillet S (2011) Dynamic applications on reconfigurable systems: from UML model design to FPGAs implementation. In: Proceedings of the conference on design, automation and test in europe (DATE) 2011, pp 1208–1211. doi:10.1109/DATE.2011.5763315
28. Weyns D, Schmerl B, Grassi V, Malek S, Mirandola R, Prehofer C, Wuttke J, Andersson J, Giese H, Goschka K (2013) Software engineering for self-adaptive systems II, lecture notes in computer science. In: Lemos R, Giese H, Müller HA, Shaw M (eds) On patterns for decentralized control in self-adaptive systems. Springer, Berlin, pp 76–107. doi:10.1007/978-3-642-35813-5_4
29. Wiegand T, Schwarz H, Joch A, Kossentini F, Sullivan GJ (2003) Rate-constrained coder control and comparison of video coding standards. IEEE Trans Circuits Syst Video Technol 13(7):688–703. doi:10.1109/TCSVT.2003.815168

Chapter 14
Split of Composite Components for Distributed Applications

**Ansgar Radermacher, Önder Gürcan, Arnaud Cuccuru,
Sébastien Gérard and Brahim Hamid**

Abstract Composite structures as in UML are a way to ease the development of complex applications. Composite classes contain sub-components that are instantiated, interconnected and configured along with the composite. Composites may also contain operations and further attributes. Their deployment on distributed platforms is not trivial, since their sub-components might be allocated to different computing nodes. In this case, the deployment implies a split of the composite. In this chapter, we will motivate why composites need to be allocated to different nodes in some cases by examining the particular case of interaction components. We will also discuss several options to achieve the separation and their advantages and disadvantages including modeling restrictions for the classes.

Keywords Unified Modeling Language (UML) · Composite stucture · Distributed application · Component-oriented approaches · Architecture Description Language (ADL) · Common Object Request Broker Architecture (CORBA) · Flex-eWare Component Model (FCM) · System modeling · Socket connector · Generic Interaction Support (GIS)

A. Radermacher (✉) · Ö. Gürcan · A. Cuccuru · S. Gérard
CEA LIST, Laboratory of Model Driven Engineering for Embedded Systems,
Point Courrier 174, 91191 Gif-sur-Yvette, France
e-mail: ansgar.radermacher@cea.fr

Ö. Gürcan
e-mail: onder.gurcan@cea.fr

A. Cuccuru
e-mail: arnaud.cuccuru@cea.fr

S. Gérard
e-mail: sebastien.gerard@cea.fr

B. Hamid
IRIT, University of Toulouse, Toulouse, France
e-mail: Brahim.Hamid@irit.fr

© Springer International Publishing Switzerland 2015
M.-M. Louërat and T. Maehne (eds.), *Languages, Design Methods,
and Tools for Electronic System Design*, Lecture Notes in Electrical Engineering 311,
DOI 10.1007/978-3-319-06317-1_14

14.1 Introduction

The basic idea behind any component-oriented approach is that elementary application pieces (i.e., components) can be composed together in order to achieve the functionality of a more complex system. Component-oriented approaches are usually grounded on a design process including component development or reuse, assembly, and deployment.

In the **component assembly** step, the system under design is itself considered as a component. It is hierarchically defined by an assembly of existing components using an Architecture Description Language (ADL) [4], where the assembly is concretely specified by connections expressed between sub-components (parts). In the context of this chapter, we focus on Unified Modeling Language (UML) [16] as modeling language. Sub-components can themselves be defined as assemblies, resulting in hierarchical systems of arbitrary depth.

In the **deployment specification** step, the target execution platform for the application is considered. The model of the execution platform usually consists, at least, of an identification of the various execution nodes, as well as available communication paths between them. The deployment specification consists of allocating the components of the application model to execution nodes of the platform (often indirectly by allocating them to processes or threads which in turn are allocated to execution nodes, but we simplify this aspect in the context of this chapter). Allocation is usually done taking into account *non-functional* requirements of the system under design, such as execution time constraints, memory footprint, communication throughput, etc.

It is sometimes necessary to allocate sub-components to different execution nodes, which requires a split of the associated composite. The next section illustrates this problem by means of a small example. Section 14.3 provides multiple options how to split composites. Section 14.4 examines how existing component frameworks split composites. An evaluation and comparison of these options is given in Sect. 14.5. Sect. 14.6 concludes this chapter.

14.2 Motivating Example

In this section, we motivate why some composites need to be split by examining interaction components.

Consider a very simple application with two components, A and B as shown in Fig. 14.1. A has a port q with a required interface I, B has a port p with a provided interface I.

Now consider that the communication between A and B is realized by a component that implements the interaction on top of the operating system's socket API. We call such a component an interaction component (also called connector in the context of the DDS-for-CCM specification [15]). On a logical level, this component is a single

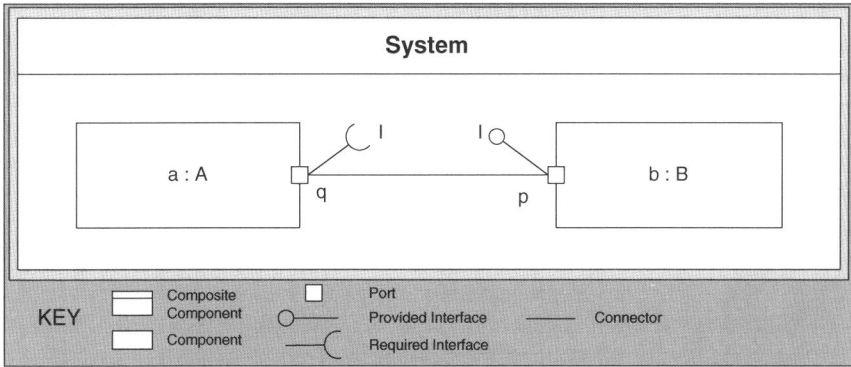

Fig. 14.1 A system with two components and uni-directional communication

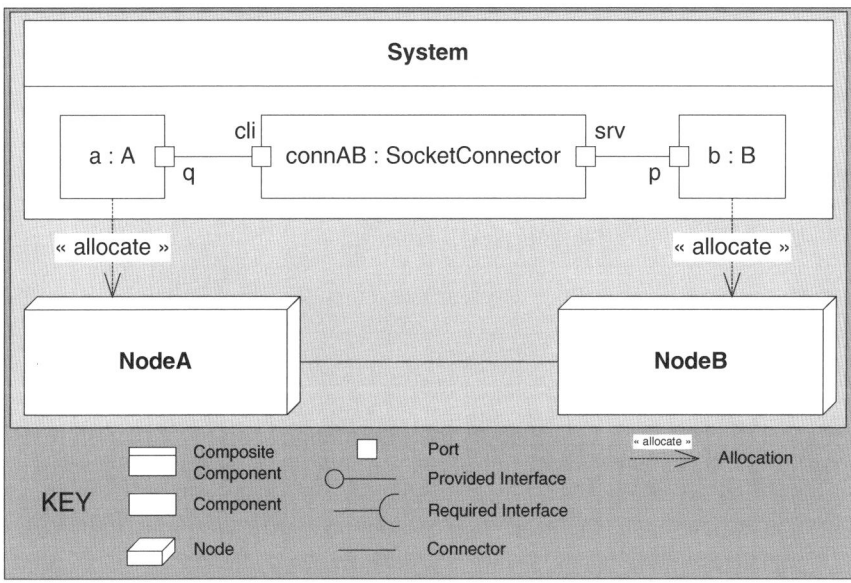

Fig. 14.2 Distribution of the system given in Fig. 14.1. The System composite component is now with sockets and allocation

entity that may contain configuration data such as a port number, connection policies or a unique identifier (object reference).

If we want to distribute the application onto two nodes, a and b are allocated to different nodes. Figure 14.2 shows the architecture of the example system. Please note that the composite structure diagram distinguishes between a role (corresponding to a kind of instance) and its type, i.e., the socket is not a nested classifier within the system but a part of the system on an instance level. Thus, the first component that is split is the component representing the system itself (System). However, the System component

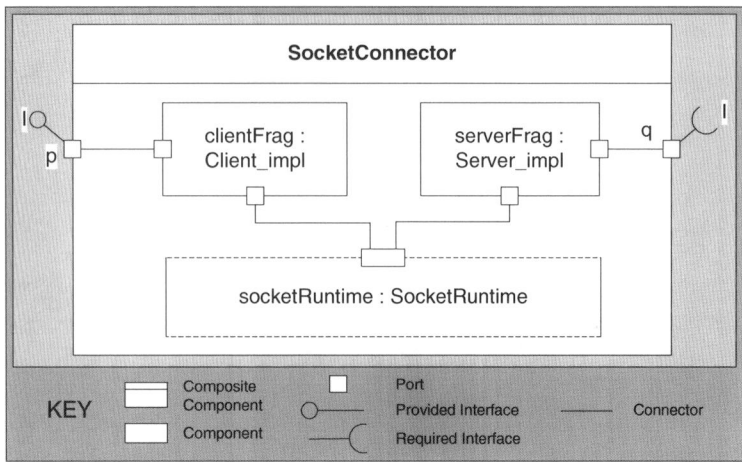

Fig. 14.3 Internal structure of the SocketConnector composite component given in Fig. 14.2

is a particular case, since there exists only one instance, it has no behavior of its own and there are no connections from the system boundary to inner parts (called delegation connectors in UML [16]). Thus, it is a pure assembly component and basically used to define the instances of a system and their interconnections.

Figure 14.3 shows the internal structure of the SocketConnector component. It consists of a client and server stub (clientFrag and serverFrag, respectively) which both access a socket run-time. The dashed outline of the latter indicates that this component is shared: it is not instantiated along with SocketConnector but exists independently. The access to a shared resource within a composite corresponds to a kind of vertical connection: the communication of the stubs with the run-time is a communication between different layers, pre-assembled within the composite.

Since the communication with the interaction component is a simple local communication, the interaction component itself needs to be separated. We can further follow local connections within the connector to determine the allocation of the internal parts of the connector. The allocations within the socket connector can thus be derived from the allocations of the application components: the client fragment of the connector needs to be co-located with A and the server fragment with B. An interesting aspect is the socket run-time that is shared by the client and by server fragments. Whereas it exists only once from a logical viewpoint, it must be present on each node and thus be allocated to NodeA and to NodeB. Figure 14.4 shows the resulting split of the SocketConnector.

Since a composite can enable distribution, its split should be authorized under the condition that *this split does not modify the component's semantics.*[1] This is the case if a composite does not have a behavior of its own (only delegation to parts) nor

---

[1] Preserving semantics of components is also important in order to be able to analyze them correctly [11].

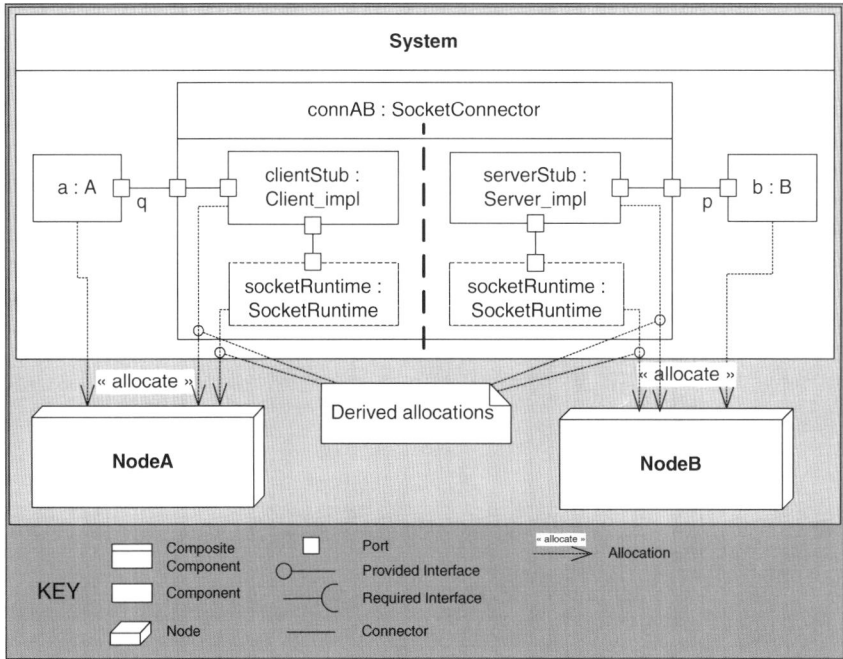

Fig. 14.4 Splitting the composite component SocketConnector during the system distribution

any configuration data. Since the latter is too strict, the composite may offer virtual configuration attributes that are effectively realized by its parts. This means that the configuration attributes of the composite are linked with configuration attributes in the parts. The same attribute might appear in multiple parts.

Now consider a slight extension of the example: B also talks to A, using the same interface, A has an additional port p, B has an additional port q and both are connected, as shown in Fig. 14.5.

In this case two parts (connAB and connBA) are typed with SocketConnector. But, the allocation of the sub-part is different for the two instances (parts):

Since a is on NodeA, the clientStub part of the instance connAB must be on NodeA as well to satisfy the co-localization constraint caused by the assumption of insepa- rable simple connections. But with the same argument, clientStub of instance connBA must be on NodeB, co-localized with b. Thus, allocation is instance based and it might happen that two different instances of a composite have different allocation specifications for their parts. Thus, the split is not trivial and we will study multiple options how to split the composite in case of the example in Sect.14.3.

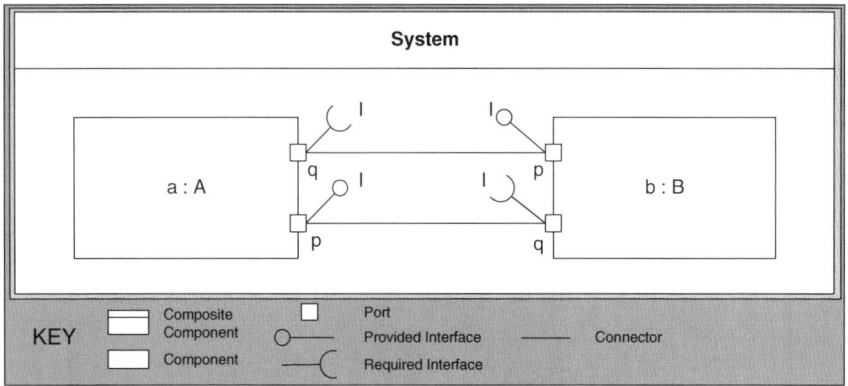

Fig. 14.5 A system with two components and bi-directional communication

14.3 Different Ways to Split Composites

In the sequel, three different options to split composites are shown by means of the simple example from Sect. 14.2.

14.3.1 Option 1—Keep Composites

The first option is to keep a modified variant of the composite that only contains the subset of parts which are deployed on a certain node. Figure 14.6 shows the result for the uni-directional variant of the example: SocketConnector' is the variant of the original SocketConnector. It contains the subset of parts that are allocated on NodeA, clientStub and socketRuntime. Note that splitting is in general not trivial, since the split must also consider super-classes. In our case, the ports of the socket are inherited by an abstract interaction component (a.k.a. connector type). Depending on how the super-class is organized, the composite only inherits from a subset of super-classes or super-classes need to be split as well, which complicates the design.

Please note that it is not the part that is allocated on a certain node, but the (sub-)instance that is associated with a part. If there is a second instance, which sub-instances are allocated in a different way, a second variant of the composite with a part subset must be created. This is shown in Fig. 14.7. The creation of multiple variants implies a certain overhead which—although small—may be non-acceptable on resource-constrained systems.

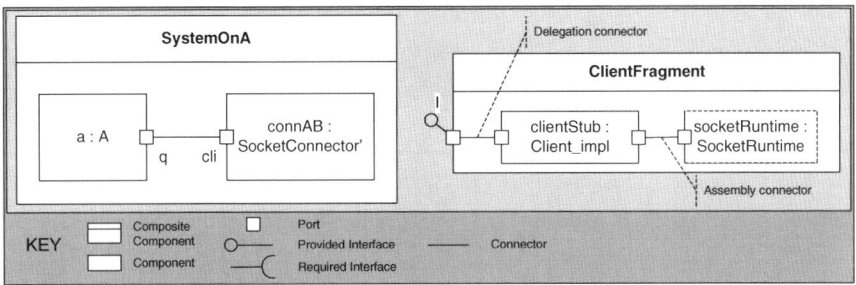

Fig. 14.6 Option 1: Splitting components in the uni-directional example given in Fig. 14.1 by *keeping* composites

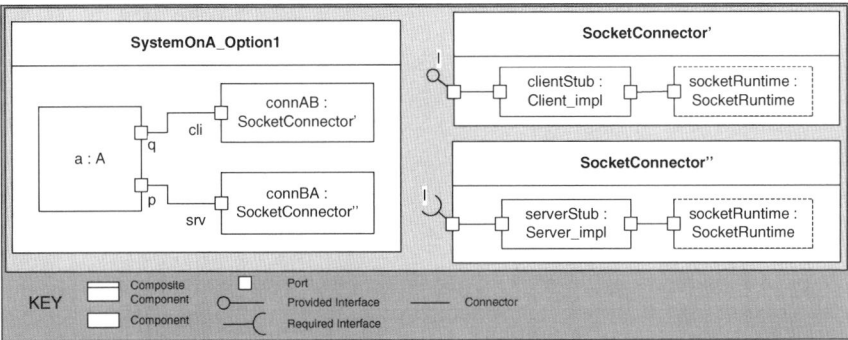

Fig. 14.7 Option 1: Splitting components in the bi-directional example given in Fig. 14.5 by *keeping* composites

14.3.2 Option 2—Flatten Composite

Flattening a composite component is a well-known approach in the literature [6, 7, 11], in which a composite component may disappear in the deployment model, i.e., it is replaced by its internal structure. The internal assembly connections of a composite become assembly connections of the containing composite (the System class in case of the example). The delegation connections[2] refine the final targets of existing assembly connections in the containing composite.

Figure 14.8 shows the example system for NodeA, in which the SocketConnector composite component has been flattened. The two parts in the system typed by a socket implementation have been replaced by parts that are directly typed with

---

[2] *Assembly connectors* are connections between inner parts; *delegation connectors* are connections from the composite to an inner part.

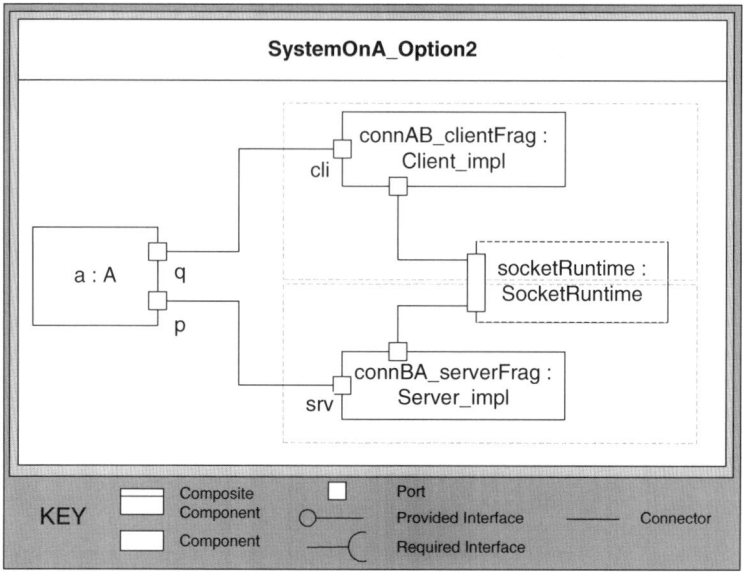

Fig. 14.8 Option 2: Splitting components in the bi-directional example given in Fig. 14.5 by *flattening* composites

elements of the socket implementation. The original composition hierarchy may still be visible via a suitable naming convention for these new parts by prefixing them with the original part name, as done in the example with the prefixes connAB and connBA.

14.3.3 Option 3—Flatten Composite, Require Explicit Fragment Sub-Components

The third option is a variation of the 2nd solution. We also flatten the SocketConnector composite component, but require that the composite must contain exclusively specific sub-components that we call *fragments*. A fragment encapsulates the parts of a composite that are allocated on the same node, conversely each fragment within a composite is typically allocated on a different node. The latter implies a restriction that is verified by a validation rule: fragments may not be connected by UML assembly connectors. The modeling of SocketConnector with fragments is shown in Fig. 14.9.

The resulting system is shown in Fig. 14.10. The composite has been flattened; the fragments have become top-level elements. The result looks very similar as the solution in Fig. 14.7, effectively the explicitly modeled fragments replace the derived subsets of the composite.

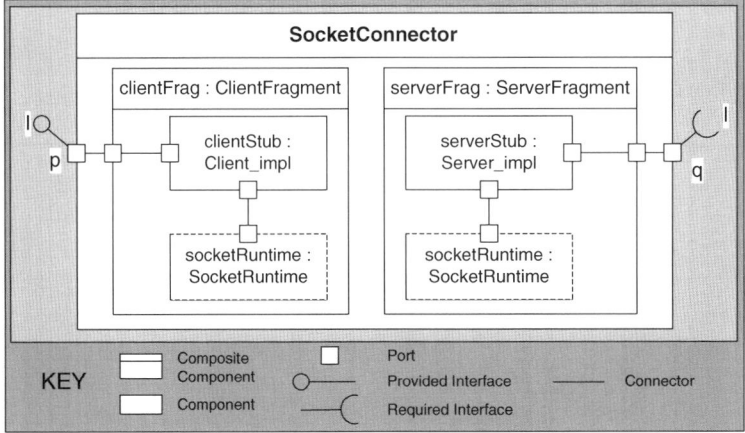

Fig. 14.9 Option 3: The SocketConnector composite component with explicit *fragments*

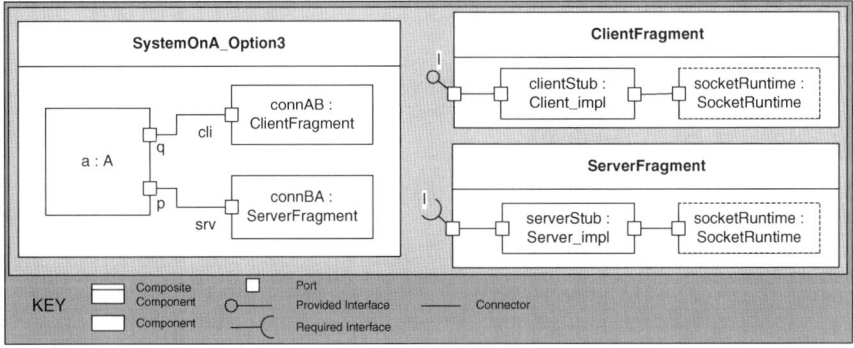

Fig. 14.10 Option 3: Splitting components in the bi-directional example given in Fig. 14.5 by *flattening* the System composite with explicit *fragments*

14.4 Support for Splitting Composites in Existing Frameworks

In the following, we sketch the existing component frameworks that have a specific support for interaction components[3] and show how these frameworks may handle composite splitting, mainly in the context of interaction components.

---

[3] Having specific support for interaction components is needed in order to be able to address the composite split in a systematic way.

14.4.1 DDS for CCM

The connector element that we have used in the motivating example is supported in multiple component models. As already mentioned, it has been standardized within the context of the OMG (Object Management Group) standard CCM (CORBA Component Model) [13]. More specifically, it is part of the DDS for CCM [15] specification, enabling component interactions via OMG's Data Distribution Service (DDS). Within this specification, the term GIS (Generic Interaction Support) is introduced. GIS will be part of the upcoming OMG Unified Component Model (UCM) [17]. The underlying connector extension for CCM has been proposed in Robert et al. [19]. Deployment with CCM is based on the specification for Deployment and Configuration (D&C) of distributed component-based applications [14]. The D&C standard describes a so-called deployment plan, a specification of instances that refer to component implementations, the interconnections between these instances, their configuration and their allocation to a node.

In the DDS for CCM specification, DDS interaction components are not identified as composites, since there are separate writer and consumer components. This is useful in case of a DDS, in which connections are implicitly created by sharing the same topic, i.e., there is no single component that represents an interaction. However, the generic interaction support enables explicit point-to-point interactions, for which composites would be useful. D&C supports two kinds of implementations of software components [14]:

- *Monolithic implementations*, where the code of the composite component is compiled as a single block.
- *Assembly (composite) implementations*, including the set of implementation of all the parts that the composite component includes. There must eventually be monolithic implementations at the "leaves" of the hierarchical implementation. Assembly allows dependent packages to be deployed on distinct target nodes, enabling flexibility in composite component instantiation.

While the D&C specification allows composites, the composites have no identity and cannot be reused. This has been analyzed in Lau and Wang [10]. In this article, the authors review and compare the ability of 13 component models of handling component composition. They identify the development with D&C as a "deposit only" repository for composites: a composite component that results from the component assembly step can be deposited in a repository but cannot be retrieved from it, because it does not have an identity of its own. In the end, only monolithic components are deployed, i.e., the component hierarchy is flattened. Note that this does not only apply to interaction components but to all composite components, even if they deployed on the same node, i.e., a stronger variant of the *flatten* option in Sect. 14.3.

14.4.2 Fractal

Connectors have also been introduced in the context of Fractal[4] [2, 5]: a binding is defined as a communication path between component interfaces. Bindings can be *primitive* or *composite*. A primitive binding (direct connector) binds one client interface and one server interface in the same address space. A composite binding is a communication path between an arbitrary number of distributed component interfaces and is represented as a set of primitive bindings and binding components. *Binding components* are called Fractal connectors, and are normal Fractal components, which role is dedicated to communication.

However, there is no support for splitting in Fractal in the sense of interaction components as shown in Sect. 14.2.

14.4.3 SOFA

In SOFA[5] [1, 3, 12], connectors are used to support transparent distribution of applications. A connector might support a transport mechanism such as CORBA[6] or low-level socket mechanisms. In this context, they are responsible for marshalling and unmarshalling as well as interfacing with the transport layer. But they can also be used for synchronization or interception. Connectors are automatically generated.

In SOFA, the connector plugging is performed after component instantiation using a split of the connector into two parts: the *server* and the *client* connector units (fragments). Whenever component interfaces query a connector reference, the corresponding *server* connector unit is returned (instead of returning a reference directly to an interface). Similarly, whenever an interface is being connected to another component, a *client* connector unit is created and bound. The connector composite specifies the parts, into which it is later split explicitly, corresponding to the *fragment* option (Option 3).

14.4.4 Qompass

The FCM [8] (Flex-eWare Component Model Flex-eWare component model) has the objective to unify the component models of Fractal and CCM. It extends the UML composite structures with dedicated interaction components—as, for instance, the socket connector presented in the motivating example (Sect. 14.2)—flexible ports and container services. This component model is supported by an add-on to the

---

[4] The Fractal Component Model, http://fractal.objectweb.org/specification/, last access on 07/02/2014.

[5] SOFA 2, http://sofa.ow2.org/, last access on 07/02/2014.

[6] Common Object Request Broker Architecture, http://www.corba.org/, last access on 07/02/2014.

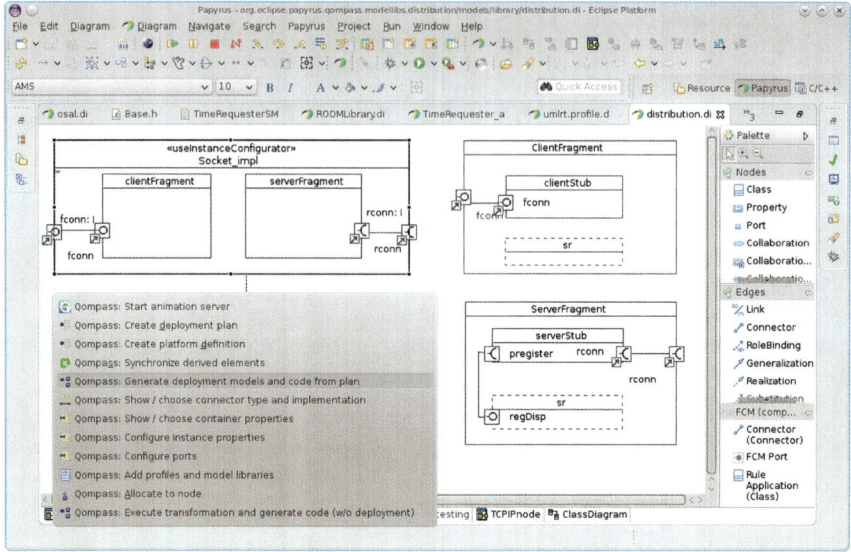

Fig. 14.11 Definition of a socket connector within Papyrus, using the client and server *fragments*

Papyrus[7] UML modeler called Qompass designer. Figure 14.11 shows the definition of the socket interaction component within a Qompass modeling library and the Qompass context menu. This add-on was first introduced as eC3M (embedded Component Container Connector Model/Middleware) [18]. Upon deployment, the tool chain executes a model transformation that replaces annotated UML connectors with the associated interaction components, as shown in the motivating example given in Sect. 14.2 (the transition from Fig. 14.1 to 14.2). This transformation includes an instantiation of the interaction component to the context in which it is used (similar to the generation of in SOFA). A further model transformation produces a model per node. During the latter, the composites within the FCM models are split. The composites that are concerned are mainly interaction components and the dedicated system component.

In Qompass, interaction components with explicitly identified fragments are flattened, i.e., the *fragment* option (Option 3). Being based on UML, Qompass must handle the specific case of a dedicated system component. Such a component is required, since connections can only be defined in the context of an enclosing composite (unlike for instance in D&C). Thus, Qompass must also split the system component, if the contained components are deployed on different nodes. The approach that has been chosen is to create a specific variant of the System component on each node, i.e., the *keep* option (Option 1). Note that it is not possible to *flatten* the system component, since the UML component model requires an enclosing composite for defining connections.

---

[7] The Papyrus UML modeler, http://www.eclipse.org/papyrus, last access on 23/01/2014.

Table 14.1 Footprint of different deployment options for the simple uni-directional (Fig. 14.1) and simple bi-directional (Fig. 14.5) systems. The results show that flattening has the smallest footprint among three deployment (splitting) options

Example system	Code size (bytes)		
	Option 1 (keep)	Option 2 (flatten)	Option 3 (fragments)
Simple uni-direct	13,904	12,233	13,936
Simple bi-direct	14,668	13,754	14,710

14.5 Discussion

Obviously, all splitting options increase the number of classes. When the composites are *kept* (Option 1), there is no need to remove additional assembly connections from the system. *Flattening* (Option 2) makes top-level composites bigger, since these composites have to incorporate the contents of a flattened component (sub-components and their connections) instead of the component itself. In *fragmentation* (Option 3), a possible split is anticipated and explicitly defined by the developer. Since the composite may not have assembly connectors, no additional connectors are added to the System class (the composite that contains the split composite). Based on this observation, to make a quantitative comparison, we measured the *footprints* associated with the different splitting options. The code size of a complete application has been measured in case of the simple uni-directional system and its bi-directional variant for splitting options 1, 2, and 3, as shown in Table 14.1. The results were obtained on a Linux machine with gcc 4.7 (optimizations disabled). As expected, *flattening* (Option 2) results in a slightly smaller footprint compared to the other two.

However, *flattening* is evidently not possible for a top-level component, since the transformation towards a model having only monolithic components and assembly connections[8] is rather straight forward. Thus, the resulting system is different, since the internal connections become visible in the system. This may be annoying, if the same composite is instantiated more than once in the original model, e.g., if we have more than one socket connector. Also note that the internal structure of an interaction component might be more complex than the simplified SocketConnector used for illustration purposes. This makes it a bit difficult to link it with the original model, for instance when debugging is done on the level of the deployed model, but fixes must be made in the original design model. Other tasks that are affected by this difference are for instance trace mechanisms (which must translate a trace specification for a composite into suitable specifications for the inner parts) and the replacement of a composite implementation with another one (e.g., in the context of different system configurations). The advantage is a slightly reduced footprint and a resolution for the splitting problem.

---

[8] In UML-like languages, connectors are always owned by a composite, i.e., a System composite must be kept.

Another important thing to consider is debugging. *Debugging* is generally defined as the process of locating and fixing or bypassing bugs in the underlying software, to achieve reliable systems. To this end, various debugging tools are developed help to identify errors at the various stages of the software development process. Especially, debugging and visualizing the behavior of component-based embedded software using models such as the UML [16] diagrams has become a reality. For instance, model-based tools such as Papyrus and the commercial tool Rhapsody[9] ("live-animation" features) enable model-based debugging of embedded software systems using sequence diagrams and state charts. In case of Papyrus, animation is based on an injected probe that communicates with the development environment. Showing the activation of a delegation connector within a composite is evidently only possible if the composition hierarchy has not been flattened. Hence, the closer the deployed model is to the original architecture, the easier it is to debug. In this sense, since *keeping* the original composition hierarchy (Option 1) has the advantage that the deployed model is closer to the original architecture, it is a bit easier to debug compared to Options 2 and 3.

In some domains (such as aerospace and electrical cars), the overall architecture of vehicles becomes very complex [20]. One possibility to tackle this complexity at run-time is the use of dynamic reconfiguration abilities. *Dynamic reconfiguration* is a process of modifying the software architecture and enact the modifications during the system's execution [9], which means making the software evolve from one configuration to another at run-time, as opposed to design-time, while introducing little or ideally no impact on the systems execution. This prevents the system to be taken off-line and/or restarted to accommodate changes. Considering the split of composite components discussed in this chapter, a dynamic reconfiguration would replace the SocketConnector component with another interaction component. In order to be able to do this seamlessly, splitting Option 1 is better suited since we do not need to remove additional assembly connections from the system.

14.6 Conclusion

We have shown that the deployment of composite instances, which are partly allocated on one node and partly on another can be tackled in several ways with different advantages and disadvantages. The choice of a suitable split option depends on properties of the composite that should be split. For instance, in Qompass designer, we keep the composite of the System component, since this particular component (no inheritance, single instance) can be split easily and since *flattening* would result in multiple top-level components. On the other hand, we flatten interaction components and require the explicit use of fragments, since we want to avoid the problems that come with multiple instances (creating potentially multiple variants of a split component). The choice depends also on the deployment goals, e.g., whether an optimized application compared to a debug-enabled application should be delivered. The

---

[9] IBM Rational Rhapsody Developer, http://www-03.ibm.com/software/products/en/ratirhap/, last access on 04/02/2014.

options are rather evident, but—to our knowledge—the task had not been examined systematically earlier.

The interest of deploying composites with complex allocation properties is not artificial: a composite definition is a suitable choice for interaction components enabling distribution. In this context, the raised issues concern principally framework and tool developers, i.e., developers of interaction components and developers of model transformations associated with the split of composites. However, the results also apply to a sub-system modeled by composite classes that need to be allocated on multiple execution nodes. In this case, system modellers or designers are concerned since they need to respect restrictions associated with the split of a composite and should know the consequences of different split options.

References

1. Bálek D (2002) Connectors in software architectures. Ph. D. thesis, Charles University Prague, Faculty of Mathematics and Physics; Department of Software Engineering
2. Bruneton E, Coupaye T, Leclercq M, Quéma V, Stefani JB (2006) The FRACTAL component model and its support in java: experiences with auto-adaptive and reconfigurable systems. Softw Pract Experience 36(11–12):1257–1284. doi:10.1002/spe.v36:11/12
3. Bureš T, Plasil F (2004) Communication style driven connector configurations. Lect Notes Comput Sci 3026:102–116
4. Clements P (1996) A survey of architecture description languages. In: Proceedings of the international workshop on software specification and design, pp 16–25. doi:10.1109/IWSSD.1996.501143
5. Coupaye T, Stefani JB (2007) Fractal component-based software engineering. In: Südholt M, Consel C (eds) Object-oriented technology. ECOOP 2006 workshop reader, Lecture notes in computer science, vol 4379. Springer, Berlin, pp 117–129. doi:10.1007/978-3-540-71774-4_ 13
6. Feher P, Meszaros T, Lengyel L, Mosterman P (2013) A novel algorithm for flattening virtual subsystems in simulink models. In: Proceedings of the international conference on system science and engineering (ICSSE) 2013, pp 369–375. doi:10.1109/ICSSE.2013.6614693
7. Huang G, Yang J, Sun Y, Mei H (2008) Quality aware flattening for hierarchical software architecture models. In: Lee R (ed) Software engineering research, management and applications, studies in computational intelligence, vol 150. Springer, Berlin, pp 73–87. doi:10.1007/978-3-540-70561-1_6
8. Jan M, Jouvray C, Kordon F, Kung A, Lalande J, Loiret F, Navas J, Pulou J, Pautet L, Radermacher A, Seinturier L (2011) Flex-eWare: a flexible model driven solution for designing and implementing embedded distributed systems. Softw Pract Experience 42(6)
9. Kramer J, Magee J (1990) The evolving philosophers problem: dynamic change management. IEEE Trans Softw Eng 16(11):1293–1306. doi:10.1109/32.60317
10. Lau KK, Wang Z (2007) Software component models. IEEE Trans Softw Eng 33(10):709–724
11. Leveque T, Carlson J, Sentilles S, Borde E (2011) Flexible semantic-preserving flattening of hierarchical component models. In: Proceedings of the EUROMICRO conference on software engineering and advanced applications (SEAA) 2011, pp 31–38. doi:10.1109/SEAA.2011.15
12. Malohlava M, Hnetynka P, Bures T (2013) SOFA 2 component framework and its ecosystem. Electronic notes in theoretical computer science. In: Proceedings of the 9th international workshop on formal engineering approaches to software components and architectures (FESCA) 2013. vol 295. pp 101–106. doi:10.1016/j.entcs.2013.04.009

13. OMG (2006a) CORBA Component Model Specification, Version 4.0. OMG, OMG Document formal/2006-04-01
14. OMG (2006b) Deployment and Configuration of Component Based Distributed Applications, v4.0. OMG, OMG document formal/2006-04-02
15. OMG (2011a) DDS for Lightweight CCM, v1.1. OMG, OMG document ptc/2011-01-14
16. OMG (2011b) Unified Modeling Language: Superstructure, Version 2.4.1. OMG, OMG Document formal/2011-08-06
17. OMG (2013) Unified Component Model for Distributed, Real-Time and Embedded Systems, Request For Proposal Draft. OMG, OMG document mars/13-05-03
18. Radermacher A, Cuccuru A, Gerard S, Terrier F (2009) Generating execution infrastructures for component-oriented specifications with a model driven toolchain—a case study for MARTE's GCM and real-time annotation. In: Proceedings of the international conference on generative programming and component engineering (GPCE) 2009, ACM press, pp 127–136
19. Robert S, Radermacher A, Seignole V, Gérard S, Watine V, Terrier F (2005) Enhancing interaction support in the CORBA component model. In: Rettberg A, Zanella MC, Rammig FJ (eds) From Specification to Embedded Systems Application, Springer, IFIP On-Line Library in Computer Science: International Embedded Systems Symposium (IESS), pp 137–146
20. Venkatesh Prasad K, Broy M, Krueger I (2010) Scanning advances in aerospace and automobile software technology. Proc IEEE 98(4):510–514. doi:10.1109/JPROC.2010.2041835

Chapter 15
ProMARTES: Performance Analysis Method and Toolkit for Real-Time Systems

Konstantinos Triantafyllidis, Egor Bondarev and Peter H.N. de With

Abstract In this chapter, we present a cycle-accurate performance analysis method for real-time systems that incorporates the following phases: 1. profiling SW components at high accuracy, 2. modeling the obtained performance measurements in MARTE-compatible models, 3. generation, scheduling analysis and simulation of a system model, 4. analysis of the obtained performance metrics, and 5. a subsequent architecture improvement. The method has been applied to a new autonomous navigation system for robots with advanced sensing capabilities, enabling validation of multiple performance analysis aspects, such as SW/HW mapping, real-time requirements and synchronization on multiprocessor schemes. The case-study has proved that the method is able to use the profiled low-level performance metrics throughout all the phases, resulting in high prediction accuracy. We have found a range of inefficient design directions leading to RT requirements failure, and recommended to robot owners a design decision set to reach an optimal solution.

Keywords Profiling · Modeling · Component-based · Simulation · Schedulability analysis · Performance analysis · Evaluation · Prediction · Assessment · Optimization · Real-time system · Modeling and Analysis of Real-Time and Embedded systems (MARTE)

15.1 Introduction

The composition of real-time systems based on the mapping of software (SW) and hardware (HW) components, has become an adopted practice, since it enables rapid prototyping and development of a system from existing blocks. The resulting real-time systems still should meet the performance requirements, such as

K. Triantafyllidis (✉) · E. Bondarev · P.H.N. de With
Eindhoven University of Technology, 5600 MB Eindhoven, The Netherlands
e-mail: k.triantafyllidis@tue.nl

E. Bondarev
e-mail: e.bondarev@tue.nl

P.H.N. de With
e-mail: p.h.n.de.with@tue.nl

© Springer International Publishing Switzerland 2015   281
M.-M. Louërat and T. Maehne (eds.), *Languages, Design Methods,*
and Tools for Electronic System Design, Lecture Notes in Electrical Engineering 311,
DOI 10.1007/978-3-319-06317-1_15

throughput, latency, etc. At the early composition phase, the architect needs reliable assessment methods in order to evaluate and predict the performance of the designed system. An incorrect performance prediction may lead to adopting an inefficient system architecture with the consequences of system re-design or re-implementation.

The challenge of performance predictions comes from the fact that at design-time the HW and SW components implementations are frequently not available. Instead, models representing abstractions of the components are used for computation and reasoning. Accuracy of such models is vital for obtaining reliable performance predictions on behavior of the future system. Component vendors, supplying the models, still face the challenges of automated generation and detailed profiling specification of models.

The analysis of the component composition should take into account the intrinsic properties of the hardware, such as cache hierarchy and dynamics, bus/network congestion, tasks floating across the processor cores and parameter-dependent execution/workload. These system aspects severely complicate the performance analysis of a composed system, even if an architect deploys detailed and accurate component models.

Another challenge comes from the limitations of analysis mechanisms, which are normally classified in two categories: analytical (formal) methods and simulation techniques. The former are not able to provide a detailed execution timeline, while the latter cannot guarantee reachability of worst cases.

In this chapter, we present a cycle-accurate performance analysis method for real-time systems (ProMARTES). Our analysis method consists of four individual phases: 1. *profiling* SW components at high accuracy, 2. *modeling* the obtained performance measurements in MARTE-compatible models, 3. *composition*, *scheduling analysis*, and *simulation* of a system model, 4. *analysis* of obtained performance metrics, and 5. a subsequent architecture improvement. The presented method is the cornerstone of our Design Space Exploration (DSE) approach [28], which is targeting automated identification of optimal architecture alternatives.

In our previous work [28], we have presented the first phase of cycle-accurate profiling and parameter-dependent MARTE-based modeling of individual components. In this chapter, we extend this phase with network utilization metrics and a detailed memory usage model, thereby addressing the above-mentioned challenges of model generation. The focus of this chapter is on the component/system composition, performance analysis and evaluation phases. The following paragraphs outline the contributions to these phases.

At the component *composition* phase, candidate SW/HW architectures and a set of workload scenarios are defined. The SW architecture is represented by composition of individual components with associated performance models. The mapping of software components on hardware nodes defines the SW/HW architecture. A set of scenarios defines the worst-case workload on a system. The instruction-level metrics of models are converted to processor-specific execution-time metrics, thereby incorporating intrinsic hardware properties.

At the system *evaluation* phase, we perform scheduling analysis and simulation of the scenarios, obtaining the usage and sharing of all involved hardware Intellectual

Property (IP) blocks. The scheduling and simulation results provide predicted performance properties, i.e., latencies, throughput and bottlenecks of the designed system. Comparing the predicted properties to the system requirements, we identify weak points of the candidate architecture, which direct us to a more efficient alternative.

To validate this method, we have performed a case study on the real-world problem of a new autonomously navigating robot with advanced sensing capabilities. The most critical performance attributes of the system are latency in the navigation control loop and throughput. We have profiled and modeled the available components, proposed a number of architectural alternatives and analyzed them with respect to the critical attributes. Based on the analysis results, we have proposed to the robot owners the optimal architecture with low HW costs that still satisfies the real-time requirements.

The sequel of this chapter is as follows. Section 15.2 records the related literature to our work. Section 15.3 explains the overall DSE methodology. Sections 15.4, 15.5 and 15.6 describe the method in detail. Section 15.7 presents the tooling developed for the method. Section 15.8 illustrates the case study used for the validation of our method. Section 15.9 describes our findings from the case study. Section 15.10 concludes the chapter.

15.2 Related Work

In the last decade, the real-time research community developed several innovative methods addressing the problems of SW/HW component modeling, predictable assembly and evaluation of real-time systems. Currently, a wide variety of modeling profiles are available for composition of (real-time) embedded systems: SysML, UML-RT, MARTE and AADL. The Systems Modeling Language (SysML) [22] is a UML profile for specifying, analyzing, and verifying complex systems that may include hardware, software, information, personnel, procedures, and facilities. However, SysML does not provide sufficient primitives for the real-time systems domain. Thereby, the OMG released the MARTE profile [23], which targets specification of real-time and embedded systems. MARTE enables the HW and SW modeling and defines specific primitives for timing and power consumption analyses. Nevertheless, MARTE lacks low-level resource modeling metrics, such as instructions, effective execution cycles and cache misses. Another alternative, AADL [24], enables modeling of SW and the HW components of real-time embedded systems. The AADL models are of high abstraction level and therefore do not provide cycle-accurate modeling primitives. Numerous composition and evaluation approaches have been proposed based on the aforementioned modeling profiles.

Cortellessa et al. [6] have proposed a comprehensive approach for modeling, composition, and mapping of SW components onto HW platforms and consequent behavior simulation. Their modeling methodology is based on UML-RT [11], which is the predecessor of the UML-MARTE. The simulation is performed by RRT [14], which is a proprietary simulator and this limits broad applicability of the approach.

However, this approach can be adopted by any modeling and simulation technique without focusing on a specific toolkit.

For automotive real-time systems, Klobedanz et al. [16] have discussed a performance analysis technique based on the AUTomotive Open System ARchitecture (AUTOSAR) [1] model. The technique targets such performance attributes as CPU load, end-to-end latency, and throughput. The AUTOSAR model lacks high-detailed performance attributes, since it is aiming at simple Electronic Control Unit (ECU)-type nodes and therefore can be applied only in the automotive domain.

For a general-purpose distributed real-time system, Määttä [21] have proposed an analysis method based on combining the UML-MARTE profile and the Ptolemy II simulator [8]. The authors target the methodology to multi-core network-on-a-chip platforms. The Ptolemy II broadly supports network communication schemes, which is crucial for performance analysis of Real Time (RT) distributed systems. Moreover, it provides the majority of the real-time scheduling policies. A limitation of this method is that the conversion of the MARTE models for the Ptolemy II tool is not fully automated, which requires a high effort of the architect during the analysis.

The UML profile is the cornerstone of many well-established performance analysis methods. Bertolino and Mirandola [3] proposed the performance analysis method CB-SPE. The components are modeled based on the UML-SPT profile, while the performance models are based on queuing networks. Another performance analysis method based on UML and queuing networks was proposed by Mos and Murphy [17] from National ICT Australia Ltd (NICTA). Unlikely to the CB-SPE approach, the one from NICTA does not distinguish the roles between the component developer and the architect of the system. Moreover, the applicability of this approach is limited, supporting only EJB applications. Another framework dealing with EJB applications is the Component Performance Assurance Solutions (COMPAS) framework, developed by Mos and Murphy [20]. The COMPAS framework deploys three individual phases: monitoring, modeling (UML), and performance analysis. Unfortunately, COMPAS does not provide any performance analysis tool.

Regarding the RT component-based embedded systems domain, the method Prediction-Enabled Component Technology (PECT) was introduced by Hissam et al. [13]. The PECT is based on the Component Composition Language (CCL) and supports synchronous and asynchronous communication, behavioral models as well as composite components. Focusing on the distributed system domain, Wu and Woodside [30] have proposed the Component-Based Modeling Language (CBML). The CBML comprises an extension of the layered queuing networks and, similarly to PECT, enables synchronous and asynchronous calls among the components. Both, PECT and CBML, approaches provide performance analysis either by formal or by simulation techniques.

Focusing on the performance evaluation of distributed systems, Becker et al. [2] proposed Palladio. Palladio incorporates its proprietary metamodel, the Palladio Component Model (PCM). During the performance evaluation, PCM is transformed to specific performance models (EQN, LQN), enabling both formal and simulation analysis techniques.

Grassi et al. [12] proposed Kernel LAnguage for PErformance and Reliability analysis (KLAPER), a method which facilitates the transformation from proprietary component structure (UML, OWL-S) to desired performance models. The performance models are expressed as Markov chains or queuing networks, enabling simulation analysis technique. KLAPER does not consider composite components and graphical modeling tools.

Following a completely different analysis approach, both, Thompson et al. [26] and Silvano et al. [25], annotate the source code of the SW components with relative performance costs of the corresponding HW platform and simulate the execution providing the performance results. Both approaches require source code of the components and cannot be considered as a conventional model-based solution.

In the domain of pure model-based techniques, Bondarev et al. [4, 5] have proposed a solution for design and performance analysis of conventional Component-Based Software Engineering (CBSE) embedded real-time systems (ROBOCOP components [15]). The Component ARchitectures Analysis Tool (CARAT) supporting the approach, synthesizes SW/HW component models, constructs a system model with corresponding scenario models and simulates the resulting models for worst, best, and average cases of CPU load, latency and throughput. CARAT supports Earliest Deadline First (EDF), Rate Monotonic (RM), and Deadline Monotonic (DM) scheduling algorithms, but it does not support network modeling and simulation. Also, CARAT does not provide a cycle-accurate profiling tool for components, which leads to less precise performance analysis.

In contrary to the above-described simulation- and scheduling-based techniques, Wandeler et al. [29] have presented a compositional method, incorporating both types of techniques for distributed embedded systems. The Modular Performance Analysis (MPA) approach models resources and their usage in a high abstraction layer, while the performance components represent the transformation of the input timing properties to the output timing properties. The modularity of the MPA approach enables the analysis and the exploration of different mapping and resource sharing strategies. As a result, the technique guarantees rapid identification of the worst-case resource load and latencies. However, intrinsic cycle-accurate execution properties cannot be incorporated during the analysis, while the task execution timeline and interleaving aspects cannot be obtained with this technique.

15.3 Overview: Architecture Analysis and Optimization

The presented performance evaluation approach is part of a larger DSE framework for real-time embedded systems, which has been developed over the last decade [4, 5, 28]. Let us first outline the framework phases, that are subdivided into three blocks: 1. Profiling and Modeling, 2. Architecture Composition, and 3. Architecture Evaluation and Optimization (Fig. 15.1).

During the *Profiling and Modeling* phase, the component developer profiles the developed SW components at cycle-accurate instruction level and generates a

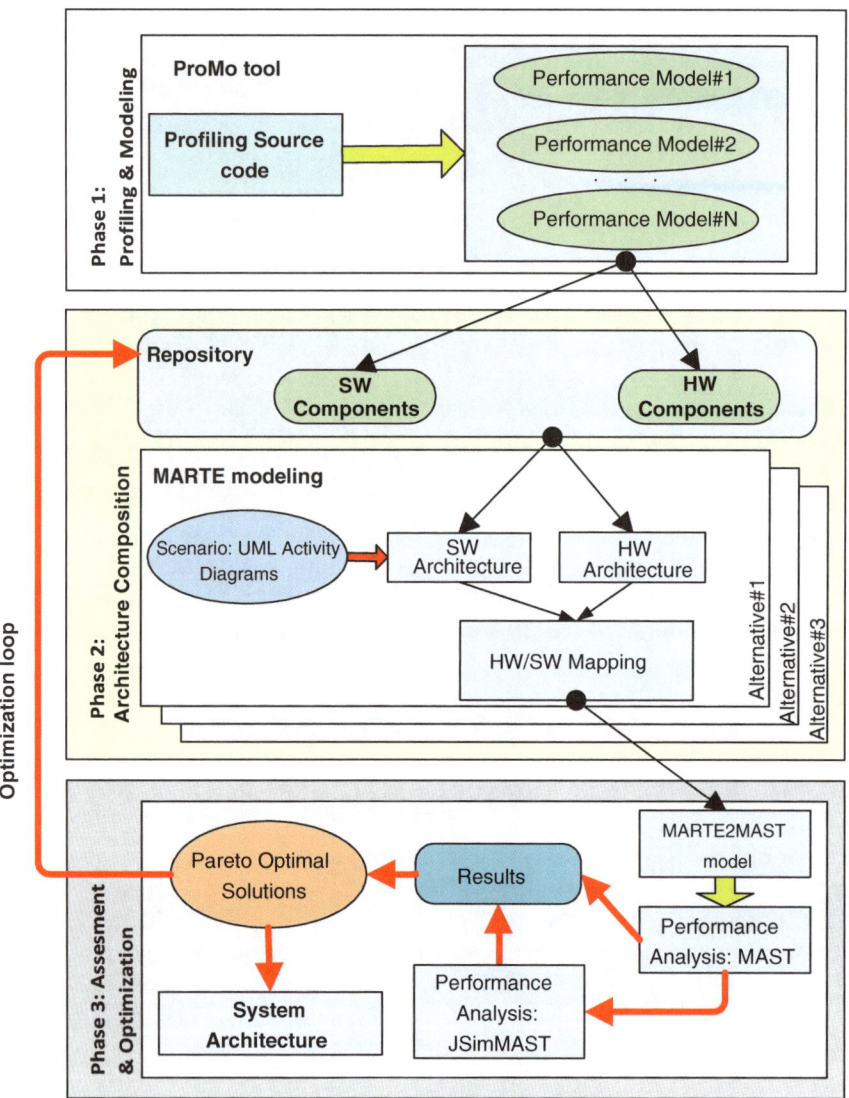

Fig. 15.1 Overview of the DSE methodology. We focus on the architecture composition, analysis and assessment phases

performance model for each individual component [28]. Each performance model may target various hardware usage aspects (CPU, bus, Random Access Memory (RAM), network, etc.) and can be specified for multiple platforms. Our tooling supports automated profiling and model specification, as well as the repository placement for the subsequent phases.

The *Architecture Composition* phase aims at selection of the required components (based on functional requirements) and the automated generation of a model of the composed system using defined workload scenarios. The composition can be performed for a number of architectural alternatives. Each alternative design includes the component instantiations and connections, as well as the mapping on a selected HW platform. Being applied onto the critical execution scenarios, the design specification is converted into a system model. Other challenges addressed are the support for multiple component architecture styles and provisioning of resulting composition models in common formats (MARTE, UML, etc.).

The *Analysis and Optimization* phase enables evaluation of system performance properties by schedulability analysis and simulation of the obtained system models. Both techniques support various hardware platforms, multiple scheduling policies and different network protocols. Performing schedulability analysis and simulation of the system model brings predicted performance properties such as latency, hardware use, and throughput. The system is validated with a comparison to the requirements, leading to consequent design iteration(s). Each iteration searches for an optimal architecture by tuning the allowed *factors of freedom* (SW component, hardware structure, SW/HW mapping and scheduling policies). In the sequel, we focus on the architecture composition and analysis blocks of the framework.

15.4 Detailed Profiling and Modeling

With the ProMo tool [28], a component developer profiles the execution behavior and hardware resource usage of each individual SW component. The ProMo tool provides the following benefits. Firstly, the profiling phase is generic, supporting the majority of the CPUs available (AMD, Intel, ARM). Secondly, the obtained performance metrics are cycle-accurate, since the measurements are directly collected from the performance monitor unit of the attached CPU. Finally, the automatically composed performance models are compatible with MARTE or AADL resource models.

The ProMo tool has been extended in order to support network utilization metrics and a more detailed memory usage model. The structural view on the model is depicted in Fig. 15.2.

We model performance attributes for each operation of the provided interfaces of the component. For each operation, the attributes are grouped into four types: `cpuUsage`, `memoryUsage`, `busUsage`, and `networkUsage`.

The Profiling and Modeling phase supports identification and specification of best-, average-, and worst-cases execution of the profiling as well as parameter-dependent usage of hardware resources.

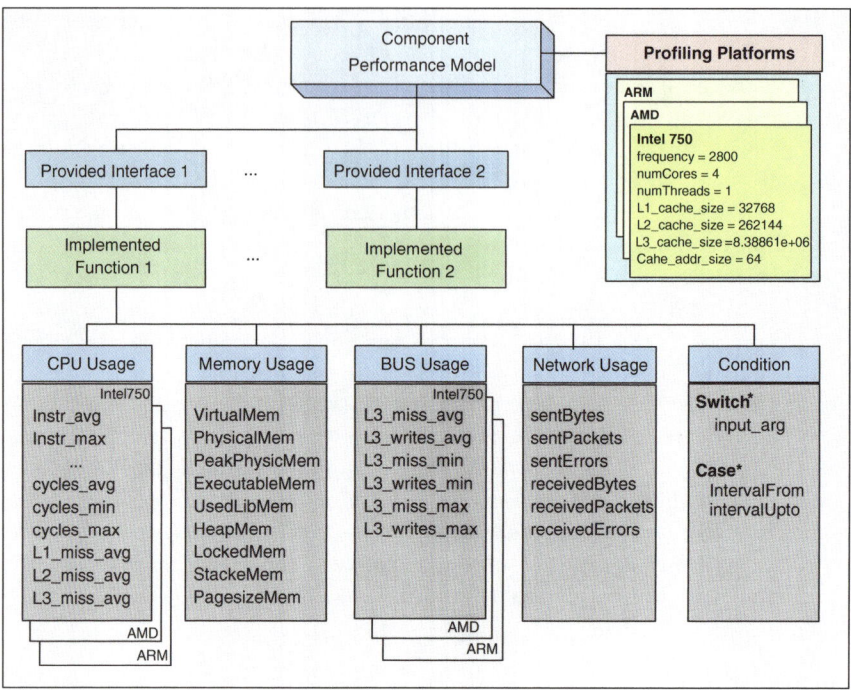

Fig. 15.2 Structure of a component performance model

15.5 Architecture Composition

At this phase, an architect selects software components that may satisfy the defined functional and extra-functional requirements and graphically specifies the component composition by instantiating and connecting the involved components. The hardware architecture can be specified in parallel, but in most of the cases, a hardware platform is already pre-specified. If not, an architect can select available hardware components from a repository and choose a specific topology, number of processing nodes, types of memory, communication means and scheduling policy.

Once the software and hardware architecture are specified, the mapping of the software components on the hardware nodes is made. The mapping shows on which processing node each software component should be executed. Efficient mapping is required to distribute the load of hardware resources in an optimal way. However, at the first mapping iteration, it is not clear how to best deploy the software components to achieve the optimal load distribution. Various mapping alternatives are possible at this stage. Each alternative represents a system architecture.

Additionally, an architect needs to define the workload on a system by means of execution scenarios, which represent either internal or external triggers for the system and the operations that are invoked by those triggers.

The creation of a system model is based on the performance models of the involved components, the scenario models and the SW/HW mapping architecture. The resulting system model represents an executable structure that can be simulated and/or analysed for performance.

15.6 System Model Analysis

The system model obtained from the previous phase is applicable to both types of evaluation techniques: schedulability analysis and simulation.

Schedulability analysis enables prediction of the best- and worst-case response latencies for each task instance, associated with a real-time deadline. This type of analysis provides guaranteed worst-case boundary conditions and can be executed within a few seconds. However, it does not provide 1. detailed behavior timeline data, 2. average-case resource usage, 3. corresponding latencies. To find the latter metrics, we also apply simulation techniques.

Simulation-based analysis deploys JsimMAST virtual schedulers that simulate the execution of the tasks specified in the system model. The selection of scheduling algorithms is dictated by 1. the types and the number of the used CPUs/HW platforms, 2. the protocols, 3. the topology of the deployed communication lines/networks, and 4. the operating system used for the composed system. While simulation techniques cannot guarantee identification of worst-case executions, they provide detailed system behavior (execution timeline of the system tasks), thereby enabling identification of possible bottlenecks already at the early design phases. However, simulation requires a substantial time span (from minutes to days), to obtain stable prediction results. Therefore, simulation can be selectively used for a detailed exploration of execution problems in the architecture, such as buffering and task-interleaving problems.

The worst-case performance properties obtained from the schedulability analysis can be further used as a guideline for next iterations of the design space exploration process, since the analytical techniques are time-efficient. The filtered candidates that meet the system requirements are further simulated to identify performance attributes which cannot be extracted by formal methods, like possible bottlenecks and average response time. Moreover, at the saturation point of identification of local or global optima, the simulation techniques facilitate the analysis of other efficient alternatives. In order to analyze the performance of the system, we have developed the ProMARTES framework, which involves several tools for profiling, modeling, both simulation and schedulability analysis and also the corresponding metamodel transformation tools. These tools are summarized in the next section.

15.7 Tooling

The ProMARTES toolchain is a combination of a number of proprietary and existing tools, which distinguish two main categories: 1. Models transformation tools, 2. Performance Analysis tools. The model transformation tools enable the conversions from various metamodels to model formats, complying with the performance analysis methods that are supported by the ProMARTES toolchain. The converted metamodels act as sources to the performance analysis tools, which analyze the performance of the system by using either formal (MAST) or simulation (JSimMast) methods.

15.7.1 Promo2Marte: Metamodel Transformation

During the phase of *Profiling and Modeling*, the generated performance models are based on the structure of the ProMo performance model and comply with the XML Metadata Interchange (XMI) framework as depicted in Fig. 15.4. The metamodel converter "ProMo2MARTE" has been implemented for the translation of the ProMo performance models to MARTE-compatible SW/HW instantiations. The ProMo2MARTE tool receives as input the ProMo performance models (generated by the ProMo tool) and it outputs MARTE-compatible SW/HW platform instantiations, ready to be placed into the repository for future utilization by the architect of the system.

15.7.2 Marte2Mast: Metamodel Transformation

In the *Architecture Composition* phase, we specify the generated system model by using the MARTE profile. In order to simulate the composed system, we use the MAST performance analysis tool [10, 23] with a proprietary model input format. Therefore, a meta-modeling tool is required for transforming the MARTE system model to the MAST format. We have performed this transformation by using the Marte2Mast tool [18, 19] and have extended the MARTE conversion to support: 1. the latest Eclipse IDE Juno and its Papyrus modeling plugin, 2. the extended MARTE profile, and 3. the MAST 1.4.0 scheduling analysis tool.

15.7.3 MAST: Schedulability Analysis Tool

The MAST analysis tool [10] provides a set of schedulability analysis methods, resulting in identification of worst-case latencies, throughputs, blocking times, and resource utilization. Furthermore, the MAST sensitivity analysis techniques enable predictions on the system robustness. In our approach, we deploy the following algorithms: Response Time Schedulability Analysis (RTA), Offset-Based Optimized RTA and Holistic RTA for fixed priorities, as well as the local and global EDF algorithms. For hierarchical scheduling, we deploy the varying priorities RTA, EDF mono-processor RTA and EDF-within-priorities RTA.

15.7.4 JSimMAST: Simulation Tool

The *JsimMAST* [7] simulation tool is used for analysis of the architectural alternatives, which have been pre-approved by the MAST tool. The *JsimMAST* simulation algorithms receive a pre-formatted MAST-2 system model as an input, and result in detailed task-execution timelines, buffer/bus/network-load timelines and task-interleaving/blocking aspects.

15.7.5 MARTE Extension

Since the MARTE profile does not support the Network Drivers of the MAST tool, we have extended the MARTE profile with the new stereotype `SaNetwork` in the *Schedulability Analysis Modeling* package. Our ProMARTES toolkit is available as an open-source distribution [27].

15.8 Case Study: Autonomously Navigating Robot

15.8.1 Introduction

We have been requested to verify real-time requirements for an advanced setup of an autonomous robot with complex navigation algorithms and propose an optimal architecture with respect to the latency of control loops and cost. We use this case study to validate our performance evaluation approach.

The autonomous robot control is normally composed of multiple hardware units: a robot with an embedded PC and a set of processing remote workstations (Fig. 15.3). The provided robot has a set of infrared laser sensors and a differential axis with two wheels (left and right). The data from infrared sensors is used to compute the map of the obstacles surrounding the robot. Based on the computed map, remote workstations suppose to send timely feedback signals to the robot for the wheel control, thereby imposing real-time requirements for specific tasks.

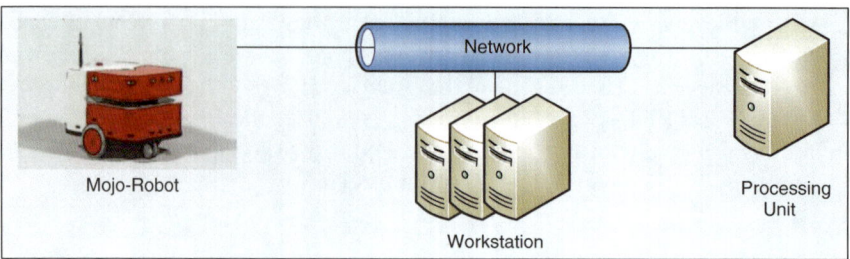

Fig. 15.3 Infrastructure for autonomous robot control

15.8.2 Component Selection

For the case study, we have selected a set of suitable components from a ROS repository [9], such that the component set enables the functionality of the autonomous control loop. Robot Operating System (ROS) is an open-source operating system for robot-based applications. It provides hardware abstraction, device drivers, message-passing and package management. The ROS is based on the publish-subscribe architectural style, where each component subscribes for, or publishes, a service wrapped into a `topic` type.

From the hardware point of view (Fig. 15.3), the system is fully distributed. The robot sends the infrared and odometry sensor data to a processing unit, which facilitates wireless communication with the robot and tunnels the packets to remote workstations, which handle major computations. The computation results are transmitted back to the processing unit to control the robot movements in real-time.

From software component point of view, the creation of surrounding map is the cornerstone of the navigation process, defining the reference model of the real world for further robot control. We have selected a GM (`Slam_Gmapping`) component for the map generation. The GM component instantly receives the laser and the odometry data and generates the actual 2-D map.

To control the robot wheels, we have selected the MB (`Move_Base`) component. This component subscribes for the actual 2D map as well as for the laser/odometry data. It also computes the local and the global plan to provide a global strategy and to publish navigation control signals.

The RVIZ component visualizes the environment (2-D map) to the control officer and allows setting the final destination goal for the robot. This goal is issued to the MB component for further wheel control. The MD (`Mojo_Driver`) component is responsible for the communication between the robot and ROS. The MD publishes the sensor data and subscribes to the wheel control data. The MF (`Mojo_Frame`) component publishes the 3-D geometrical representation of the robot with the exact position of the infrared sensors in space.

Table 15.1 HW platforms used for profiling

Model	CPU			BUS	RAM	
	Frequency (GHz)	Cache (MB)	Cores/ threads	Transfer rate (GT/s)	Size (GB)	Frequency (MHz/dual)
Intel i5-750	2.8	8	4/4	2.5	4	1,333
Intel i5-2520	2.5	3	2/4	5.0	8	1,066

```
SW component = MB
function = ExecuteCycles
  property = cpuUsage
  CPU = Intel(R) Core(TM) i5-2520M CPU@2.50GHz
    instructions_avg, min, max = 1.81985e+07
    cycles_avg, min, max = 1.364+07, 2.129+06, 2.183+07
    L2_miss_avg, min, max = 55068.9, 6592, 88475
    L3_miss_avg, min, max = 16698, 2200, 35827

  CPU = Intel(R) Core(TM) i5 CPU 750@2.67GHz
    instructions_avg, min, max = 1.93538e+07
    cycles_avg, min, max = 1.6003+07, 2.117+06, 2.4029+07
    L2_miss_avg, min, max = 39541.2, 6864, 69087
    L3_miss_avg, min, max = 6308.96, 391, 30626

  property = busUsage
  BUS = bus_Intel(R) Core(TM) i5-2520M CPU@2.50GHz
    L3_miss_avg, min, max = 16698, 2200, 35827
    L3_cache_wr_avg, min, max = 8960.51, 335, 15039

  BUS = bus_Intel(R) Core(TM) i5 CPU 750@2.67GHz
    L3_miss_avg, min, max = 6308.96, 391, 30626
    L3_cache_wr_avg, min, max = 7323.12, 246, 14497

  property = memoryUsage
        averageClaim, averageRelease = 966224
        minClaim, minRelease = 966224
        maxClaim, maxRelease = 966224
END ExecuteCycles
```

Fig. 15.4 An example of a ProMo performance model for the `MoveBase` (MB) component

15.8.3 Component Profiling and Modeling

We have profiled the selected components for two different platforms: i5-750 CPU (1st generation) and i5-2520 Mobile CPU (2nd generation). The specifications of the HW platforms are summarized in Table 15.1. During the profiling process, the ProMo tool automatically generates a performance model for each individual SW component. An example of the performance model of the MB component is depicted in Fig. 15.4.

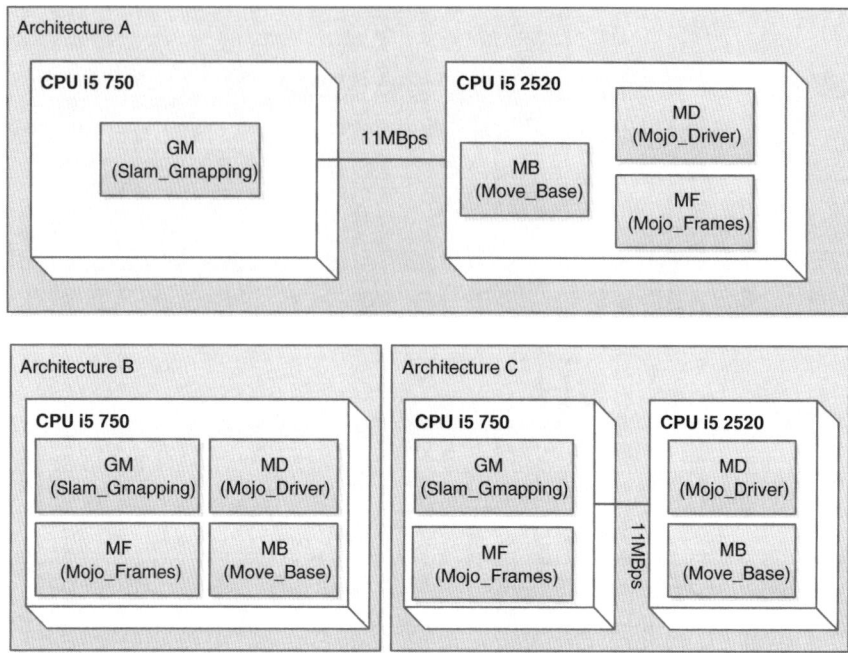

Fig. 15.5 Three initial architecture alternatives of the Autonomously Navigating Robot

15.8.4 System Composition Phase

At the system composition phase, we have decided to use three different architecture alternatives varying in hardware topologies and SW/HW mapping (Fig. 15.5). In Architecture A, we deploy 2 processing nodes. The most computationally expensive component, GM, is mapped on the i5-750 CPU node, while the rest of navigation components are mapped on the i5-2520 node. In Architecture B, all SW components are mapped on the single i5-750 CPU processing node. Architecture C balances the workload of the system between two nodes: the i5-750 CPU processing node executes the GM and the MF components, while the i5-2520 CPU processing node executes the MD and the MB components.

15.8.5 Scenario Definition

The analysis of predicted system behavior has shown that the workload on the system can be characterized by nine execution scenarios (Fig. 15.6 and Table 15.2). In this section, we describe the functionality, the trigger period and the deadline for each individual scenario. The triggering periods and real-time requirements (deadlines)

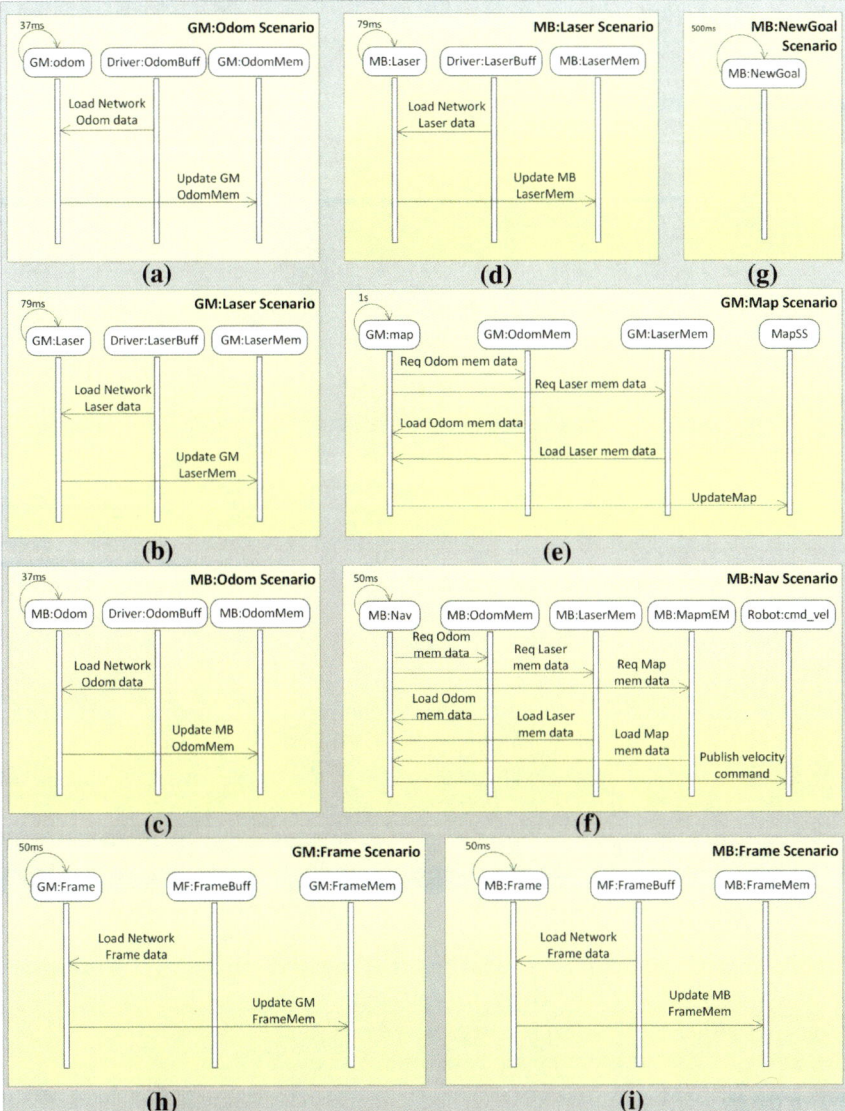

Fig. 15.6 Specification of scenarios in a message sequence chart diagram style

for each scenario are presented in Table 15.2. The deadlines are not scoped by periods and are defined to maintain the system's stability only, here the freshness of the data is not critical for each period.

This paragraph defines infrastructural scenarios with no hard real-time deadlines and is optional for reading. The scenario GM : Odom describes the robot odometry data transmission to the GM component. In this scenario, the odometry data is transmitted

Table 15.2 Periods and
deadlines for the RT scenarios

Scenario Name	Period (ms)	Deadline (ms)
GM:Odom	37	1,000
GM:Laser	79	1,000
MB:Odom	37	1,000
MB:Laser	79	1,000
GM:Frame	1,000	1,000
MB:NewGoal	500	500
MB:Frame	1,000	1,000
GM:Map	1,000	1,000
MB:Nav	50	150

every 37 ms and the GM component stores these data to an internal buffer (memory). Similarly, during the scenario GM:Laser, the infrared sensor data is transmitted to the GM every 79 ms for internal storage. Commonly, the scenarios MB:Odom and MB:Laser are triggered every 37 ms and 79 ms, respectively, transferring the odometry and the laser data from the robot node to the MB component for internal storage. The scenarios GM:Frame and MB:Frame describe transmission of the infrared sensor positions to the GM and MB components. The scenario is iterative with 50 ms period and a missing deadline does not lead to system failure.

The scenario GM:Map describes generation of the 2-D map of the robot environment. The Map operation is the core of the GM component. The operation is triggered with a 1,000 ms period. During iteration, Map loads the laser and odometry data from the internal buffer and updates/generates the actual 2-D map. The completion deadline of the task instance is set to 1,000 ms.

The scenario MB:Nav is the main scenario of interest, since it is computationally expensive and has a hard real-time deadline at a very low time span (150 ms). The scenario describes the feedback control loop for the robot wheels. The Nav operation of the MB component is triggered every 50 ms. It loads the odometry, laser, and the actual 2-D map data from the internal buffers, creates the global/local planners and sends the control signal for the engine of the robot wheels.

The scenario MB:NewGoal represents the workload on the system when a new goal arrives from RVIZ during the navigation process. For worst-case analysis, we set it to be periodic with 500 ms period.

The deadlines for the scenarios have been defined to guarantee that the autonomous navigation system: 1. avoids collisions with newly appeared obstacles and 2. does not introduce a noticeable response in delay in setting a new goal.

In this case study, we strived for identification of an architecture alternative able to satisfy the two hard real-time requirements at a low cost of materials.

Table 15.3 Performance evaluation of three architecture alternatives

			Architecture A	Architecture B	Architecture C
CPU Util.	i5-2520	Core 1	7.62%	–	48.74%
		Core 2	27.24%	–	2.62%
	i5-750	Core 1	53.93%	4.80%	1.67%
		Core 2	–	21.06%	48.74%
		Core 3	–	38.76%	–
		Core 4	–	93.35%	–
Network Utilization/Slack			10.43%	–	10.39%
Events	GM:Odom Scenario		595.213 ms	915.925 ms	575.751 ms
	GM:Laser Scenario		604.806 ms	929.603 ms	576.429 ms
	MB:Odom Scenario		111.186 ms	19.726 ms	280.717 ms
	MB:Laser Scenario		112.963 ms	19.726 ms	280.967 ms
	GM:Frame Scenario		602.565 ms	985.737 ms	576.023 ms
	MB:NewGoal		5.158 ms	176.959 ms	188.964 ms
	MB:Frame Scenario		113.653 ms	176.404 ms	280.775 ms
	GM:Map Scenario		609.131 ms	947.877 ms	577.534 ms
	MB:Nav Scenario		113.903 ms	18.635 ms	270.594 ms

15.8.6 Performance Evaluation Phase

The evaluation of the three alternative system architectures has been performed using both schedulability analysis MAST and the simulation JsimMAST techniques. First, we perform schedulability analysis for all parallel tasks in the identified scenarios and check whether their real-time requirements are satisfied. The alternatives featuring all met deadlines were applied to a simulation analysis, providing detailed behavior predictions. Finally, we compare the predicted performance results of the alternatives and propose the optimal alternative with respect to the response latency in critical scenarios, robustness, resource load and cost. In the following paragraphs, we present the analysis results (see Table 15.3).

The Architecture A meets all the real-time requirements specified in Table 15.2. However, the worst-case response time of the task in the MB:nav scenario (113.9 ms) is close to the deadline. For this task, unexpected CPU overload may lead to a missing deadline and to an undesirable collision, making the alternative sensitive to higher load conditions (low robustness).

The Architecture B, where all the SW components are mapped on a single node, satisfies all hard real-time requirements (Table 15.2 and 15.3). However, due to the mapping of all SW components on the same node, the CPU load is high. This influences the end-to-end execution time of the GM:Map scenario (947 ms with a 1,000 ms deadline), thereby substantially reducing the robustness of the map generation task under overload conditions.

Table 15.4 Performance validation of schedulability versus simulation versus actual execution for the crucial RT events GM:MapScenario and MB:NavScenario of the autonomously navigating robot

Model	GM:Map Scenario (ms)	MB:Nav Scenario (ms)
WCET schedulability analysis	947.877	19.726
WCET simulation analysis	874.818	11.565
WCET in reality	899.123	18.833
ACET simulation analysis	559.331	3.807
ACET in reality	565.983	3.687

Architecture C fails to satisfy the hard real-time requirement (150 ms) for the MB : Nav scenario task, accounting to 270 ms. The introduced network delay increases the task latency resulting in the RT requirement failure.

In the optimal architecture selection phase, the reasoning is as follows. Both, architectures A and B, satisfy the real-time requirements, but have low robustness under overload conditions. The task in MB : nav is not sufficiently robust in Architecture A, while the task in GM : Map is not robust in Architecture B. The former task has more severe consequences when missing a deadline (potential collision), therefore this task has higher importance for deadline fulfillment. Therefore, the Architecture B is a first choice to be considered optimal among alternatives. Moreover, the Architecture B deploys only one processing node, reducing the system cost. Finally, deployment of more than one HW node (platform, network) increases the possibility of HW or communication failures. Therefore, the Architecture B has been advised to robot owners as an optimal solution among the alternatives.

15.8.7 Validation of the Predictions on the Implemented System

In order to validate our method, we have implemented the optimal Architecture B, then measured the latencies of the most critical scenarios, MapScenario and NavScenario, and compared them to the predicted latencies. For the worst-case, the predicted vs. actual latency deviations have shown to be within a 6 % range, while for the average case it is even lower, as presented in Table 15.4. In order to compute the actual worst-case and average-case latency of the two scenarios, we let the system navigate for more than 1 h, while recording the latencies to log files. However, for the actual measurements, we cannot guarantee that the system has reached its worst-case latency.

By comparing the delays that are depicted in Table 15.3, a number of interesting observations can be made. Firstly, the Worst-Case Execution Time (WCET) that the simulation analysis predicts is lower than the predicted WCET of the schedulability analysis. This can be explained by the nature of these two performance analysis methods, since that the simulation analysis cannot guarantee the reachability of the WCET, in contrast to the schedulability analysis where WCET is predicted due to its formal substance. Secondly, the actual WCET is lower than the WCET predicted by the schedulability analysis, indicating that during the actual execution of the system, the worst-case scenario has not been reached. Thirdly and finally, we notice that the simulation prediction delays are lower than the actual worst-case scenarios, highlighting that during simulation, the system did not execute a scenario, which imposes a delay close to the WCET that both the schedulability analysis and the actual execution have reached. This means that if we were building our navigating robot system based only on the simulation predictions, our prototype would not meet the real-time requirements, leading to a failure. Thus, the simulation WCET cannot act as a reliable source for building a hard real-time system and should be combined with the schedulability analysis predictions, in order to sufficiently predict 1. the worst- and average-case execution scenarios and 2. the identifications of possible bottlenecks.

15.9 Case Study Findings and Lessons Learned

During the schedulability analysis we have noticed that the mapping of tasks on CPU cores a plays a definitive role for the system performance. Two computationally expensive tasks mapped on the same core may introduce high task interleaving, therefore increasing WCET of both tasks. An advised strategy is to map the identified heavy tasks on separate CPU cores. This will reduce latencies and increase robustness for each task, even at the expense of under-use of each CPU core.

We have also observed that the architecture, where the critical tasks are data- and execution-independent from performance of all other tasks, is very robust under overload conditions. All our tasks in the scenarios execute independently of the success/failure of other tasks, by taking the data from internal buffers. Failure of neighboring task to store actual data to the internal buffer does not influence the critical task delay, but only decreases the operation quality.

Another interesting conclusion is that for low-latency tasks, the mapping of involved components on different nodes led to disproportionally high increase in latency due to the added communication delay. One example is the low-latency MB:nav task rendering 18 ms delay in single-node mapping and 113 ms delay in multiple-node mapping case. In opposite, heavy but high-latency tasks improve on execution speed when being mapped on several nodes, e.g., the GM:Map task with 947 ms delay on a single node and 609 ms delay in two-node mapping.

The case study revealed a number of limitations of our approach. We were not able to analyze the internal memory-CPU bus usage and cache behavior. For this case study, it was not critical, but it would be of high importance for data-intensive systems. Besides this, we noticed an influence of Operating System (OS) tasks on the latency values during the performance validation on implemented system, while these tasks were not taken into account during the analysis. Finally, the migration of the tasks over different cores within one processing node is extremely difficult to predict, since it is dynamically defined by an OS scheduler at run time.

15.10 Conclusions

In this chapter, we have proposed an accurate performance analysis method for real-time systems. We have evaluated and validated our method on a real-time autonomous navigation robot system. The method incorporates 1. profiling and modelling of SW components at cycle-accurate level, 2. automated generation of system performance model from the models of individual components, and 3. evaluation of the obtained system model by schedulability analysis and simulation techniques resulting in predicted latencies, throughput, resource usage, and robustness. The presented component and system models are MARTE-compatible.

The method features multiple advantages for profiling, modeling and evaluating real-time systems. Firstly, the profiling provides cycle-accurate performance measurements collected directly by the Performance Monitor Unit (PMU) of the attached CPU. Secondly, the performance models are compatible with the commonly used UML-MARTE profile. Thirdly, the established pipeline generating models at different analysis phases automates the analysis process and carries the profiled low-level metrics of the components through all phases till the overall system performance is predicted. This brings high accuracy in predictions, as it was shown by the case study (6 % error range). Fourthly, the method deploys both types of techniques: schedulability analysis and simulation enabling predictions of both guaranteed worst-case executions and detailed behavior of tasks. Finally, we integrate the Eclipse Papyrus IDE into our tooling pipeline, so that an architect can easily design the SW/HW architectures graphically and automatically convert the design into models.

Our method has a number of limitations that require further research. Firstly, the component performance models can be obtained only for Linux-based operating systems, and the component profiling requires availability of actual HW platforms. Secondly, we do not provide reliable solution for generation of behavior models for component operations, leaving this notorious task to a component developer. Thirdly, we do not fully take into account the influence of the memory, internal bus and cache behavior on the performance of the system. Similarly, we do not provide support for GPU-based and OS tasks, which may decrease the prediction accuracy.

Let us briefly outline our future research steps. Since our performance analysis method does not incorporate the memory and the CPU cache structures, we plan to integrate a cycle-accurate platform simulator executing cache behavior. Moreover, due to the increasing popularity of applications that can be executed on a GPU, it is vital to support the modeling and the evaluation of such systems. Last but not least, the architecture alternatives are composed manually, which bounds the DSE process to a limited number of alternatives. We are developing an architecture generation method to automate the creation and assessment of optimal alternatives. The latest open source version of the ProMARTES toolchain is available online [27].

References

1. AUTOSAR (2003) AUTOSAR: AUtomotive Open System Architecture. AUTOSAR Consortium, http://www.autosar.org/
2. Becker S, Koziolek H, Reussner R (2007) Model-based performance prediction with the palladio component model. In: Proceedings of the 6th international workshop on software and performance (WOSP) 2007, ACM, pp 54–65. doi:10.1145/1216993.1217006
3. Bertolino A, Mirandola R (2004) CB-SPE tool: putting component-based performance engineering into practice. In: Crnkovic I, Stafford JA, Schmidt HW, Wallnau K (eds) Component-based software engineering, vol 3054, Springer, Berlin and Heidelberg, Germany, pp 233–248. doi:10.1007/978-3-540-24774-6_21
4. Bondarev E, Chaudron M, de With P (2006) A process for resolving performance trade-offs in component-based architectures. In: Gorton I, Heineman GT, Crnković I, Schmidt HW, Stafford JA, Szyperski C, Wallnau K (eds) Component-based software engineering—9th international symposium, CBSE 2006, Västerås, Sweden, Lect Notes Comput Sci, vol 4063, Springer, Berlin, Germany, pp 254–269. doi:10.1007/11783565_18
5. Bondarev E, Chaudron M, de With P (2007) CARAT: a toolkit for design and performance analysis of component-based embedded systems. In: Proceedings of the design, automation test in Europe conference exhibition (DATE) 2007, pp 1–6. doi:10.1109/DATE.2007.364428
6. Cortellessa V, Pierini P, Rossi D (2007) Integrating software models and platform models for performance analysis. IEEE Trans Softw Eng 33(6):385–401. doi:10.1109/TSE.2007.1014
7. Cuesta C (2010) JsimMAST: The performance analysis simulator for real-time. http://mast.unican.es/jsimmast/
8. Eker J, Janneck JW, Lee EA, Liu J, Liu X, Ludvig J, Neuendorffer S, Sachs S, Xiong Y (2003) Taming heterogeneity—the Ptolemy approach. Proc IEEE 91(1):127–144. doi:10.1109/JPROC.2002.805829
9. Garage W (2007) ROS: the robot operating system. Open Source Robotics Foundation. http://www.ros.org/wiki/
10. Gonzalez Harbour M, Gutierrez Garcia J, Palencia Gutierrez J, Drake Moyano J (2001) MAST: modeling and analysis suite for real time applications. In: Proceedings of the 13th Euromicro conference on real-time systems 2001, pp 125–134. doi:10.1109/EMRTS.2001.934015
11. Graf S, Ober I, Ober I (2006) A real-time profile for UML. Int J Softw Tools Technol Transf 8(2):113–127. doi:10.1007/s10009-005-0213-x
12. Grassi V, Mirandola R, Sabetta A (2005) From design to analysis models: a kernel language for performance and reliability analysis of component-based systems. In: Proceedings of the 5th international workshop on software and performance (WOSP) 2005, ACM, New York, NY, USA, pp 25–36.doi:10.1145/1071021.1071024
13. Hissam S, Moreno G, Stafford J, Wallnau K (2003) Enabling predictable assembly. J Syst Softw 65(3):185–198. doi:10.1016/S0164-1212(02)00038-9 component-Based Software Engineering
14. IBM (2003) Rationale rose: a modeling environment. IBM. http://www-01.ibm.com/software/rational/
15. ITEA (2000) ROBOCOP: robust open component based software architecture for configurable devices project. ITEA. http://www.hitech-projects.com/euprojects/robocop/
16. Klobedanz K, Kuznik C, Thuy A, Mueller W (2010) Timing modeling and analysis for AUTOSAR-based software development—a case study. In: Proceedings of design, automation test in Europe conference exhibition (DATE) 2010, pp 642–645. doi:10.1109/DATE.2010.5457125
17. Liu Y, Fekete A, Gorton I (2005) Design-level performance prediction of component-based applications. IEEE Trans Softw Eng 31(11):928–941. doi:10.1109/TSE.2005.127
18. Medina JL, Cuesta AG (2011) MAST: modeling and analysis suite for real-time applications. http://mast.unican.es

19. Medina JL, Garcia Cuesta A (2011) Model-based analysis and design of real-time distributed systems with Ada and the UML profile for MARTE. In: Reliable software technologies—Ada-Europe 2011, Lect Notes Comput Sci, vol 6652, Springer, Berlin and Heidelberg, Germany, pp 89–102.doi:10.1007/978-3-642-21338-0_7

20. Mos A, Murphy J (2002) A framework for performance monitoring, modelling and prediction of component oriented distributed systems. In: Proceedings of the 3rd international workshop on software and performance (WOSP) 2002, ACM, New York, NY, USA, pp 235–36. doi:10.1145/584369.584403

21. Määttä S, Indrusiak LS, Ost L, Möller L, Glesner M, Moraes FG, Nurmi J (2010) Model based approach for heterogeneous application modelling for real time embedded systems. In: Proceedings of the third international workshop on model based architecting and construction of embedded systems (ACES-MB 2010), held as part of the 2010 international conference on model driven engineering languages and systems (MoDELS'10), Oslo, Norway. http://www.inf.pucrs.br/moraes/my_pubs/papers/2010/2010_ACESMB_moller.pdf

22. OMG (2007) SysML: systems modeling language. OMG. http://www.omgsysml.org/

23. OMG (2009) MARTE: modeling and analysis of real time and embedded systems. OMG. http://www.omgmarte.org

24. SAE (2000) Architecture analysis and design language. SAE. http://www.aadl.info/aadl/currentsite

25. Silvano C, Fornaciari W, Palermo G, Zaccaria V, Castro F, Martinez M, Bocchio S, Zafalon R, Avasare P, Vanmeerbeeck G, Ykman-Couvreur C, Wouters M, Kavka C, Onesti L, Turco A, Bondik U, Marianik G, Posadas H, Villar E, Wu C, Dongrui F, Hao Z, Shibin T (2010) MULTICUBE: multi-objective design space exploration of multi-core architectures. In: Proceedings of the 2010 IEEE Annu Symp VLSI (ISVLSI), IEEE Comput Soc, Washington, DC, USA, pp 488–493. doi:10.1109/ISVLSI.2010.67

26. Thompson M, Polstra S, Erbas C, Pimentel AD (2008) Calibration of abstract performance models for system-level design space exploration. J Signal Proc Syst 50(2):99–114. doi:10.1007/s11265-007-0085-2

27. Triantafyllidis K (2013) ProMARTES: profiling, modeling, analysis of real-time embedded systems. http://vca.ele.tue.nl/demos/ProMARTES

28. Triantafyllidis K, Bondarev E, de With P (2012) Low-level profiling and MARTE-compatible modeling of software components for real-time systems. In: Proceedings of the 38th EUROMICRO conference on software engineering and advanced applications (SEAA) 2012, pp 216–223.doi:10.1109/SEAA.2012.25

29. Wandeler E, Thiele L, Verhoef M, Lieverse P (2006) System architecture evaluation using modular performance analysis: a case study. Int J Softw Tools Technol Transfer 8(6):649–667. doi:10.1007/s10009-006-0019-5

30. Wu X, Woodside M (2004) Performance modeling from software components. In: Proceedings of the 4th international workshop on software and performance (WOSP) 2004, ACM, New York, NY, USA, pp 290–301. doi:10.1145/974044.974089

Index

© Springer International Publishing Switzerland 2015
M.-M. Louërat and T. Maehne (eds.), *Languages, Design Methods, and Tools for Electronic System Design*, Lecture Notes in Electrical Engineering 311, DOI 10.1007/978-3-319-06317-1

Printed by Printforce, the Netherlands